Chromatography-the Ultimate Analytical Tool

Chromatography-the Ultimate Analytical Tool

Editors

Victoria Samanidou
Natasa Kalogiouri

MDPI • Basel • Beijing • Wuhan • Barcelona • Belgrade • Manchester • Tokyo • Cluj • Tianjin

Editors
Victoria Samanidou
Chemistry
Aristotle University of Thessaloniki
Thessaloniki
Greece

Natasa Kalogiouri
Chemistry
Aristotle University of Thessaloniki
Thessaloniki
Greece

Editorial Office
MDPI
St. Alban-Anlage 66
4052 Basel, Switzerland

This is a reprint of articles from the Special Issue published online in the open access journal *Molecules* (ISSN 1420-3049) (available at: www.mdpi.com/journal/molecules/special_issues/chromatography_analytical_tool).

For citation purposes, cite each article independently as indicated on the article page online and as indicated below:

LastName, A.A.; LastName, B.B.; LastName, C.C. Article Title. *Journal Name* **Year**, *Volume Number*, Page Range.

ISBN 978-3-0365-4078-8 (Hbk)
ISBN 978-3-0365-4077-1 (PDF)

Cover image courtesy of Victoria Samanidou

© 2022 by the authors. Articles in this book are Open Access and distributed under the Creative Commons Attribution (CC BY) license, which allows users to download, copy and build upon published articles, as long as the author and publisher are properly credited, which ensures maximum dissemination and a wider impact of our publications.

The book as a whole is distributed by MDPI under the terms and conditions of the Creative Commons license CC BY-NC-ND.

Contents

About the Editors vii

Preface to "Chromatography-the Ultimate Analytical Tool" ix

Natasa P. Kalogiouri and Victoria F. Samanidou
A Validated Ultrasound-Assisted Extraction Coupled with SPE-HPLC-DAD for the Determination of Flavonoids in By-Products of Plant Origin: An Application Study for the Valorization of the Walnut Septum Membrane
Reprinted from: *Molecules* **2021**, *26*, 6418, doi:10.3390/molecules26216418 1

Natasa P. Kalogiouri, Petros D. Mitsikaris, Dimitris Klaoudatos, Athanasios N. Papadopoulos and Victoria F. Samanidou
A Rapid HPLC-UV Protocol Coupled to Chemometric Analysis for the Determination of the Major Phenolic Constituents and Tocopherol Content in Almonds and the Discrimination of the Geographical Origin
Reprinted from: *Molecules* **2021**, *26*, 5433, doi:10.3390/molecules26185433 19

Aswini Panigrahi, Julius Benicky, Renhuizi Wei, Jaeil Ahn, Radoslav Goldman and Miloslav Sanda
A Rapid LC-MS/MS-PRM Assay for Serologic Quantification of Sialylated O-HPX Glycoforms in Patients with Liver Fibrosis
Reprinted from: *Molecules* **2022**, *27*, 2213, doi:10.3390/molecules27072213 33

Prawez Alam, Faiyaz Shakeel, Abuzer Ali, Mohammed H. Alqarni, Ahmed I. Foudah and Tariq M. Aljarba et al.
Simultaneous Determination of Caffeine and Paracetamol in Commercial Formulations Using Greener Normal-Phase and Reversed-Phase HPTLC Methods: A Contrast of Validation Parameters
Reprinted from: *Molecules* **2022**, *27*, 405, doi:10.3390/molecules27020405 43

Arman Kulyyassov
Application of Skyline for Analysis of Protein–Protein Interactions In Vivo
Reprinted from: *Molecules* **2021**, *26*, 7170, doi:10.3390/molecules26237170 59

Sunanta Wangkarn, Kate Grudpan, Chartchai Khanongnuch, Thanawat Pattananandecha, Sutasinee Apichai and Chalermpong Saenjum
Development of HPLC Method for Catechins and Related Compounds Determination and Standardization in Miang (Traditional Lanna Fermented Tea Leaf in Northern Thailand)
Reprinted from: *Molecules* **2021**, *26*, 6052, doi:10.3390/molecules26196052 71

Joanna Sobiak, Matylda Resztak, Maria Chrzanowska, Jacek Zachwieja and Danuta Ostalska-Nowicka
The Evaluation of Multiple Linear Regression–Based Limited Sampling Strategies for Mycophenolic Acid in Children with Nephrotic Syndrome
Reprinted from: *Molecules* **2021**, *26*, 3723, doi:10.3390/molecules26123723 83

Alisa Pautova, Natalia Burnakova and Alexander Revelsky
Metabolic Profiling and Quantitative Analysis of Cerebrospinal Fluid Using Gas Chromatography–Mass Spectrometry: Current Methods and Future Perspectives
Reprinted from: *Molecules* **2021**, *26*, 3597, doi:10.3390/molecules26123597 103

Camille Keisha Mahendra, Syafiq Asnawi Zainal Abidin, Thet Thet Htar, Lay-Hong Chuah, Shafi Ullah Khan and Long Chiau Ming et al.
Counteracting the Ramifications of UVB Irradiation and Photoaging with *Swietenia macrophylla* King Seed
Reprinted from: *Molecules* **2021**, *26*, 2000, doi:10.3390/molecules26072000 . . . . **133**

Yu Ra Lee, Ji Won Lee, Jongki Hong and Bong Chul Chung
Simultaneous Determination of Polyamines and Steroids in Human Serum from Breast Cancer Patients Using Liquid Chromatography–Tandem Mass Spectrometry
Reprinted from: *Molecules* **2021**, *26*, 1153, doi:10.3390/molecules26041153 . . . . **175**

About the Editors

Victoria Samanidou

Dr Victoria Samanidou is a Full Professor and Director of the Laboratory of Analytical Chemistry in the Department of Chemistry of Aristotle University of Thessaloniki, Greece. Her research interests focus on the development of sample preparation methods using novel sorptive extraction approaches prior to liquid and/or gas chromatographic analysis. She has co-authored more than 250 original research articles and reviews in peer-reviewed journals, 65 editorials/in view and 51 chapters in scientific books (h-index 40). She is an editorial board member of more than 30 scientific journals and a guest editor in more than 29 Special Issues. She has peer reviewed ca 700 manuscripts for ca 150 scientific journals.

In 2016, she was included in top 50 power list of women in Analytical Science (https://theanalyticalscientist.com/power-list/the-power-list-2016). In 2021, she was included in the "The Analytical Scientist"2021 Power List of top 100 influential people in analytical science. https://theanalyticalscientist.com/power-list/2021. In 2020, she was included in the 2% top world scientists in the field of Analytical Chemistry (career—2019, as well as single year 2019) published in PLOS Biology based on citations from SCOPUS. In 2021, she was again included in the list of World Top 2% Scientists, for her whole career up to now as well as for the single year 2020. The list is prepared by the Stanford University (USA) and it is based on standardized citation indicators. She is the Leader of Working Group 1 Science and Fundamentals of EuChemS-DAC SamplePreparation Study Group and Network, since 2021. Since 2016, she has been the President of the Steering Committee of the Association of Greek Chemists-Regional Division of Central & Western Macedonia.

http://users.auth.gr/samanidu
https://www.researchgate.net/profile/Victoria_Samanidou
http://orcid.org/0000-0002-8493-1106
Scopus Author ID 7003896015
https://sciprofiles.com/profile/152347

Natasa Kalogiouri

Natasa Kalogiouri is a Postdoctoral Researcher at the Laboratory of Analytical Chemistry of the Department of Chemistry, at the Aristotle University of Thessaloniki. She received her Bachelor's degree from the Department of Chemistry of the Aristotle University of Thessaloniki in 2011. She continued her studies in the same Department, from where she received a Master's Degree in 2013 on "Advanced Chemical Analysis". In 2017, she was awarded a Doctorate Title by the Department of Chemistry of the National and Kapodistrian University of Athens, where she has worked on the development of high-resolution mass spectrometric methods for the investigation of food authenticity. As a Postdoctoral Researcher she received a grant from the State Scholarships Foundation (IKY), and her research focused on the development of novel analytical methods with the use of green sample preparation techniques combined with advanced chemometric tools to guarantee food quality and authenticity. She has published 36 scientific papers and 2 Chapter Books. She has served as a Guest Editor in four Special Issues in MDPI journals. She is a member of the Association of Greek Chemists and Member of European Chemical Society-Division of Analytical Chemistry (EuChemS-DAC). Her research interests focus on microextraction sample preparation techniques, high-pressure liquid chromatography, high-resolution mass spectrometry, gas chromatography, targeted and nontargeted screening determinations, and chemometrics.

ORCHID: 0000-0001-6183-7857

Links

https://scholar.google.com/citations?user=sQvK1rkAAAAJ&hl=el
https://www.linkedin.com/in/natasa-kalogiouri-757a5253/
https://www.scopus.com/authid/detail.uri?authorId=55795994000
https://www.researchgate.net/profile/NatasaKalogiouri/research

Preface to "Chromatography-the Ultimate Analytical Tool"

Since its early introduction by the Russian botanist Mikhail Semyonovich Tsvet, chromatography has been undoubtedly the most powerful analytical tool in analytical chemistry.

Separation, qualitative analysis, and quantitative analysis can be achieved by choosing the right conditions. Thus, numerous gas chromatographic, liquid chromatographic, and supercritical fluid chromatographic methods have been developed and applied for most types of samples and most kinds of analytes. Additionally, older varieties such as paper chromatography and thin-layer chromatography were pioneer analytical techniques in many laboratories. Especially when hyphenated to spectrometric techniques, chromatography also allows the identification of separated analytes in a single run. Highly sophisticated equipment can answer all analytical problems very quickly. Chromatographers cooperate with many scientific fields and give their lights to medical doctors, veterinarians, food scientists, biologists, dentists, archaeologists, etc. In this Special Issue, analytical chemists were invited to prove that chromatography-based separation techniques are the ultimate analytical tool and their significant contribution is reflected in ten interesting articles.

Victoria Samanidou and Natasa Kalogiouri
Editors

Article

A Validated Ultrasound-Assisted Extraction Coupled with SPE-HPLC-DAD for the Determination of Flavonoids in By-Products of Plant Origin: An Application Study for the Valorization of the Walnut Septum Membrane

Natasa P. Kalogiouri and Victoria F. Samanidou *

Laboratory of Analytical Chemistry, Department of Chemistry, Aristotle University of Thessaloniki, 54124 Thessaloniki, Greece; kalogiourin@chem.auth.gr
* Correspondence: samanidu@chem.auth.gr

Abstract: Walnut byproducts have been shown to exert functional properties, but the literature on their bioactive content is still scarce. Among walnut byproducts, walnut septum is a dry ligneous diaphragm tissue that divides the two halves of the kernel, exhibiting nutritional and medicinal properties. These functional properties are owing to its flavonoid content, and in order to explore the flavonoid fraction, an ultrasound-assisted (UAE) protocol was combined with solid phase extraction (SPE) and coupled to high-performance liquid chromatography with diode array detection (HPLC-DAD) for the determination of flavonoids in Greek walnut septa membranes belonging to Chandler, Vina, and Franquette varieties. The proposed UAE-SPE-HPLC-DAD method was validated and the relative standard deviations (RSD%) of the within-day and between-day assays were lower than 6.2 and 8.5, respectively, showing good precision, and high accuracy ranging from 90.8 (apigenin) to 97.5% (catechin) for within-day assay, and from 88.5 (myricetin) to 96.2% (catechin) for between-day assay. Overall, seven flavonoids were determined (catechin, rutin, myricetin, luteolin, quercetin, apigenin, and kaempferol) suggesting that the walnut septum is a rich source of bioactive constituents. The quantification results were further processed using ANOVA analysis to examine if there are statistically significant differences between the concentration of each flavonoid and the variety of the walnut septum.

Keywords: walnut septum; UAE; SPE; flavonoids; functional; HPLC-DAD

1. Introduction

The development of analytical methodologies to characterize natural ingredients and produce high added value food products is a field of research that has attracted the interest of the scientific community, industry, and consumers, as well. During the last decades, there has been an increasing demand for foods that not only have high nutritional and sensorial quality but also deliver health promoting benefits through certain ingredients, namely "bioactive" or "functional" [1]. Moreover, emphasis is given on the valorization of food byproducts that combine nutritional value, flavor, and health benefits, as well. Towards this direction and in order to protect both public health and environment agri-food byproducts, generated in large amounts worldwide, can be exploited as a promising source of valuable compounds for novel food applications.

Walnuts (*Juglans regia* L.) are a valuable nutritional source with pleasant taste, consumed on a global scale. Walnut has been cultivated since 1000 BC, and it has naturally diverged to several cultivars, such as Chander, Vina, Franquette, Mellanaise, Lara, Marbot, Hartley Mayette, Serr, Tulare, Sorento, etc. [2]. Apart from the nutritional benefits, walnuts exert health-promoting properties, such as antioxidant, anti-inflammatory, antimicrobial, antidiabetic, hepatoprotective, anticancer, and cardioprotective, among others [3,4]. These benefits are owing to the presence of phytonutrients, and specifically to the phenolic compounds [5–7]. Even though the nutritional importance of the walnut is mainly related to the kernel, walnut byproducts, such as walnut husks, leaves, etc. [8,9], have been shown to exert functional properties, as well. Among walnut byproducts, walnut septum is a dry ligneous diaphragm tissue that divides the two halves of the kernel. Traditionally, the walnut septum has been used as a nutraceutical and medicinal material, and has been documented in the Chinese Pharmacopoeia [10]. The septum contains bioactive constituents, and it has been shown to exhibit antitumor, antioxidant, and immunoenhancement effects in vitro [11,12]. Phenolic compounds, and particularly flavonoids, are an important class of bioactive constituents that might be related to the functional properties of the walnut septum. Despite the fact that flavonoids are responsible for several pharmacological activities [13], the literature on the flavonoid content of walnut septum is still scarce.

Thus, the phenolic fraction of this unexplored byproduct has to be assessed with analytical methodologies. The generic analytical procedure for the determination of flavonoids in agricultural products involves first sample preparation, and then separation, detection, and quantification. Several traditional extraction techniques have been proposed for the extraction of flavonoids from natural products, including percolation, maceration, hydro-distillation, boiling, reflux, Soxhlet [14,15]. These techniques, however, present several disadvantages such as a lot of labor and time, large amounts of organic solvents, low selectivity, low extraction yield, and high cost [16]. For this reason, advanced green microextraction techniques are continually being developed to overcome these limitations. Green extraction techniques involve ultrasound-assisted extraction (UAE), microwave assisted extraction (MAE), supercritical fluid extraction (SFE), and solid phase extraction (SPE), among others. The current state-of-the art on the green extraction of natural constituents from food products has already been critically reviewed [4,17].

The main analytical technique used for the separation and determination of phenolic compounds is traditionally liquid chromatography (LC) coupled to UV, diode array detection (DAD), or mass spectrometric detectors (MS) [18–21]. The recent trends in analytical determination also involve the development of high-resolution mass spectrometric (HRMS) techniques, which provide separation efficiency and high accuracy in identifications, as it has already been reviewed [4,15,22,23]. Among these techniques, HPLC-DAD is a rapid analytical technique that enables the separation, identification, and quantification of flavonoids, offering several advantages in terms of sensitivity, specificity, ruggedness, and low cost of analysis. The accurate determination of the flavonoids and the further quantification of such analytes that exist in trace levels in food matrices is a challenging task. The further processing of the quantification results with statistical tools enhances the conclusions derived from the experimental data, allowing the discovery of trends and behaviors among the samples.

In this work, a green UAE-SPE-HPLC-DAD methodology was developed and validated for the determination of flavonoids in walnut septum belonging to different varieties (Chandler, Vina, and Franquette) cultivated in Greece. The determined analytes were quantified, and the results were further analyzed by one-way ANOVA to examine if there are statistically significant differences between the analytes' concentrations and the walnut variety. To the best of our knowledge, this is the first report of assessing the flavonoid profile of Greek walnut septa.

2. Results and Discussion
2.1. Method Development and Validation

An HPLC-DAD methodology was developed and validated to assess the flavonoid profile of walnut septum and all the analytical parameters, including the calibration curves, linear range, the determined coefficients (r^2), accuracy and precision, limits of detection (LODs), and limits of quantification (LOQs) are presented in Table 1. The analytical curves presented an adequate fit when submitted to the lack-of-fit test ($F_{calculated}$ was less than $F_{tabulated}$ in all cases), with r^2 above 0.99, proving that they can be used for the quantification of the flavonoids. The LOQs were found to range between 0.30 µg/g and 0.90 µg/g, while the LODs were calculated over the range 0.10 µg/g to 0.30 µg/g. The RSD% of the within-day ($n = 6$) and between-day assays ($n = 3 \times 3$) were lower than 6.2, and 8.5, respectively, showing adequate precision. The accuracy was assessed by means of relative percentage of recovery (%R) at low, medium, and maximum concentration levels of 1, 5, and 10 µg/g, and the results were acceptable, ranging from 90.8 (apigenin, at 10 µg/g concentration level) to 97.5% (catechin, at 10 µg/g concentration level) for within-day assay ($n = 6$) (Table 2), and from 88.5 (myricetin, at 1 µg/g concentration level) to 96.2% (catechin, at 5 µg/g concentration level) for between-day assay ($n = 3 \times 3$) (Table 3).

Table 1. HPLC-DAD method analytical parameters.

Compound	Calibration Equation y = (a ± Sa) + (b ± Sb)x (Linear Range: 1–10 µg/g)	r^2	F_{calc}	F_{tab}	LOD (µg/g)	LOQ (µg/g)
catechin	y = (1095 ± 1115) + (11808 ± 305)x	0.997	7.9×10^{-9}	0.2334	0.31	0.94
rutin	y = (389 ± 1200) + (19857 ± 204)x	0.995	1.9×10^{-9}	0.2334	0.20	0.60
myricetin	y = (989 ± 1450) + (20005 ± 424)x	0.993	5.6×10^{-9}	0.2334	0.24	0.72
luteolin	y = (1017 ± 1608) + (17008 ± 440)x	0.995	2.9×10^{-9}	0.2334	0.20	0.60
quercetin	y = (−1032 ± 1128) + (18404 ± 153)x	0.993	6.5×10^{-9}	0.2334	0.20	0.60
apigenin	y = (1732 ± 152) + (1745 ± 665)x	0.994	4.6×10^{-7}	0.2334	0.29	0.87
kaempferol	y = (1710 ± 54.3) + (19045 ± 685)x	0.996	1.7×10^{-9}	0.2334	0.29	0.90

Ftab: Ftabulated, Fcalc: Fcalculated, LOD: limit of detection, LOQ: limit of quantitation.

Table 2. %Recoveries (%R, $n = 6$) for the evaluation of repeatability.

Compound	%R Low Conc. Level (1 µg/g)	%RSD	%R Medium Conc. Level (5 µg/g)	%RSD	%R Maximum Conc. Level (10 µg/g)	%RSD
catechin	97.1	5.3	96.4	6.2	97.5	5.3
rutin	93.5	5.1	92.5	4.5	98.4	2.5
myricetin	91.8	4.3	94.4	5.2	91.2	1.7
luteolin	94.2	3.8	95.6	4.6	93.6	3.9
quercetin	93.6	3.4	98.8	4.2	94.1	4.2
apigenin	94.4	5.4	91.7	6.1	90.8	5.1
kaempferol	92.1	2.8	93.5	3.2	93.5	2.9

Conc.: Concentration.

Table 3. %Recoveries (%R, $n = 3 \times 3$) for the evaluation of intermediate precision.

Compound	%R Low Conc. Level (1 µg/g)	%RSD	%R Medium Conc. Level (5 µg/g)	%RSD	%R Maximum Conc. Level (10 µg/g)	%RSD
catechin	95.1	5.2	96.2	4.8	92.1	6.4
rutin	93.5	7.1	95.7	5.9	95.2	5.2
myricetin	88.5	6.2	93.4	6.5	94.4	6.1
luteolin	91.8	5.8	89.2	7.5	93.5	8.5
quercetin	94.5	7.8	90.3	8.1	92.3	6.9
apigenin	93.7	6.3	91.1	7.4	94.2	5.4
kaempferol	95.5	5.8	88.9	6.6	95.4	7.3

Conc.: Concentration.

2.2. Walnut Septum Analysis

Twenty-four walnut septum membranes belonging to the varieties Chandler, Vina, and Franquette were analyzed in triplicate and the flavonoids: catechin, rutin, myricetin, luteolin, quercetin, apigenin, and kaempferol were determined. The chromatographic identification results, including the retention times (R_{ts}) of the analytes, and their respective maximum absorption wavelengths (λ, nm) are presented in Table 4. Figure 1 illustrates the chromatographic separation of the flavonoids in a walnut septum extract that was monitored at 280 nm.

Table 4. Retention time and maximum absorption wavelength of the determined flavonoids.

Compound	Chemical Structure	Rt	λ (nm)
catechin		5.8	278
rutin		10.1	353
myricetin		16.5	370
luteolin		21.1	356

Table 4. Cont.

Compound	Chemical Structure	Rt	λ (nm)
quercetin		22.7	378
apigenin		24.5	360
kaempferol		26.1	360

Rt: retention time.

2.3. Quantitative Analysis of Flavonoids

The walnut septa were analyzed in triplicate ($n = 3$). The identified flavonoids (catechin, rutin, myricetin, luteolin, quercetin, apigenin, and kaempferol) were quantified on the basis of their maximum absorption wavelengths. The presence of the determined flavonoids was linked to certain positive health effects and other bioactive functions that have already been reported in the literature to highlight the potential functional activity of the analyzed byproduct [4]. Table 5 presents the quantification ranges and mean values (±SD) of the determined flavonoids in the walnut septa belonging to Chandler, Vina, and Franquette variety. Box and Whisker plots were constructed to graphically illustrate the concentrations of the flavonoids and present the distributional characteristics of each variety (Chandler, Vina, Franquette). The quantification results were further analyzed with ANOVA to examine if there are statistically significant differences between the concentrations of the determined flavonoids and the varieties of the walnut septa.

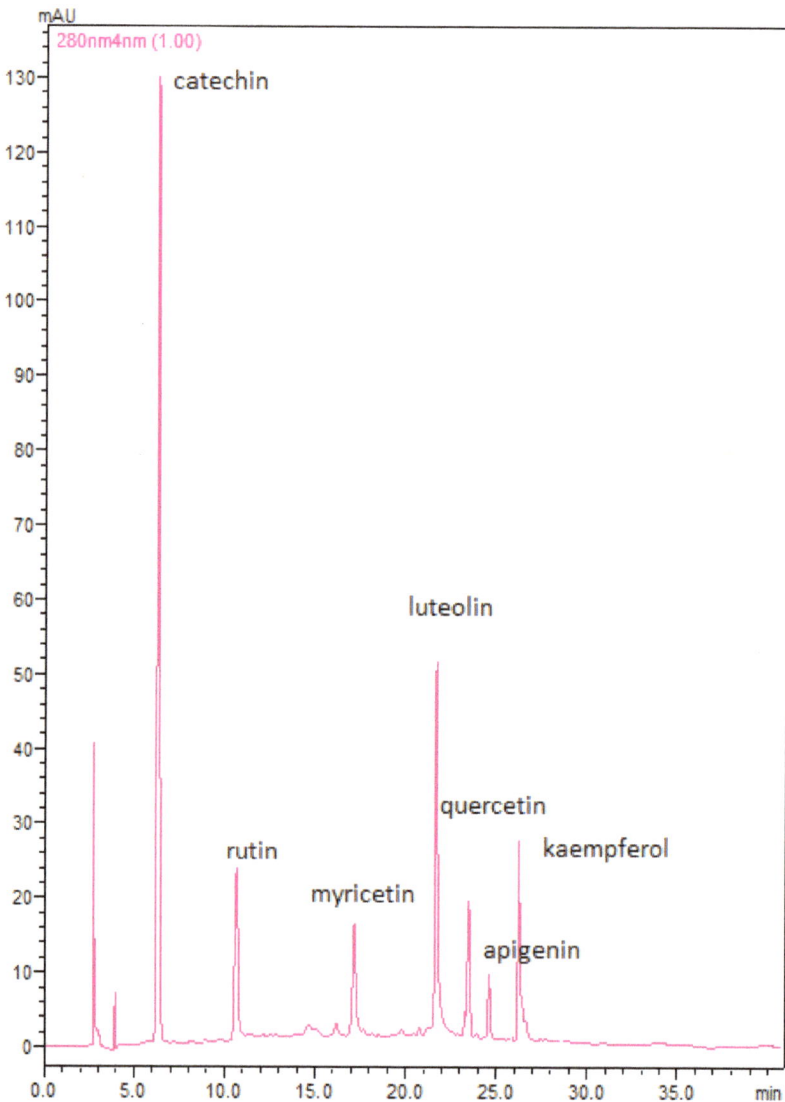

Figure 1. Characteristic chromatogram of a walnut septum extract; monitored at 280 nm.

2.3.1. Catechin

Catechin was the flavonoid detected in higher concentration in all the analyzed septa. Specifically, the highest mean concentration was determined in Vina septa (47 μg/g), and then Chandler (32 μg/g), and Franquette (31 μg/g) varieties followed (Figure 2). The ANOVA analysis exhibited a statistically significant difference in catechin concentration among all of the walnut septa of the different varieties with $p = 0.04$. The concentration ranges indicate that the walnut septum is rich in catechin, and it could be effectively used as a food additive to increase the antioxidant potential of food products, enrich feeds, while it could also be used as a functional ingredient of novel functional food products, including soft beverages, teas, infusions, etc. Several health properties have been associated with catechins, as they have proven to be promising protective agents against diabetes, arterial hypertension and ischemic stroke, obesity, metabolic syndrome [24]. Moreover, catechins demonstrate significant biological activities against oral cancer, breast cancer, Alzheimer's disease, Parkinson's disease [24]. Furthermore, catechin displays unique features responsible for several pharmacological and biological properties, as it possesses the ability to block ROS-induced chain reactions, acting as a promising antioxidant [25]. It also demonstrates significant antidiabetic function, via hepatoprotective, antineurodegenarative effects, insulin-mimetic effects, hindering amyloid formation [24].

Table 5. Quantification results of the flavonoids determined in walnuts septa belong to Chandler, Vina, and Franquette varieties (samples analyzed in triplicate, n = 3).

Variety	Chandler		Vina		Franquette	
Compound	Concentration Range (µg/g)	Mean Value (µg/g ± SD)	Concentration Range (µg/g)	Mean Value (µg/g ± SD)	Concentration Range (µg/g)	Mean Value (µg/g ± SD)
catechin	30–33	32 ± 5	41–53	47 ± 3	25–37	31 ± 1
rutin	1–3	2.4 ± 0.2	3–6	5 ± 2	3–6	4 ± 2
myricetin	1–6	3.8 ± 0.8	3–11	8 ± 2	5–9	7 ± 2
luteolin	1–4	2.6 ± 0.2	2–5	3.4 ± 0.9	1–3	2.4 ± 0.3
quercetin	5–14	9 ± 1	9–16	12 ± 4	4–8	6 ± 2
apigenin	1–6	4 ± 1	4–9	6 ± 2	2–10	5.5 ± 0.7
kaempferol	2–9	5 ± 3	2–8	6 ± 1	2–5	3.7 ± 0.9

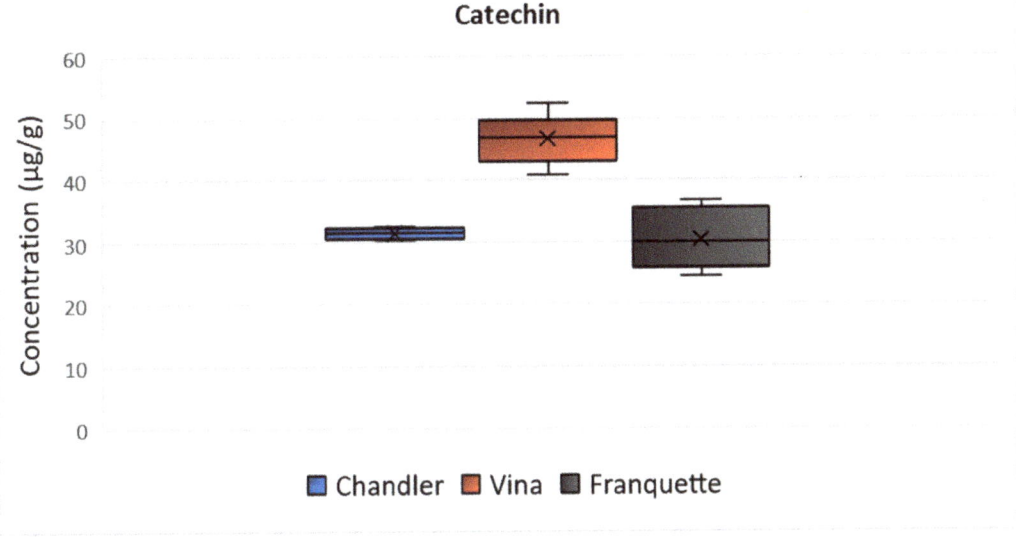

Figure 2. Box and Whisker plot for the concentration of catechin among walnut septa, belonging to the Chandler, Vina, and Franquette varieties.

2.3.2. Rutin

Rutin was determined over the range 1–3 µg/g in Chandler walnut septa, and over the ranges 4–6 µg/g and 3–6 µg/g in Vina and Franquette walnut septa, respectively (Figure 3). A statistically significant difference was observed for the concentration of rutin among the septa of the different varieties ($p = 0.002$). Rutin possesses promising antioxidant potential and exhibits significant biological properties, thus, playing an essential role in the human body's numerous physiological functions [26]. According to the literature, rutin may provide a wide variety of therapeutic effects, such as antiallergic, antiviral, antihypertensive, vasoactive, cytoprotective, anti-inflammatory, antiprotozoal, hypolipidemic, antispasmodic, anticarcinogenic, antibacterial, and antiplatelet activities [27]. Furthermore, rutin contributes to the strengthening of the blood vessels capillaries, due to its high radical scavenging capacity, thereby preventing fragility-associated hemorrhagic disorders in humans [28].

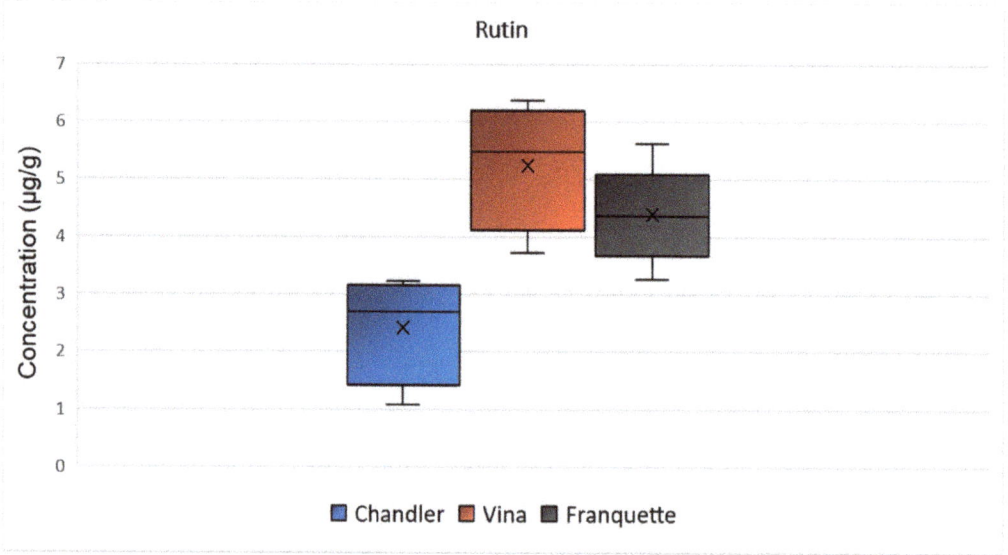

Figure 3. Box and Whisker plot for the concentration of rutin among walnut septa belonging to the Chandler, Vina, and Franquette varieties.

2.3.3. Myricetin

The mean concentration values of myricetin were equal to 4 µg/g, 8 µg/g, and 7 µg/g in Chander, Vina, and Franquette walnut septa, respectively (Figure 4). The ANOVA analysis showed that there is significant statistically difference among the septa of the different varieties ($p = 0.002$). Great scientific interest has been raised in myricetin, which has been shown to exhibit antioxidant, anti-inflammatory, antiviral, and anticarcinogen activities. It is functioning as an antineoplastic agent in human patients, since it has demonstrated strong suppressive effects on the activities of several types of cancer cells (e.g., cancer cell invasion or metastasis), thus, regulating apoptosis, and inhibitory properties on their proliferation [29–32].

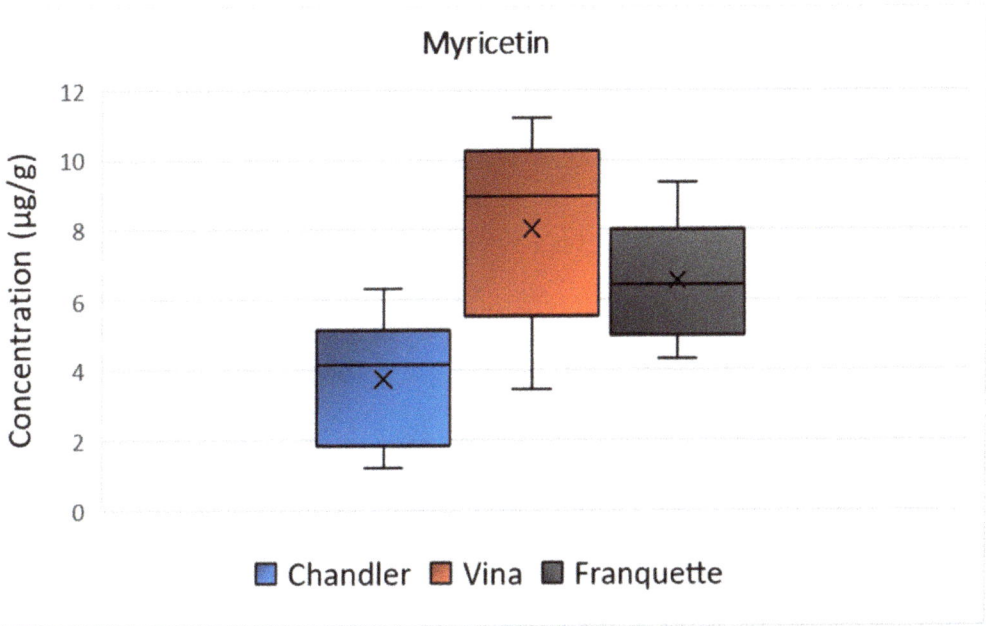

Figure 4. Box and Whisker plot for the concentration of myricetin among walnut septa belonging to the Chandler, Vina, and Franquette varieties.

2.3.4. Luteolin

The mean concentration values of luteolin were equal to 3 µg/g, 3 µg/g, and 2 µg/g in Chandler, Vina, and Franquette walnut septa, respectively (Figure 5). Statistically significant differences were observed among the luteolin concentrations determined in septa of the different varieties ($p = 0.039$). As for its biological action, luteolin is a naturally occurring flavonoid with a yellow crystalline appearance, and scientific research has indicated that it may display multiple cellular effects, hence, favorably affecting human health. It may exhibit antioxidant properties, protecting cells from ROS induced damage or act as an antineoplastic, anti-inflammatory, antimicrobial, or estrogenic regulatory compound and prevent liver diseases [33].

2.3.5. Quercetin

Quercetin was determined at 9 µg/g, in Chandler walnut septa. The mean concentration of quercetin was higher in walnut septa belonging to the Vina variety (12 µg/g), and lower for those belonging to the Franquette variety (6 µg/g), as it is presented in Figure 6. The ANOVA analysis showed that there is statistically significant difference among the quercetin concentration and the analyzed varieties ($p = 0.0008$). According to the literature, quercetin is responsible for the bitter flavor of foods, and in synergy with myricetin and other phenolic compounds has been recently used as food additives to protect meat products against bacteria's development [34]. It demonstrates several significant health-promoting functions, including cardioprotective, anti-ulcer, antidiabetic, antioxidant properties and chemopreventive potential [35]. It may also produce anti-inflammatory and anti-allergy effects through the inhibition of the lipoxygenase and cyclooxygenase pathways, thereby reducing the production of pro-inflammatory or pro-oxidant mediators [35].

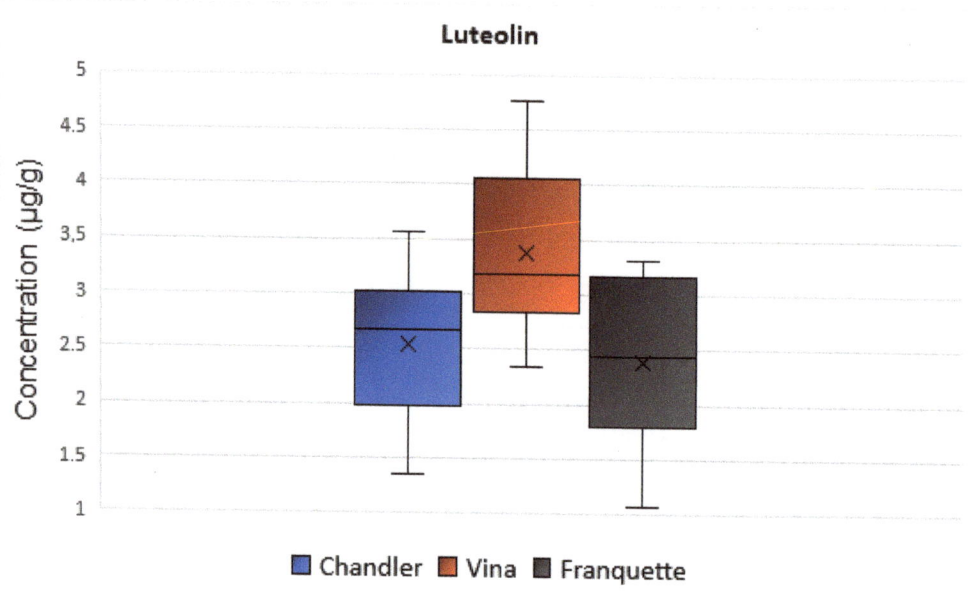

Figure 5. Box and Whisker plot for the concentration of luteolin among walnut septa belonging to the Chandler, Vina, and Franquette varieties.

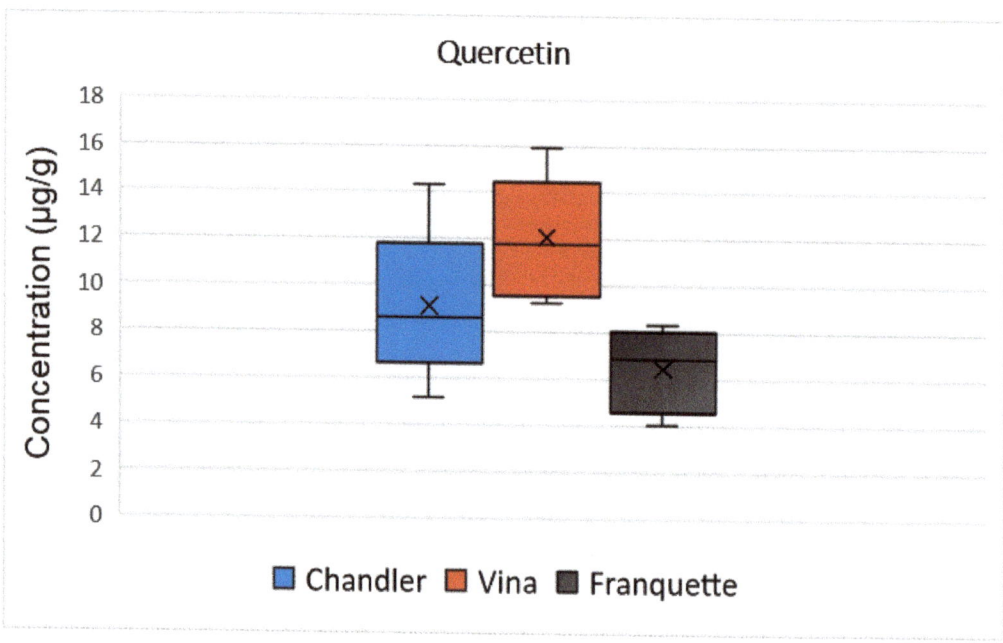

Figure 6. Box and Whisker plot for the concentration of quercetin among walnut septa belonging to the Chandler, Vina, and Franquette varieties.

2.3.6. Apigenin

Apigenin was determined at mean concentrations of 4 µg/g, 6 µg/g, and 5 µg/g in Chandler, Vina, and Franquette walnut septa, respectively (Figure 7). The ANOVA analysis showed that there was no statistically significant difference among the apigenin concentrations in the analyzed septa of the different varieties ($p = 0.147$). Traditionally, extracts, oils, and teas from plants with a high content of apigenin were used for its soothing qualities as a sedative, mild analgesic and sleep medication [36], reinforcing the idea that walnut septa could be used in infusion making. Moreover, in the food industry, apigenin possesses a role as a flavoring or adjuvant agent, enhancing the human body's response to antigens [35]. It may yield antiproliferative and antimetastatic effects, suppressing the formation and inducing malignant tumor cells [28]. Furthermore, it could prevent skin or colon cancer and acts as an anti-inflammatory, antioxidant, antiallergic, antimicrobial, antiviral, cardioprotective, and neuroprotective agent [28].

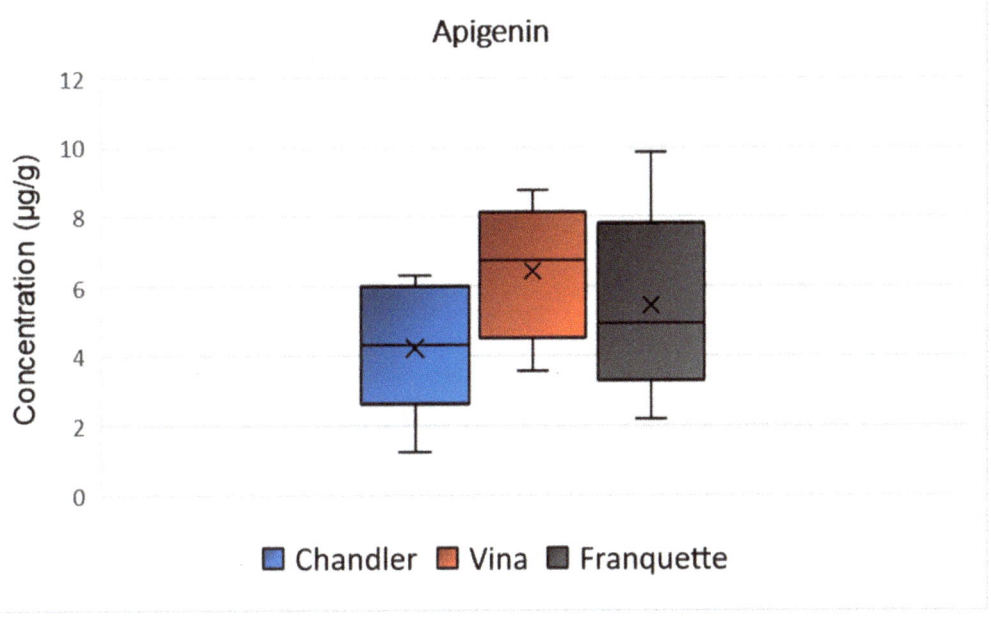

Figure 7. Box and Whisker plot for the concentration of apigenin among walnut septa belonging to the Chandler, Vina, and Franquette varieties.

2.3.7. Kaempferol

The highest mean concentrations of kaempferol were observed in the Vina variety (5 µg/g), while Chandler variety followed with 6 µg/g, and Franquette exhibited the lowest mean concentration of 4 µg/g (Figure 8). No statistically significant difference was observed for the concentration of kaempferol among the septa of the different varieties ($p = 0.06$). As for its biological action, kaempferol is a natural dietary flavonoid, potentially acting as chemopreventive agent [37], protecting against oxidative stress and inflammatory chronic disorders.

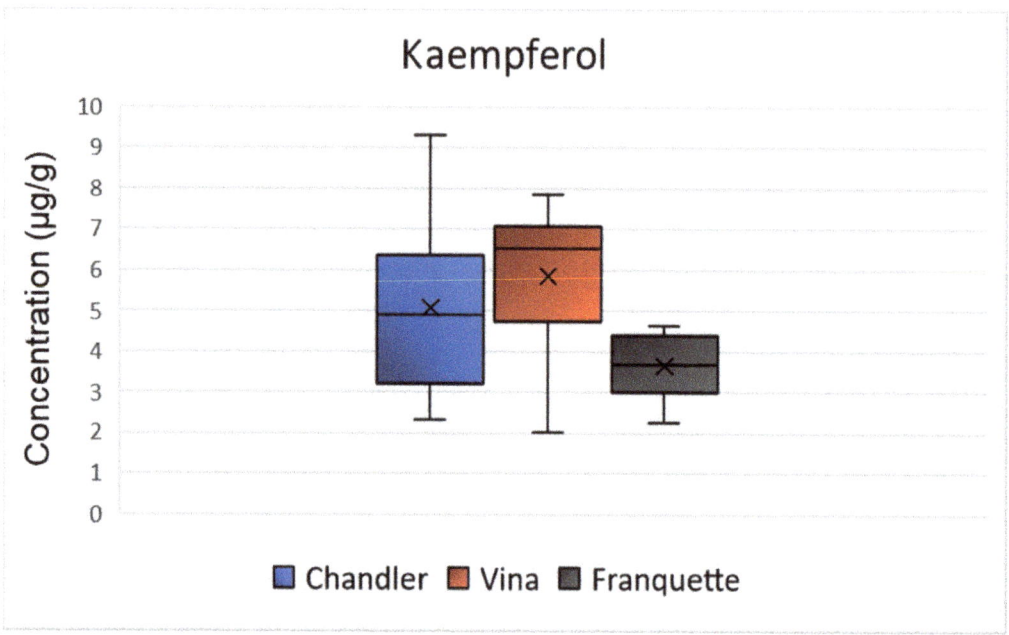

Figure 8. Box and Whisker plot for the concentration of kaempferol among walnut septa belonging to the Chandler, Vina, and Franquette varieties.

2.4. Effect of the Variety on the Phenolic Content

The determination of the seven flavonoids clearly indicates that walnut septum is a byproduct that is rich in flavonoids, and, thus, could be characterized as a "nutraceutical" raw ingredient. This is also supported by the quantification ranges of each individual flavonoid identified. This fruit septum of walnuts could be utilized as a substance for medicinal purposes, while it could be also further exploited for the development of novel functional foods and beverages, and for the enrichment of feeds, as well.

Except for the functional properties that have been attributed to each individual analyte, and were discussed in detail in Section 2.3, the findings of this research also demonstrated that the flavonoid content of the walnut septum is affected by the variety. The sum of the individually quantified flavonoids showed that the walnut septum of the Vina variety exhibited the highest flavonoid content (88 ± 15 μg/g). The flavonoid contents of the septa belonging to Chandler variety, and the Franquette variety were approximately similar, with calculated concentrations of 59 ± 12 μg/g, and 60 ± 9 μg/g, respectively. The graph in Figure 9 presents the sum of the mean values of the individual concentrations of the flavonoids determined in the walnut septa belonging to Chandler, Vina, and Franquette varieties. Considering that the analyzed samples originated from the same geographical origin (Thessaly, Greece), the variation in the concentration of the flavonoids could be linked to the genetic fingerprint of each variety.

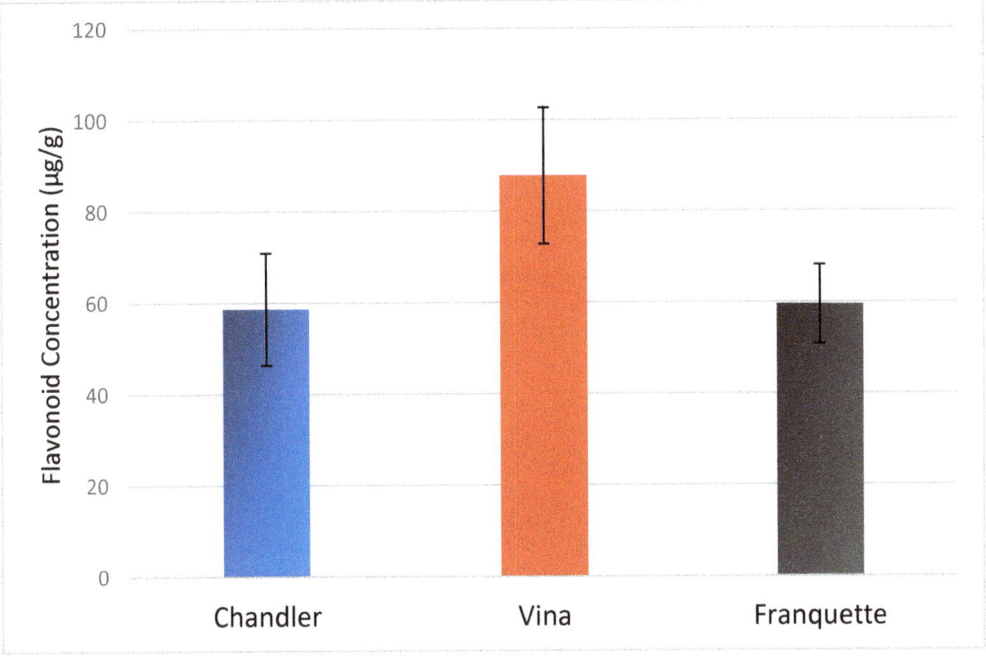

Figure 9. Average flavonoid concentration of walnut septum belonging to Chandler, Vina, and Franquette varieties.

Furthermore, the ANOVA analysis showed that there is statistically significant difference between the concentrations of catechin, rutin, myricetin, luteolin, quercetin, and the walnut variety. Furthermore, PCA analysis was employed to investigate the similarities in the flavonoid content of the septa according to their variety, and the first two PCs explained the 70% of variance. The PCA score plot and loading plot are presented in the Supplementary Material in Figures S1 and S2, respectively. Even though several works have already associated the bioactive content of the walnut kernel with the geographical origin of walnuts [18,38], and the variety [39,40], these are the first reports relating the flavonoid content of the walnut septum with the variety.

3. Materials and Methods

3.1. Chemicals and Reagents

Methanol (MeOH) and acetonitrile (ACN), HPLC grade, were acquired from Merck (Darmstadt, Germany). Acetic acid 99% and trifluoroacetic acid (TFA) 99% were purchased from Sigma-Aldrich (Steinheim, Germany). The LiChrolut RP-18 (C18, 3 mL, 500 mg) SPE cartridges used were supplied by Merck (Darmstadt, Germany). Ultrapure water was provided by a Milli-Q® purification system (Millipore, Bedford, MA, USA). The flavonoids catechin 98%, rutin 98%, myricetin 98%, luteolin 98%, quercetin 98%, kaempferol 98%, and apigenin 98% were supplied by Sigma-Aldrich (Steinheim, Germany). Stock standard solutions at 1000 mg/L concentration level were prepared and stored in dark brown glass bottles at −20 °C. Working standard solutions were prepared in MeOH after appropriate dilution of the stock solutions every laboratory day, before analysis.

3.2. Instrumentation

A quaternary low-pressure gradient HPLC–DAD system by Shimadzu (Kyoto, Japan) was used for analysis. The HPLC system consisted of: (a) an FCV-10ALVP mixing system, (b) a Rheodyne 7725i injection valve (Rheodyne, Cotati, CA, USA), and a 20 µL loop for sample injection, (c) an LC- 10ADVP pump equipped with a Shimadzu SCL-10ALVP System Controller, (d) an SPD-M10AVP photodiode array detector. Real time analysis monitoring and post run processing were carried out using the software Lab Solutions-LC solutions, supplied by Shimadzu. A glass vacuum filtration apparatus, acquired by Alltech Associates (Deereld, IL, USA), and nylon 0.2 μm membrane Filters (Alltech Associates, Chicago, IL, USA) were utilized for the filtration of the mobile phase, and a DGU-10B de-gassing unit with helium was used for degassing. A vortexer purchased from FALC Instruments (Treviglio (BG), Italy) was used for sample agitation. Centrifugation was carried out using a HermLe centrifuge, model Z-230 (B. HermLe, Gosheim, Germany). An ultrasonic bath (MRC: DC-150H) by MRC (Essex, UK) was used for sample preparation. For evaporation, after SPE extraction, a ReactiVap 9-port evaporator model 18,780 by Pierce (Rockford, IL, USA) was utilized. For sample filtration, prior to the injection in the chromatographic system, Q-Max RR syringe filters (0.45 μm nylon membrane) were purchased from Frisenette ApS (Knebel, Denmark).

3.3. Chromatographic Separation and Analysis

The chromatographic separation of the flavonoids was achieved on a C18 UniverSil column (250 mm × 4.6 mm, 5 μm), supplied by Fortis Technologies Ltd. (Neston, UK). A reverse-phase HPLC assay was carried out using a gradient system with 1 mL/min flow rate, thermostated at 30 °C. The mobile phase consisted of (A) 1% acetic acid in water, and (B) ACN. The gradient elution program begun with 80:20, *v/v* (A: B), gradually increasing to 50:50, *v/v* (A: B), in the following 25 min, and then remaining constant for the next 5 min. The initial conditions were restored for 10 min, prior to the next injection. The injection volume was 20 μL of solution and the total run time was less than 25 min for each injection. For peak identification, the R_{ts} of the peaks of the real samples were compared with the R_{ts} of the standard compounds, along with the spectral information provided by the DAD detector that operated over the range 280–400 nm. Peak monitoring and quantitation were performed at the maximum wavelength of each analyte.

3.4. Sample Collection

Twenty-four walnut septa samples were created after crushing walnuts in the laboratory using a wooden hammer, and carefully removing the walnut septa. Each walnut septum sample was a bulk sample that consisted of ten walnut septa originating from the same tree. In this way, eight bulk walnut septum samples were created in the laboratory for each variety (Chandler, Vina, Franquette). All the walnut samples were collected during the harvesting period of November 2020 from Kokkinogi, in Thessaly, Greece (40°2′2.62″ N 22°9′54.00″ E).

3.5. Sample Preparation

The samples were homogenized in a porcelain mortar and stored at −20 °C until analysis. For sample preparation, 50 mg of each homogenized bulk sample was weighted in 2-mL Eppendorf tubes, and then, 0.5 mL of 0.05% TFA in methanol: water at 60:40 ratio (*v/v*), was added [18]. The mixture was vortexed for 1 min and, then ultra-sound assisted extraction took place in an ultrasonic bath for 10 min at 25 °C. Each sample was centrifuged for 10 min at 10,000 rpm and, then, the supernatant was collected, according to Kalogiouri et al. [18]. The extract was further diluted with water at a final volume of 2 mL. The diluent was purified using a modified version of the SPE protocol proposed by Bajkacz et al. [41]. The LiChrolut RP-18 (C18, 3 mL, 500 mg) SPE cartridges were used for this purpose. First, the C18 column was conditioned with 2 mL MeOH, followed by 2 mL of water. Then, the diluted sample extract was passed through the sorbent at a flow rate of approximately 1 mL/min. The analytes were eluted with 3 mL MeOH and the eluates were evaporated to dryness with nitrogen. The residues were dissolved in 1 mL MeOH, and filtered through 0.22 μm nylon syringe filters. Finally, 20 μL was injected into the chromatographic system.

3.6. Method validation and Quantification

Linearity, selectivity, LODs and LOQs, within-day, and between-day accuracy and precision were evaluated. Linearity was examined by testing the lack-of-fit of the calibration calibration curves over the range 1–10 μg/g. The slopes, intercepts, and the determination coefficients of each analyte were calculated using last square linear regression analysis. LODs and LOQs were calculated by the equations:

$$LOD = 3.3 \times Sa/b \qquad (1)$$

$$LOQ = 10 \times Sa/b \qquad (2)$$

where, Sa is the standard error of the intercept α; and b is the slope of the calibration curve [42]. Accuracy was evaluated after spiking a bulk sample at 1, 5, and 10 mg/kg concentration level, and analysis was performed in triplicate. Accuracy was expressed as relative recovery, and precision was expressed as relative standard deviation (RSD%). Repeatability, expressed as within-day precision, was assessed in six replicates ($n = 6$), and reproducibility, expressed as between-days precision, was assessed after analyzing the spiked bulk samples within three consecutive days ($n = 3 \times 3$). The analytes were quantified using the corresponding calibration curves of the standards. For the quantification of the analytes with high concentrations that exceeded beyond the linear range, such as catechin, the extracts were further diluted with MeOH and re-injected in the chromatographic system to ensure that their calculated concentration was within the linear range of the curves.

3.7. Chemometric Analysis

The quantification results were processed with one-way analysis of variance (ANOVA), using the data analysis tool of Microsoft Excel (Microsoft, Redmond, WA, USA). ANOVA was applied to examine potential statistically significant differences among the flavonoids' concentrations and the different varieties of the walnut septa (Chandler, Vina, Franquette). A *p*-value was used for a confidence level of 95% to evaluate the results. If the *p*-value was calculated higher than 0.05, there was no statistically significant difference, and in the cases that the *p*-value was lower than 0.05, there was statistically significant difference among the samples. Principal Component Analysis (PCA) was employed to investigate the interrelationships among the determined flavonoids and the samples belonging to the different varieties (Chandler, Vina, and Franquette). PCA was created in R using the MetaboAnalystR package [43].

4. Conclusions

A UAE-SPE-HPLC-DAD analytical method was developed and validated to determine flavonoids in twenty-four walnut septa belonging to Chandler, Vina, and Franquette varieties cultivated in Greece. Overall, seven flavonoids were determined (catechin, rutin, myricetin, luteolin, quercetin, apigenin, and kaempferol), indicating that the walnut septum is rich in flavonoids, and it could be further exploited and utilized in the pharmaceutical industry, and in the food and feed industry, as well. The calculated concentrations of all the individual flavonoids were further analyzed with ANOVA, and the results indicated that there is statistically significant difference among the concentrations of catechin, rutin, myricetin, luteolin, quercetin, and the variety of the analyzed septa, demonstrating that the flavonoid content of the walnut byproduct is affected by the variety. PCA analysis was employed to investigate the similarities in the flavonoid content of the septa according to their variety, and the first two PCs explained the 70% of variance. Overall, the findings of this work suggest that the walnut septum is a promising raw ingredient with functional properties and several potential uses.

Supplementary Materials: The following are available online. Figure S1: PCA 3D score plot in color presenting pairwise correlation between PCs in the clustering between walnut septa belonging to Chandler, Vina, and Franquette variety; Figure S2: PCA loading plot showing the projection of the data set in PC1 × PC2 plane.

Author Contributions: Conceptualization, N.P.K. and V.F.S.; methodology, N.P.K.; validation, N.P.K.; formal analysis, N.P.K.; investigation, N.P.K.; resources, V.F.S.; data curation, N.P.K.; writing—original draft preparation, N.P.K.; writing—review and editing, V.F.S.; visualization, N.P.K.; supervision, V.F.S.; project administration, N.P.K. and V.F.S. All authors have read and agreed to the published version of the manuscript.

Funding: This research was co-financed by Greece and the European Union (European Social Fund, ESF) through the Operational Programme "Human Resources Development, Education and Lifelong Learning" in the context of the project "Reinforcement of Postdoctoral Researchers—2nd Cycle" (MIS5033021), implemented by the State Scholarships Foundation (IKY) with grant no. 2019-050-0503-17749.

Acknowledgments: The authors would like to thank Terra Rossa di Olimpo and Ioannis Tsirogiannis for supporting the sampling of walnuts originating from Kokkinogi in Thessaly, Greece used in this study.

Conflicts of Interest: The authors declare no conflict of interest.

Sample Availability: Samples of the compounds are available from the authors.

References

1. Kaderides, K.; Kyriakoudi, A.; Mourtzinos, I.; Goula, A.M. Potential of pomegranate peel extract as a natural additive in foods. *Trends Food Sci. Technol.* **2021**, *115*, 380–390. [CrossRef]
2. Martínez, M.L.; Mattea, M.A.; Maestri, D.M. Varietal and crop year effects on lipid composition of walnut (Juglans regia) genotypes. *JAOCS, J. Am. Oil Chem. Soc.* **2006**, *83*, 791–796. [CrossRef]
3. Abdallah, I.B.; Tlili, N.; Martinez-Force, E.; Rubio, A.G.P.; Perez-Camino, M.C.; Albouchi, A.; Boukhchina, S. Content of carotenoids, tocopherols, sterols, triterpenic and aliphatic alcohols, and volatile compounds in six walnuts (Juglans regia L.) varieties. *Food Chem.* **2015**, *173*, 972–978. [CrossRef]
4. Kalogiouri, N.P.; Manousi, N.; Rosenberg, E.; Zachariadis, G.A.; Samanidou, V.F. Advances in the Chromatographic Separation and Determination of Bioactive Compounds for Assessing the Nutrient Profile of Nuts. *Curr. Anal. Chem.* **2020**, *16*, 1–17.
5. Medic, A.; Jakopic, J.; Hudina, M.; Solar, A.; Veberic, R. Identification and quantification of the major phenolic constituents in Juglans regia L. peeled kernels and pellicles, using HPLC–MS/MS. *Food Chem.* **2021**, *352*, 129404. [CrossRef] [PubMed]
6. Slatnar, A.; Mikulic-Petkovsek, M.; Stampar, F.; Veberic, R.; Solar, A. Identification and quantification of phenolic compounds in kernels, oil and bagasse pellets of common walnut (Juglans regia L.). *Food Res. Int.* **2015**, *67*, 255–263. [CrossRef]
7. Pereira, J.A.; Oliveira, I.; Sousa, A.; Ferreira, I.C.F.R.; Bento, A.; Estevinho, L. Bioactive properties and chemical composition of six walnut (Juglans regia L.) cultivars. *Food Chem. Toxicol.* **2008**, *46*, 2103–2111. [CrossRef] [PubMed]
8. Medic, A.; Jakopic, J.; Solar, A.; Hudina, M.; Veberic, R. Walnut (J. regia) agro-residues as a rich source of phenolic compounds. *Biology (Basel)*. **2021**, *10*, 1–24.
9. Pereira, J.A.; Oliveira, I.; Sousa, A.; Valentão, P.; Andrade, P.B.; Ferreira, I.C.F.R.; Ferreres, F.; Bento, A.; Seabra, R.; Estevinho, L. Walnut (Juglans regia L.) leaves: Phenolic compounds, antibacterial activity and antioxidant potential of different cultivars. *Food Chem. Toxicol.* **2007**, *45*, 2287–2295. [CrossRef]
10. Liu, P.; Li, L.; Song, L.; Sun, X.; Yan, S.; Huang, W. Characterisation of phenolics in fruit septum of Juglans regia Linn. by ultra performance liquid chromatography coupled with Orbitrap mass spectrometer. *Food Chem.* **2019**, *286*, 669–677. [CrossRef]
11. Fizeșan, I.; Rusu, M.E.; Georgiu, C.; Pop, A.; Ștefan, M.G.; Muntean, D.M.; Mirel, S.; Vostinaru, O.; Kiss, B.; Popa, D.S. Antitussive, antioxidant, and anti-inflammatory effects of a walnut (Juglans regia L.) septum extract rich in bioactive compounds. *Antioxidants* **2021**, *10*, 119. [CrossRef] [PubMed]
12. Meng, Q.; Wang, Y.; Chen, F.; Xiao, T.; Zhang, L. Polysaccharides from Diaphragma juglandis fructus: Extraction optimization, antitumor, and immune-enhancement effects. *Int. J. Biol. Macromol.* **2018**, *115*, 835–845. [CrossRef] [PubMed]
13. Scarano, A.; Chieppa, M.; Santino, A. Looking at flavonoid biodiversity in horticultural crops: A colored mine with nutritional benefits. *Plants* **2018**, *7*, 98. [CrossRef] [PubMed]
14. Alara, O.R.; Abdurahman, N.H.; Ukaegbu, C.I. Soxhlet extraction of phenolic compounds from Vernonia cinerea leaves and its antioxidant activity. *J. Appl. Res. Med. Aromat. Plants* **2018**, *11*, 12–17. [CrossRef]
15. de Albuquerque Mendes, M.K.; dos Santos Oliveira, C.B.; Veras, M.D.A.; Araujo, B.Q.; Dantas, C.; Chaves, M.H.; Júnior, C.A.L.; Vieira, E.C. Application of multivariate optimization for the selective extraction of phenolic compounds in cashew nuts (Anacardium occidentale L.). *Talanta* **2019**, *205*, 120100. [CrossRef]

16. Manousi, N.; Raber, G.; Papadoyannis, I. Recent Advances in Microextraction Techniques of Antipsychotics in Biological Fluids Prior to Liquid Chromatography Analysis. *Separations* **2017**, *4*, 18. [CrossRef]
17. Bodoira, R.; Maestri, D. Phenolic Compounds from Nuts: Extraction, Chemical Profiles, and Bioactivity. *J. Agric. Food Chem.* **2020**, *68*, 927–942. [CrossRef] [PubMed]
18. Kalogiouri, N.P.; Samanidou, V.F. HPLC Fingerprints for the Characterization of Walnuts and the Detection of Fraudulent Incidents. *Foods* **2021**, *10*, 2145. [CrossRef]
19. Motilva, M.J.; Serra, A.; Macià, A. Analysis of food polyphenols by ultra high-performance liquid chromatography coupled to mass spectrometry: An overview. *J. Chromatogr. A* **2013**, *1292*, 66–82. [CrossRef]
20. Delgado-Zamarreño, M.M.; Fernández-Prieto, C.; Bustamante-Rangel, M.; Pérez-Martín, L. Determination of tocopherols and sitosterols in seeds and nuts by QuEChERS-liquid chromatography. *Food Chem.* **2016**, *192*, 825–830. [CrossRef] [PubMed]
21. Bolling, B.W.; Chen, C.Y.O.; McKay, D.L.; Blumberg, J.B. Tree nut phytochemicals: Composition, antioxidant capacity, bioactivity, impact factors. A systematic review of almonds, Brazils, cashews, hazelnuts, macadamias, pecans, pine nuts, pistachios and walnuts. *Nutr. Res. Rev.* **2011**, *24*, 244–275. [CrossRef] [PubMed]
22. Pyrzynska, K.; Sentkowska, A. Recent Developments in the HPLC Separation of Phenolic Food Compounds. *Crit. Rev. Anal. Chem.* **2015**, *45*, 41–51. [CrossRef]
23. Ghisoni, S.; Lucini, L.; Rocchetti, G.; Chiodelli, G.; Farinelli, D.; Tombesi, S.; Trevisan, M. Untargeted metabolomics with multivariate analysis to discriminate hazelnut (*Corylus avellana* L.) cultivars and their geographical origin. *J. Sci. Food Agric.* **2020**, *100*, 500–508. [CrossRef] [PubMed]
24. Bahadori, M.B.; Zengin, G.; Dinparast, L.; Eskandani, M. The health benefits of three Hedgenettle herbal teas (Stachys byzantina, Stachys inflata, and Stachys lavandulifolia) - profiling phenolic and antioxidant activities. *Eur. J. Integr. Med.* **2020**, *36*, 101134. [CrossRef]
25. Fusi, F.; Trezza, A.; Tramaglino, M.; Sgaragli, G.; Saponara, S.; Spiga, O. The beneficial health effects of flavonoids on the cardiovascular system: Focus on K+ channels. *Pharmacol. Res.* **2020**, *152*, 104625. [CrossRef]
26. Kritikou, E.; Kalogiouri, N.P.; Kolyvira, L.; Thomaidis, N.S. Target and Suspect HRMS Metabolomics for the Determination of Functional Ingredients in 13 Varieties of Olive Leaves and Drupes from Greece. *Molecules* **2020**, *25*, 4889. [CrossRef] [PubMed]
27. Babou, L.; Hadidi, L.; Grosso, C.; Zaidi, F.; Valentão, P.; Andrade, P.B. Study of phenolic composition and antioxidant activity of myrtle leaves and fruits as a function of maturation. *Eur. Food Res. Technol.* **2016**, *242*, 1447–1457. [CrossRef]
28. Zheng, J.; Lu, B.; Xu, B. An update on the health benefits promoted by edible flowers and involved mechanisms. *Food Chem.* **2021**, *340*, 127940. [CrossRef]
29. Alañón, M.E.; Pérez-Coello, M.S.; Marina, M.L. Wine science in the metabolomics era. *TrAC - Trends Anal. Chem.* **2015**, *74*, 1–20. [CrossRef]
30. Kalogiouri, N.P.; Samanidou, V.F. Liquid chromatographic methods coupled to chemometrics: A short review to present the key workflow for the investigation of wine phenolic composition as it is affected by environmental factors. *Environ. Sci. Pollut. Res.* **2020**. [CrossRef]
31. Arceusz, A.; Wesolowski, M.; Ulewicz-Magulska, B. Flavonoids and Phenolic Acids in Methanolic Extracts, Infusions and Tinctures from Commercial Samples of Lemon Balm. *Nat. Prod. Commun.* **2015**, *10*, 8–12. [CrossRef]
32. Martínez-Poveda, B.; Torres-Vargas, J.A.; Ocaña, M.D.C.; García-Caballero, M.; Medina, M.Á.; Quesada, A.R. The mediterranean diet, a rich source of angiopreventive compounds in cancer. *Nutrients* **2019**, *11*, 2036. [CrossRef] [PubMed]
33. Jegal, K.H.; Kim, E.O.; Kim, J.K.; Park, S.M.; Jung, D.H.; Lee, G.H.; Ki, S.H.; Byun, S.H.; Ku, S.K.; Cho, I.J.; et al. Luteolin prevents liver from tunicamycin-induced endoplasmic reticulum stress via nuclear factor erythroid 2-related factor 2-dependent sestrin 2 induction. *Toxicol. Appl. Pharmacol.* **2020**, *399*, 115036. [CrossRef]
34. Tamkutė, L.; Gil, B.M.; Carballido, J.R.; Pukalskienė, M.; Venskutonis, P.R. Effect of cranberry pomace extracts isolated by pressurized ethanol and water on the inhibition of food pathogenic/spoilage bacteria and the quality of pork products. *Food Res. Int.* **2019**, *120*, 38–51. [CrossRef]
35. Nour, V.; Trandafir, I.; Cosmulescu, S. Bioactive compounds, antioxidant activity and nutritional quality of different culinary aromatic herbs. *Not. Bot. Horti Agrobot. Cluj-Napoca* **2017**, *45*, 179–184. [CrossRef]
36. Srivastava, J.K.; Shankar, E.; Gupta, S. Chamomile: A herbal medicine of the past with a bright future (review). *Mol. Med. Rep.* **2010**, *3*, 895–901.
37. Dormán, G.; Flachner, B.; Hajdú, I.; András, C.D. Target identification and polypharmacology of nutraceuticals. *Nutraceuticals Effic. Saf. Toxic.* **2016**, 263–286.
38. Esteki, M.; Farajmand, B.; Amanifar, S.; Barkhordari, R.; Ahadiyan, Z.; Dashtaki, E.; Mohammadlou, M.; Vander Heyden, Y. Classification and authentication of Iranian walnuts according to their geographical origin based on gas chromatographic fatty acid fingerprint analysis using pattern recognition methods. *Chemom. Intell. Lab. Syst.* **2017**, *171*, 251–258. [CrossRef]
39. Vu, D.C.; Vo, P.H.; Coggeshall, M.V.; Lin, C.H. Identification and Characterization of Phenolic Compounds in Black Walnut Kernels. *J. Agric. Food Chem.* **2018**, *66*, 4503–4511. [CrossRef]
40. Jaćimović, V.; Adakalić, M.; Ercisli, S.; Božović, D.; Bujdoso, G. Fruit quality properties of walnut (Juglans regia l.) genetic resources in Montenegro. *Sustain.* **2020**, *12*, 1–19.
41. Bajkacz, S.; Baranowska, I.; Buszewski, B.; Kowalski, B.; Ligor, M. Determination of Flavonoids and Phenolic Acids in Plant Materials Using SLE-SPE-UHPLC-MS/MS Method. *Food Anal. Methods* **2018**, *11*, 3563–3575. [CrossRef]

42. Zakeri-Milani, P.; Islambulchilar, Z.; Majidpour, F.; Jannatabadi, E.; Lotfpour, F.; Valizadeh, H. A study on enhanced intestinal permeability of clarithromycin nanoparticles. *Brazilian J. Pharm. Sci.* **2014**, *50*, 121–129. [CrossRef]
43. Pang, Z.; Chong, J.; Zhou, G.; de Lima Morais, D.A.; Chang, L.; Barrette, M.; Gauthier, C.; Jacques, P.-É.; Li, S.; Xia, J. MetaboAnalyst 5.0: Narrowing the gap between raw spectra and functional insights. *Nucleic Acids Res.* **2021**, *49*, 388–396. [CrossRef]

Article

A Rapid HPLC-UV Protocol Coupled to Chemometric Analysis for the Determination of the Major Phenolic Constituents and Tocopherol Content in Almonds and the Discrimination of the Geographical Origin

Natasa P. Kalogiouri [1,*], Petros D. Mitsikaris [2], Dimitris Klaoudatos [3], Athanasios N. Papadopoulos [2] and Victoria F. Samanidou [1]

[1] Laboratory of Analytical Chemistry, Department of Chemistry, Aristotle University of Thessaloniki, 54124 Thessaloniki, Greece; samanidu@chem.auth.gr

[2] Laboratory of Chemical Biology, Department of Nutritional Sciences and Dietetics, International Hellenic University, Sindos, 57400 Thessaloniki, Greece; petrosmitsikaris@gmail.com (P.D.M.); papadnas@ihu.gr (A.N.P.)

[3] Laboratory of Oceanography, Department of Ichthyology and Aquatic Environment, School of Agricultural Sciences, University of Thessaly, 38446 Volos, Greece; dklaoud@uth.gr

* Correspondence: kalogiourin@chem.auth.gr

Abstract: Reversed phase-high-pressure liquid chromatographic methodologies equipped with UV detector (RP-HPLC-UV) were developed for the determination of phenolic compounds and tocopherols in almonds. Nineteen samples of *Texas* almonds originating from USA and Greece were analyzed and 7 phenolic acids, 7 flavonoids, and tocopherols ($-\alpha$, $-\beta + \gamma$) were determined. The analytical methodologies were validated and presented excellent linearity ($r^2 > 0.99$), high recoveries over the range between 83.1 (syringic acid) to 95.5% (ferulic acid) for within-day assay ($n = 6$), and between 90.2 (diosmin) to 103.4% (rosmarinic acid) for between-day assay ($n = 3 \times 3$), for phenolic compounds, and between 95.1 and 100.4% for within-day assay ($n = 6$), and between 93.2–96.2% for between-day assay ($n = 3 \times 3$) for tocopherols. The analytes were further quantified, and the results were analyzed by principal component analysis (PCA), and agglomerative hierarchical clustering (AHC) to investigate potential differences between the bioactive content of almonds and the geographical origin. A decision tree (DT) was developed for the prediction of the geographical origin of almonds proposing a characteristic marker with a concentration threshold, proving to be a promising and reliable tool for the guarantee of the authenticity of the almonds.

Keywords: almonds; HPLC; authenticity; PCA; tocopherols; phenolics

1. Introduction

The current trend in nutrition is following the Mediterranean diet, as it is considered one of the healthiest dietary patterns. Nuts are a highly nutritious food with unique taste and beneficial health properties deriving from their unique molecular composition. Popular tree nuts comprise almonds (*Prunus amygdalus* Batsch or *P. dulcis*), walnuts (*Juglans regia* L.), hazelnuts (*Corylus avellane* L.), and pistachios (*Pistachia vera* L.), among others. Almonds are one of the most popular and widely harvested culinary nuts in the world. Apart from their unique taste and texture, they have been proven to possess a wide variety of beneficial health properties deriving from their unique molecular composition. Thus, they are now considered as an important component of a healthy and highly nutritious diet [1–5].

Numerous studies have shown that various pharmacological activities can be attributed to regular consumption of almonds. A meta-analysis observed a significant reduction in LDL-C levels with almond consumption [6]. Additionally, a systematic review conducted by Kalita et al. [7] suggests that eating almonds leads to a significant reduction in total cholesterol, LDL-C, and triglycerides levels, whilst the impact on HDL-C levels is minor. In a randomized, controlled crossover study that took place over a time period of six weeks, individuals that consumed 45 g of almonds per day showed reduced LDL-C and non-HDL-C levels and, at the same time, maintained their HDL-C levels [8]. The study also demonstrated that almond intake reduced abdominal fat which is of very high significance, considering the fact that high amounts of abdominal fat are a major factor in metabolic syndrome. Furthermore, studies suggest that apart from reducing the risk of cardiovascular disease, almonds exhibit anti-inflammatory and anti-carcinogenic effects [9]. From these studies accrues the conclusion that almonds can be an effective diet tool in the process of trying to decrease an individual's cholesterol levels, hence reducing his risk of coming across any type of cardiovascular disease.

The beneficial health effects are mainly owed to their favorable phytochemical composition. Almonds are rich in bioactive constituents, mainly in phenolics and tocopherols. These compounds are defined as secondary plant metabolites and originate from carbohydrates through the shikimate and phenyl propanoid pathways [2,10]. Their chemical structure is characterized by one or more aromatic rings bearing at least one hydroxyl group [2]. Tocopherols are one of two subgroups that comprise vitamin E, with the other one being tocotrienols. Tocopherols are constituted by four derivatives: alpha, beta, gamma, and delta [11–13]. It is suggested that phenolic compounds that are found in almond skins act synergistically with vitamins C and E and protect the LDL particles from oxidation, resulting in the overall enhancement of the individual's antioxidant capacity [14].

Although polyphenols and tocopherols are ubiquitous in nuts, and particularly in almonds, their content, distribution and bioavailability vary depending on genetics, location, plant structure, pre- and post-harvest factors and climate conditions [15–17]. In this context, the analysis of almonds' phenolic content could provide useful information, making the evaluation process of different almond cultivars produced in different countries more accurate. The authentication process of various almond cultivars also contributes to the assessment of overall almond quality. However, traditional methods of doing so depend largely on environmental and production factors, making the differentiation between cultivars, geographical origin, and type of farming a difficult task to tackle [15,18–22]. Hence, the need arises to develop specific analytical methodologies and protocols that are applicable to a wide variety of nut types, with the end goal of differentiating them based on their phenolic content.

The determination of small bioactive molecules from food matrices involves the examination of several distinct aspects of the analytical methodology. Separation of phenolic compounds and tocopherols is mainly achieved with high pressure liquid chromatography (HPLC) coupled to UV [12,23,24], photodiode array (DAD) [25,26], or mass spectrometric (MS) detectors [27,28]. The most crucial step of the analytical methodology is sample preparation. Several laborious and time-consuming protocols have been proposed, suggesting the use of large volumes of organic solvents and Soxhlet-type apparatus [29,30]. The objective is to eliminate the use of organic solvents, minimize extraction times and select techniques that are suitable for the rapid determination of bioactive constituents [23]. The further processing of the results with chemometric tools increases the extensiveness of the analysis, enlightening the reliability of the conclusions derived from the experimental data. Data mining and the development of chemometric models are widely used in food authenticity studies for the investigation of several issues such as the discrimination of botanical origin, geographical origin, farming type, etc. [31–33].

The objective of this research was to develop two rapid HPLC-UV methodologies for the determination of the major phenolic compounds and tocopherols in almonds of the *Texas* variety originating from Greece and the USA. The quantification results were further analyzed with agglomerative hierarchical clustering (AHC) and principal component analysis (PCA) to investigate similarities between samples of the same geographical origin. A decision tree (DT) was developed for the classification of almonds, proving to be a promising and reliable tool for verifying the geographical origin on the basis of their phenolic profile and bioactive content.

2. Results

2.1. Analytical Performance

2.1.1. RP-HPLC-UV Method for the Determination of Phenolic Compounds

The analytical parameters of the HPLC-UV methodology for the determination of phenolic compounds, including the calibration curves, the linear range, the determination coefficients, the limits of detection (LODs) and limits of quantification (LOQs), precision and accuracy are summarized in Table S1. As it can be observed, the coefficients of determination ranged between 0.991 and 0.999, showing good linearity for all the phenolic analytes. The LOQs were found to range between 0.24 (rosmarinic acid) to 1.80 µg/g (diosmin), while the LODs were calculated equal to 0.08 (rosmarinic acid) to −0.60 µg/g (vanillin). The RSD% of the within-day ($n = 6$) and between-day assays ($n = 3 \times 3$) was lower than 6.1 and 10.3, respectively, presenting adequate precision. The accuracy was assessed by means of relative percentage of recovery (%R) at three concentration levels (0.5, 5, 10 µg/g) and ranged between 83.1 (syringic acid at 10 µg/g concentration level) to 95.5% (ferulic acid at 0.5 µg/g concentration level) for within-day assay ($n = 6$) (Table S2), and between 90.2 (diosmin) to 103.4% (rosmarinic acid) for between-day assay ($n = 3 \times 3$) (Table S3).

2.1.2. RP-HPLC-UV for the Determination of Tocopherols

The analytical parameters of the RP-HPLC-UV methodology for the determination of tocopherols are presented in Table S4. The LOQs were found to range between 0.36 (γ-tocopherol) to −0.99 µg/g (α-tocopherol), while the LODs were calculated equal to 0.12 (γ-tocopherol) to −0.33 µg/g (α-tocopherol). The RSD% of the within-day ($n = 6$) and between-day assays ($n = 3 \times 3$) was lower than 5.5 and 8.1, respectively, presenting adequate precision. The accuracy was assessed by means of relative percentage of recovery at three concentration levels (0.5, 5, 10 µg/g) and ranged between 95.1 and 100.4% for within-day assay ($n = 6$) (Table S5), and between 93.2–96.2% for between-day assay ($n = 3 \times 3$) (Table S6).

2.2. Real Samples' Application

2.2.1. Determination of Phenolic Compounds

Nineteen almond samples of the *Texas* variety from USA and Greece were analyzed. In total, fourteen phenolic compounds were determined. Gallic acid, ferulic acid, sinapic acid, rosmarinic acid, vanillic acid, p-coumaric, and caffeic acid were determined from the class of phenolic acids. Diosmin, catechin, epicatechin, quercetin, luteolin, apigenin, and kaempferol were determined from the class of flavonoids. A characteristic chromatogram, of a spiked sample at 5 µg/g is presented in the Supplementary Materials Figure S1. The retention times of the identified phenolic analytes are presented in Table S7. All samples were analyzed in triplicate and the concentration ranges as well as the mean values (±SD) are presented in Table 1.

The results are in accordance with Coric et al. [26] and Boiling [34]. Specifically, vanillic acid ranged between 1.37 to 4.25 µg/g in Greek almonds and between 1.03 to 2.23 µg/g in American almonds, similarly to Coric et al. [26] who reported a range of 0.38–2.84 µg/g. Caffeic acid ranged between 1.18 to 1.85 µg/g in Greek almonds and between 0.82 to 1.90 µg/g in American almonds, slightly higher concentrations compared to Coric et al. [26] who reported concentrations up to 1.48 µg/g. Sinapic acid ranged between 1.25 to 4.48 µg/g in Greek almonds, and between 1.02 to 3.65 µg/g in American almonds, correspondingly to Coric et al. [26] who reported concentrations up to 3.50 µg/g. Syringic acid was not detected in any of the samples, while p-coumaric acid was detected below the LOQ. Furthermore, rosmarinic acid ranged between 1.03 to 1.84 µg/g in Greek almonds and between 2.51 to 4.19 µg/g in American almonds. The detected concentrations of rosmarinic acid are higher than those reported previously by Keser et al. [35]. Significantly high concentrations up to 4.56 µg/g in Greek almonds and up to 1.81 µg/g in American almonds were detected for gallic acid, as well, compared to the literature [26,34,36].

Table 1. Phenolic compounds' quantification results in Greek and American almonds (samples analyzed in triplicate, $n = 3$).

Phenolic Compounds	Greek Almonds		American Almonds	
	Concentration Range (µg/g)	Mean Value (±SD, µg/g)	Concentration Range (µg/g)	Mean Value (±SD, µg/g)
apigenin	4.65–8.65	7.04 ± 1.35	LOQ–3.21	1.77 ± 0.13
caffeic acid	1.18–1.85	1.53 ± 0.12	0.82–1.90	1.35 ± 0.14
catechin	11.7–27.8	21.3 ± 1.35	14.5–25.2	20.2 ± 1.11
diosmin	LOQ–29.0	3.91 ± 0.08	4.32–15.6	8.06 ± 0.54
epicatechin	3.21–6.01	5.13 ± 2.39	1.01–2.21	1.25 ± 0.08
ferulic acid	1.65–2.98	2.25 ± 0.23	LOQ–1.45	1.06 ± 0.09
gallic acid	LOQ–4.56	3.40 ± 0.18	1.14–1.81	1.51 ± 0.12
kaempferol	2.19–3.21	2.55 ± 0.66	LOQ–2.87	1.30 ± 0.19
luteolin	LOQ–0.65	0.59 ± 0.06	<LOQ	–
p-coumaric acid	<LOQ	–	<LOQ	–
quercetin	LOQ–0.68	0.53 ± 0.04	<LOQ	–
rosmarinic acid	1.03–1.84	1.34 ± 0.25	2.51–4.19	3.56 ± 0.75
sinapic acid	1.25–4.48	3.26 ± 0.85	1.02–3.65	2.18 ± 0.63
syringic acid	–	–	–	–
vanillic acid	1.37–4.25	3.31 ± 0.31	1.03–2.23	1.36 ± 1.22

As far as flavonoids are concerned, catechin was the dominant phenolic compound with similar mean values of 21.3 µg/g for Greek and 20.2 µg/g for American, respectively. The second most abundant flavonoid was diosmin with a higher mean value of 8.06 µg/g in American almonds, compared to Greek almonds (3.91 µg/g). Higher concentrations of apigenin were detected in Greek almonds over the range 4.65 to 8.65 µg/g compared to American almonds (up to 3.21 µg/g). The mean concentration of luteolin in Greek almonds was found equal to 0.59 µg/g, while it was not detected in American almonds. Epicatechin ranged between 3.21–6.01 µg/g in Greek almonds and between 1.02 to 1.21 µg/g in American almonds. Kaempferol was detected in Greek almonds at a higher concentration with a mean value of 2.53 µg/g compared to 1.30 µg/g that was detected in American almonds, similarly to Coric et al. [26] who reported concentrations up to 2.63 µg/g. Finally, quercetin was detected at a mean concentration of 0.53 µg/g in Greek almonds and was not detected in American almonds, since according to the literature [28,34], the glucoside is mainly dominant in almonds and not its aglycone form.

2.2.2. Determination of Tocopherols

The separation of tocopherols was achieved within 15 min. The gradient elution program performed separation of α-tocopherol at (Rt = 9.2 min) and δ-tocopherol (Rt = 12.1 min), while β + γ tocopherols co-eluted (Rt = 10.6 min) and were analyzed as a sum according to Gliszczyńska-Świgło et al. [9]. A representative chromatogram of the 10 µg/g standard solution mixture is shown in the Supplementary Materials Figure S2. The analysis of tocopherols proved that almonds constitute a great source of α-tocopherol which ranged between 502 to 802 µg/g and between 221 to 326 µg/g in American almonds. γ-Tocopherol was measured as the sum of β- and γ-tocopherol, since these tocopherols are isomers and co-elute in RP chromatographic systems. δ-Tocopherol was not detected in any of the analyzed samples. All samples were analyzed in triplicate and quantification ranges and the mean values (±SD) are presented in Table 2.

Table 2. Tocopherols' quantification results in Greek and American almonds (samples analyzed in triplicate, $n = 3$).

Tocopherol	Greek Almonds		American Almonds	
	Concentration Range (µg/g)	Mean Value (±SD, µg/g)	Concentration Range (µg/g)	Mean Value (±SD, µg/g)
α-tocopherol	502–802	643 ± 31	221–326	267 ± 18
sum of β- and γ-tocopherol	13.6–89.3	72.3 ± 5.7	61.2–81.2	68.2 ± 6.4

2.3. Chemometric Analysis

The quantification results of the determined phenolic compounds and tocopherols (Sections 2.2.1 and 2.2.2) were further processed with chemometric tools to examine if the samples can be classified according to their phenolic composition and tocopherol content.

2.3.1. PCA

PCA was applied in the analysis of nineteen different samples of almonds originating from Greece and the USA. The data matrix consisted of sixteen features (quantification results of phenolics and tocopherols) and was normalized using the auto-scaling function of the MetaboAnalyst package [37]. Figure 1 presents the score plot and the clustering of the almonds into two individual groups, according to the geographical origin. Almonds originating from Greece are marked in red and almonds originating from the USA are marked in green. The first two Principal Components (PCs) explained 66.8% of the variance, presenting appropriate groups of samples of the same variety and geographical origin. The PCA biplot in Figure S3 presents the influence of the variables in each PC.

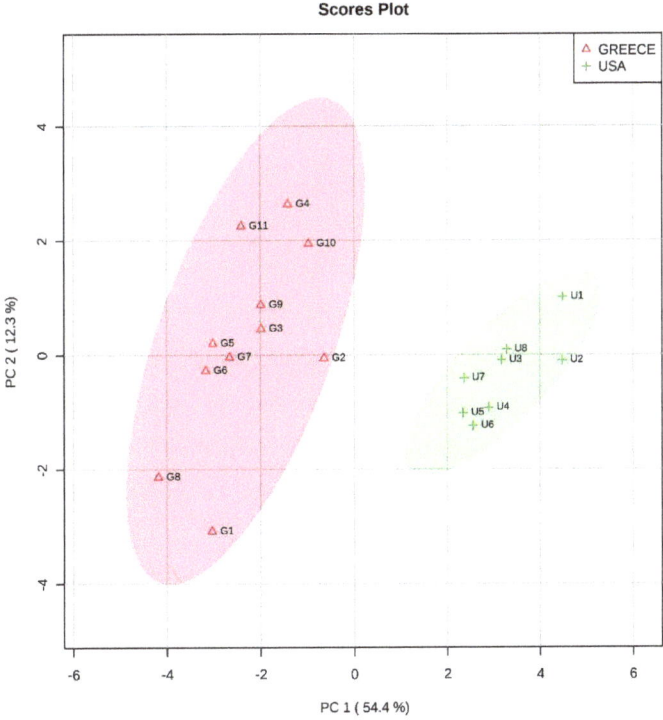

Figure 1. PCA Score plot showing the pairwise correlation between PCs in the clustering between Greek and American almonds.

2.3.2. Agglomerative Hierarchical Clustering

Cluster analysis was employed to divide the matrix into homogeneous groups measuring the distance between each pair of objects and without previous knowledge about the structure of the groups. A tree diagram was built with AHC to identify the groups that present high similarity. Each object is considered a singleton cluster (leaf) by the algorithm. Subsequently, the pairs of clusters are merged until all of them end up into a large cluster that contains all the objects [38], resulting in a tree-based representation, the so-called dendrogram.

Figure 2 presents the dendrogram of the eleven Greek and eight American almonds' clustering in two major groups according to the place of origin.

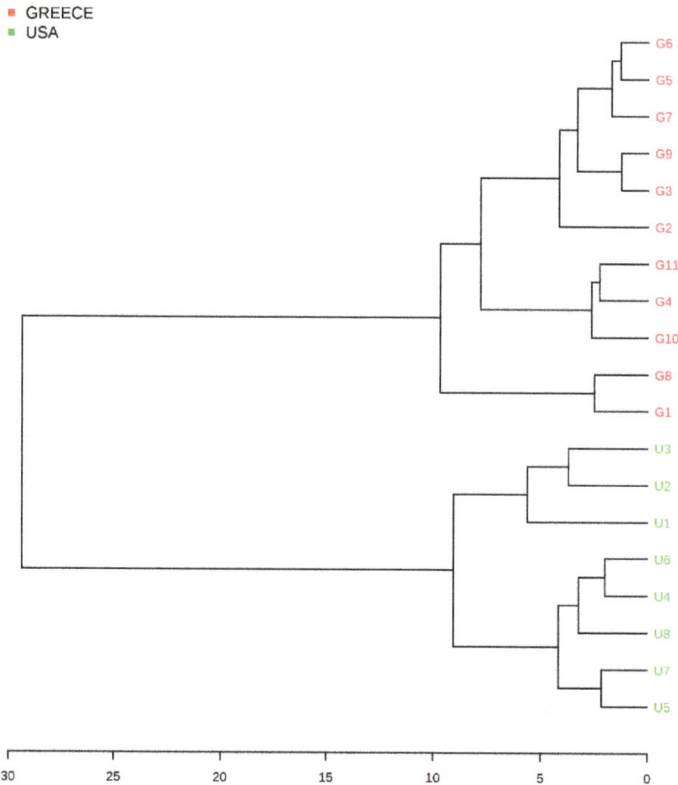

Figure 2. HCA dendrogram of Greek (G1–G11) and American (U1–U8) almonds' clustering in two major groups.

The heatmap in Figure 3 presents the data matrix showing pairwise correlations between the Greek (G1–G11) and American almonds (U1–U8). Each one of the colored cells corresponds to a concentration value; the samples are represented in the columns and the compounds in the rows.

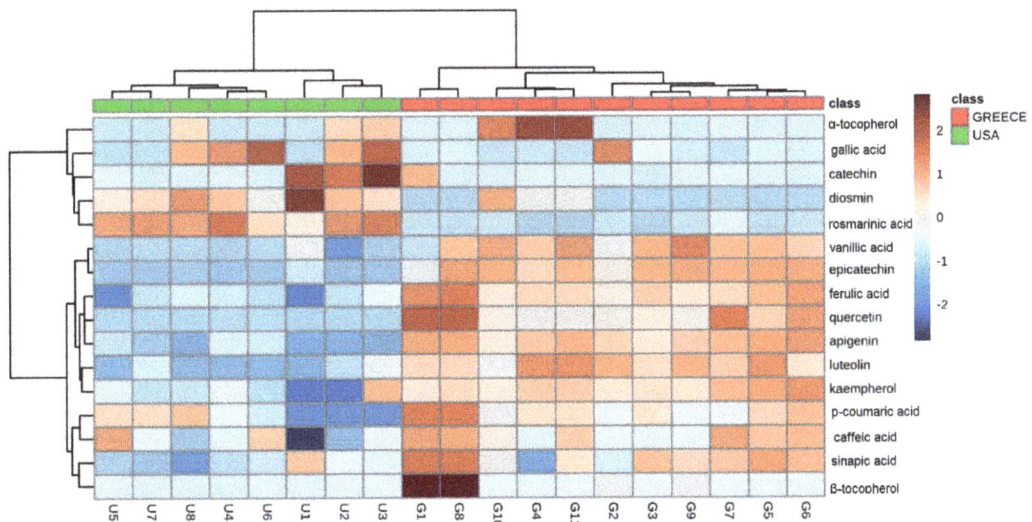

Figure 3. Clustered image map acquired by HCA dendrogram showing pairwise correlation between almonds produced in USA (U1–U8, in green color on the left-hand side of the heatmap) and Greece (G1–G11, in red color on the right-hand of the heatmap). The red blocks of the heatmap indicate positive correlations, whereas the blue blocks indicate negative correlations for the clustering of the samples. The lighter shades indicate smaller correlation values.

2.3.3. Decision Tree

The DT algorithm was built to develop a prediction model by splitting the data repeatedly into two discrete subsets according to the numerical value (i.e., concentration threshold) of the selected explanatory variable. The model selects the most significant variable that minimizes the model's total error. The initial dataset was split into a training and a test set. Twelve samples were used as training set and seven as test set. The developed DT suggested that ferulic acid could be used as a characteristic marker for the discrimination between Greek and American almonds and succeeded in classifying the samples with zero error, resulting in two terminal nodes and setting the concentration threshold of 1.54 µg/g. The developed DT was validated with a receiver operating characteristics (ROC) plot for each class of almonds with 1-specificity and zero error (Figure S4).

According to Figure 4, almonds with calculated concentrations lower or equal to 1.54 µg/g were produced in the USA, while those with higher concentrations than 1.54 µg/g were produced in Greece.

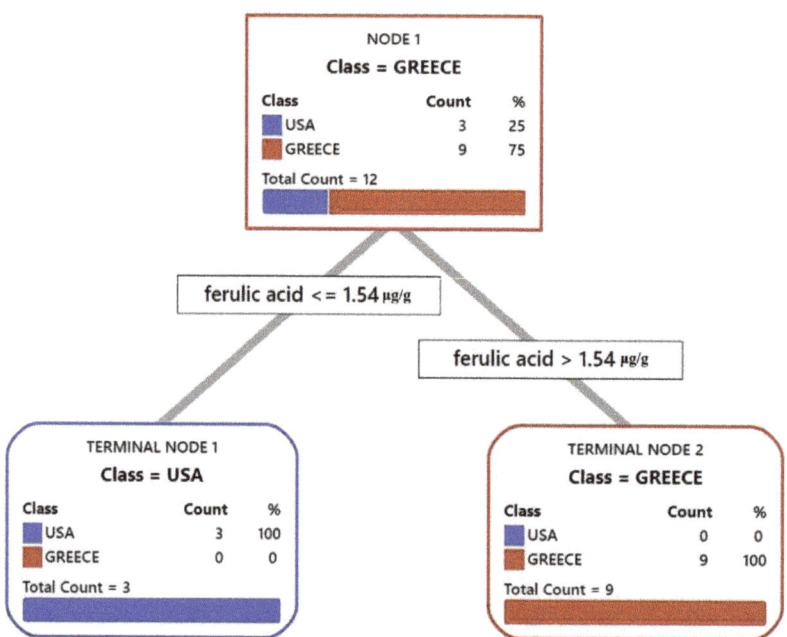

Figure 4. Optimal decision tree diagram using predictive analysis for phenolic and tocopherol concentration of almond samples originating from Greece and USA.

3. Materials and Methods

3.1. Chemicals and Standards

Acetonitrile (ACN) and 2-propanol (IPA), HPLC grade, were purchased by Panreac—AppliChem (Darmstadt, Germany). Methanol (MeOH), HPLC grade, was acquired by Carl Roth (Carlsruhe, Germany). Hexane reagent grade 99% and acetic acid 99% were purchased by Sigma-Aldrich (Steinheim, Germany). Ultrapure water was provided by a Milli-Q purification system (Millipore, Bedford, MA, USA).

Sinapic acid 95%, gallic acid 98%, ferulic acid 98%, rosmarinic acid 98%, catechin 98%, epicatechin 97%, p-coumaric 98%, quercetin 98%, diosmin 97%, kaempferol 97%, vanillic acid 97%, and caffeic acid 98% were purchased from Sigma-Aldrich (Steinheim, Germany). Luteolin 98% was acquired from Santa Cruz Biotechnologies. Apigenin 97% was purchased from Alfa Aesar (Karlsruhe, Germany). α-Tocopherol 96%, β-tocopherol 96%, γ-tocopherol 96%, and δ-tocopherol 96% were purchased by Sigma-Aldrich (Steinheim, Germany). Stock standard solutions of each analyte (1000 mg/L) were solubilized in methanol and stored at $-20\ ^\circ$C in dark brown glass bottles.

3.2. Collection of Samples

Eleven Greek almond samples belonging to the variety *Texas* were acquired from different producers, originating from different territories around Greece (Evia, Trikala, Vergina, Katerini, Adendro, Elassona, Mouzaki, Aridaia, Veroia, Drama, Larissa), and eight almond samples of the *Texas* variety originating from California and available in the Greek market were acquired from eight different traders.

3.3. Instrumentation

The chromatographic analysis of the analytes was performed in an Agilent (Santa Clara, CA, USA) 1220 Infinity HPLC-UV, using gradient elution methods. The HPLC system consisted of the following: manual injector, column oven, degasser, and lastly, a UV

Detector. In order to monitor the analysis, the Agilent Open Lab software and the package Method and Run Control were used. For data processing, the Data Analysis software package was used to identify and integrate the peaks. A glass vacuum-filtration apparatus, produced by Alltech Associates (Deerfeld, IL, USA), in combination with cellulose nitrate 0.22 µm nylon filters (Whatman Laboratory Division, Maidstone, UK) were utilized for the filtration of the aqueous and organic phase, respectively. QMax RR syringe filters (0.22 µm nylon membrane) were purchased from Frisenette ApS (Knebel, Denmark) and used for filtering the real samples prior to analysis. An ultrasonic bath (MRC: DC-150H) by MRC (Essex, UK) was utilized to remove the template from the MIP as well as for sample preparation. A vortex mixer from VELP Scientifica (Usmate Velate, Italy) was used for the agitation of the samples. A centrifuge system 3–16PK by Sigma (Osterode am Harz, Germany) was operated for centrifugation.

3.4. Chromatographic Conditions

A Nucleosil RP-18 analytical column (250 mm × 4.6 mm, 5 µm particle size), supplied by Macherey–Nagel (Düren, Nordrhein-Westfalen, Germany) was used for the analysis of polyphenols at 280 nm. The mobile phase consisted of a mixture of acetic acid in ultrapure water 2% v/v (A) and acetic acid in ACN 0.5% v/v (B). The system operated in a gradient mode at 28 °C. At the beginning of the analysis, the mixture was 100% A, gradually dropping to 20% A at the 60 min mark. The flow rate was set at 1 mL/min. Each chromatographic run lasted for 60 min. The peaks were identified by comparing the retention time of the standard compound with the peaks detected in real samples.

A Kromasil RP-18 analytical column (125 mm × 4.6 mm, 5 µm particle size), purchased from Macherey–Nagel, was used for the analysis of tocopherols at 295 nm. The mobile phase consisted of a mixture of methanol (A) and ACN (B). The system operated in a gradient mode at 28 °C. At the beginning of the analysis, the mixture was stable at 50% A and at the 7 min mark, it gradually increased to 100% A until the 12 min mark. The flow rate was set at 1 mL/min. Each chromatographic run lasted for 15 min. The peaks were identified by comparing the retention time of the standard compound with the peaks detected in real samples.

3.5. Sample Preparation

For the extraction of the phenolic compounds, a modified extraction protocol was applied, as previously suggested by Kritikou et al. [39]. In brief, 0.5 g of each homogenized sample was weighted in 2 mL Eppendorf tubes and 1 mL of MeOH: H_2O (80:20, v/v) was added. The samples were vortexed for 1 min and then they were centrifuged for 5 min at 8000 rpm. The extract was collected and filtered through 0.45 µm nylon filters and 20 µL were injected in the chromatographic system. As for the extraction of tocopherols, 1 g of homogenized samples were weighted in 15 mL falcon tubes and 10 mL of hexane were added to extract the lipid fraction. The samples were vortexed for 1 min and they were then placed in an ultra-sound bath at 40 °C for 10 min. In a further step, the falcon tubes were centrifuged at 8000 rpm for 10 min. The organic layer was transferred and evaporated in a rotary evaporator under vacuum. The almond oil product was collected and stored in dark brown vials at −20 °C. Prior to analysis, 20 mg of oil was weighed and dissolved in 500 µL 2-propanol, according to Martakos et al. [40]. The mixture was filtered through nylon 0.45 µm syringe filters and an aliquot of 20 µL was injected into the HPLC system.

3.6. Method Validation

Method validation was performed for both methodologies to estimate selectivity, linearity, LODs and limits of quantification (LOQs), within-day, and between-day accuracy and precision, respectively. Linearity studies were performed in triplicate and covered the working range of 0.5–20 µg/g which was selected for the phenolic compounds, and the working range of 5–50 µg/g was selected for tocopherols. Linearity was assessed by constructing calibration curves for each analyte using standard solutions. Eight point calibration curves were constructed by plotting the peak area versus concentration. LODs and LOQs were calculated on the basis of the S/N of the analyte until an S/N ratio of 3:1 (LOD) and 10:1 (LOQ) was reached. [41].

Accuracy and precision were studied for both methods using a pool sample spiked at three different concentrations: 0.5 µg/g—10 µg/g—20 µg/g for phenolics and, 5 µg/g—25 µg/g—50 µg/g for tocopherols, all analyzed in triplicate. Relative recoveries (R%) were calculated by means of recovery percentage, by comparing the found and added concentrations of the examined analytes (mean concentration found/concentration*100, R%), expressing accuracy. The precision of the method was expressed in terms of relative standard deviation (RSD%). Following this approach, within-day precision (repeatability) was assessed in six replicates (n = 6), while between-day precision (reproducibility) were assessed by performing triplicate analysis for spiked samples within three consecutive days (n = 3 × 3) [41]. Five blank matrices were used to assess selectivity and no interferences were observed in the same chromatographic window for both methodologies.

3.7. Chemometric Analysis

PCA was used as a mathematical tool to represent the variation in the dataset of nineteen samples and sixteen features (phenolic compounds and tocopherols). PCA is an unsupervised chemometric method used for exploratory data analysis [42]. PCA selects the most important components to reduce data dimension and retain the variation of the data with the Principal Components (PCs) which are linear combinations of the variables of the dataset. The first PC explains the largest variance, the second PC presents the second largest variance, and so on [43]. HCA was also used to represent and visualize the classes of almonds, explore the similarities of the analyzed samples, and discover patterns among them [44]. A DT was developed in an attempt to discover patterns in the quantitative data and predict the geographical origin of the analyzed samples by assigning a numerical value. PCA and HCA were created in R using the MetaboAnalyst package [37]. The DT was created in Minitab 19 software (Minitab, PA, USA).

4. Conclusions

This work presents an innovative approach for assessing the bioactive content of almonds with the development of two RP-HPLC-UV methodologies for the determination of phenolic compounds and tocopherols, respectively. Nineteen samples of almonds originating from USA and Greece were analyzed, and gallic acid, ferulic acid, sinapic acid, caffeic acid, vanillic acid, p-coumaric acid, and rosmarinic acid were determined from the class of phenolic acids. Catechin, epicatechin, diosmin, quercetin, apigenin, luteolin, and kaempferol were determined from the class of flavonoids. Furthermore, from the group of tocopherols, α-tocopherol and the sum of (β + γ)-tocopherols were determined as well. The quantification results were further processed with chemometrics. PCA analysis quantitatively showed the distribution of the almonds on the score plot and the clear formation of two separate groups on the basis of their geographical origin (Greece or USA), with the first two PCs explaining the 66.7% of variance. An HCA dendrogram was built, as well, showing the clustering of two major groups according to the origin of production. Finally, a DT was developed for the prediction of the country of origin suggesting ferulic acid as a characteristic marker and proposing a concentration value of 1.54 µg/g.

The findings of this research have made progress towards the characterization of almonds that belong to the *Texas* variety, showing that the geographical origin affects the phenolic composition and tocopherol content, as well as showing that these bioactive constituents could be used for the authentication of almonds that are commercially available in the Greek market.

Supplementary Materials: The following are available online, Figure S1: Characteristic chromatogram of an almond sample spiked at 5 µg/g. Figure S2: Characteristic chromatogram of a standard mixture of tocopherols at 10 µg/g. Figure S3: PCA biplot presenting the projection of the data set in PC1 and PC2. The red vectors show the influence to each PC (atoc: α-tocopherol; bctoc: β

+ γ-tocopherol). Figure S4: Prediction model performance characteristics. Table S1: Chromatographic retention times of the phenolic compounds determined in almonds. Table S2: Recoveries (%R) for the evaluation of repeatability. Table S3: Recoveries (%R) for the evaluation of intermediate precision. Table S4: Recoveries (%R) for the evaluation of repeatability. Table S5: Recoveries (%R) for the evaluation of repeatability. Table S6: Recoveries (%R) for the evaluation of intermediate precision. Table S7. Chromatographic retention times of the phenolic compounds determined in almonds.

Author Contributions: Conceptualization, N.P.K.; methodology, N.P.K.; validation, N.P.K.; formal analysis, N.P.K. and P.D.M.; investigation, N.P.K., P.D.M. and D.K.; resources, A.N.P.; data curation, N.P.K. and P.D.M.; writing—original draft preparation, N.P.K. and P.D.M.; writing—review and editing, N.P.K., D.K., A.N.P. and V.F.S.; visualization, N.P.K. and D.K.; supervision, A.N.P. and V.F.S.; project administration, N.P.K. and V.F.S. All authors have read and agreed to the published version of the manuscript.

Funding: This research was co-financed by Greece and the European Union (European Social Fund, ESF) through the Operational Program "Human Resources Development, Education and Lifelong Learning" in the context of the project "Reinforcement of Postdoctoral Researchers—2nd Cycle" (MIS-5033021), implemented by the State Scholarships Foundation (IKY) with grant number 2019-050-0503-17749.

Data Availability Statement: Not applicable.

Conflicts of Interest: The authors declare no conflict of interest.

Sample Availability: Samples of the compounds are available from the authors.

References

1. Maguire, L.S.; O'Sullivan, S.M.; Galvin, K.; O'Connor, T.P.; O'Brien, N.M. Fatty acid profile, tocopherol, squalene and phytosterol content of walnuts, almonds, peanuts, hazelnuts and the macadamia nut. *Int. J. Food Sci. Nutr.* **2004**, *55*, 171–178. [CrossRef] [PubMed]
2. Kalogiouri, N.P.; Manousi, N.; Rosenberg, E.; Zachariadis, G.A.; Samanidou, V.F. Advances in the Chromatographic Separation and Determination of Bioactive Compounds for Assessing the Nutrient Profile of Nuts. *Curr. Anal. Chem.* **2020**, *16*, 1–17. [CrossRef]
3. Ryan, E.; Galvin, K.; O'Connor, T.P.; Maguire, A.R.; O'Brien, N.M. Fatty acid profile, tocopherol, squalene and phytosterol content of brazil, pecan, pine, pistachio and cashew nuts. *Int. J. Food Sci. Nutr.* **2006**, *57*, 219–228. [CrossRef]
4. Cardoso, B.R.; Duarte, G.B.S.; Reis, B.Z.; Cozzolino, S.M.F. Brazil nuts: Nutritional composition, health benefits and safety aspects. *Food Res. Int.* **2017**, *100*, 9–18. [CrossRef]
5. Rodríguez-Bencomo, J.J.; Kelebek, H.; Sonmezdag, A.S.; Rodríguez-Alcalá, L.M.; Fontecha, J.; Selli, S. Characterization of the Aroma-Active, Phenolic, and Lipid Profiles of the Pistachio (*Pistacia vera* L.) Nut as Affected by the Single and Double Roasting Process. *J. Agric. Food Chem.* **2015**, *63*, 7830–7839. [CrossRef]
6. Lee-Bravatti, M.A.; Wang, J.; Avendano, E.E.; King, L.; Johnson, E.J.; Raman, G. Almond Consumption and Risk Factors for Cardiovascular Disease: A Systematic Review and Meta-analysis of Randomized Controlled Trials. *Adv. Nutr.* **2019**, *10*, 1076–1088. [CrossRef]
7. Kalita, S.; Khandelwal, S.; Madan, J.; Pandya, H.; Sesikeran, B.; Krishnaswamy, K. Almonds and cardiovascular health: A review. *Nutrients* **2018**, *10*, 468. [CrossRef]
8. Dreher, M.L. A comprehensive review of almond clinical trials on weight measures, metabolic health biomarkers and outcomes, and the gut microbiota. *Nutrients* **2021**, *13*, 1968. [CrossRef]
9. Rajaram, S.; Connell, K.M.; Sabaté, J. Effect of almond-enriched high-monounsaturated fat diet on selected markers of inflammation: A randomised, controlled, crossover study. *Br. J. Nutr.* **2010**, *103*, 907–912. [CrossRef]
10. Marchiosi, R.; dos Santos, W.D.; Constantin, R.P.; de Lima, R.B.; Soares, A.R.; Finger-Teixeira, A.; Mota, T.R.; de Oliveira, D.M.; de Paiva Foletto-Felipe, M.; Abrahão, J.; et al. Biosynthesis and Metabolic Actions of Simple Phenolic Acids in Plants. *Phytochem. Rev.* **2020**, *19*, 865–906. [CrossRef]
11. Kornsteiner, M.; Wagner, K.H.; Elmadfa, I. Tocopherols and total phenolics in 10 different nut types. *Food Chem.* **2006**, *98*, 381–387. [CrossRef]
12. Hejtmánková, A.; Táborský, J.; Kudelová, V.; Kratochvílová, K. Contents of tocols in different types of dry shell fruits. *Agron. Res.* **2018**, *16*, 1373–1382. [CrossRef]
13. Delgado-Zamarreño, M.M.; Fernández-Prieto, C.; Bustamante-Rangel, M.; Pérez-Martín, L. Determination of tocopherols and sitosterols in seeds and nuts by QuEChERS-liquid chromatography. *Food Chem.* **2016**, *192*, 825–830. [CrossRef] [PubMed]
14. Che, C.Y.; Milbury, P.E.; Lapsley, K.; Blumberg, J.B. Flavonoids from almond skins are bioavailable and act synergistically with vitamins C and E to enhance hamster and human LDL resistance to oxidation. *J. Nutr.* **2005**, *135*, 1366–1373. [CrossRef]

15. Ghisoni, S.; Lucini, L.; Rocchetti, G.; Chiodelli, G.; Farinelli, D.; Tombesi, S.; Trevisan, M. Untargeted metabolomics with multivariate analysis to discriminate hazelnut (*Corylus avellana* L.) cultivars and their geographical origin. *J. Sci. Food Agric.* **2020**, *100*, 500–508. [CrossRef]
16. Locatelli, M.; Coïsson, J.D.; Travaglia, F.; Cereti, E.; Garino, C.; D'Andrea, M.; Martelli, A.; Arlorio, M. Chemotype and genotype chemometrical evaluation applied to authentication and traceability of "tonda Gentile Trilobata" hazelnuts from Piedmont (Italy). *Food Chem.* **2011**, *129*, 1865–1873. [CrossRef]
17. Bolling, B.W.; Chen, C.Y.O.; McKay, D.L.; Blumberg, J.B. Tree nut phytochemicals: Composition, antioxidant capacity, bioactivity, impact factors. A systematic review of almonds, Brazils, cashews, hazelnuts, macadamias, pecans, pine nuts, pistachios and walnuts. *Nutr. Res. Rev.* **2011**, *24*, 244–275. [CrossRef]
18. Kalogiouri, N.P.; Manousi, N.; Klaoudatos, D.; Spanos, T.; Topi, V.; Zachariadis, G.A. Rare Earths as Authenticity Markers for the Discrimination of Greek and Turkish Pistachios Using Elemental Metabolomics and Chemometrics. *Foods* **2021**, *10*, 349. [CrossRef]
19. Kalogiouri, N.P.; Manousi, N.; Rosenberg, E.; Zachariadis, G.A.; Paraskevopoulou, A.; Samanidou, V. Exploring the volatile metabolome of conventional and organic walnut oils by solid-phase microextraction and analysis by GC-MS combined with chemometrics. *Food Chem.* **2021**, *363*, 130331. [CrossRef] [PubMed]
20. Esteki, M.; Farajmand, B.; Amanifar, S.; Barkhordari, R.; Ahadiyan, Z.; Dashtaki, E.; Mohammadlou, M.; Vander Heyden, Y. Classification and authentication of Iranian walnuts according to their geographical origin based on gas chromatographic fatty acid fingerprint analysis using pattern recognition methods. *Chemom. Intell. Lab. Syst.* **2017**, *171*, 251–258. [CrossRef]
21. Kalogiouri, N.P.; Aalizadeh, R.; Thomaidis, N.S. Investigating the organic and conventional production type of olive oil with target and suspect screening by LC-QTOF-MS, a novel semi-quantification method using chemical similarity and advanced chemometrics. *Anal. Bioanal. Chem.* **2017**, *409*, 5413–5426. [CrossRef]
22. Alasalvar, C.; Shahidi, F.; Cadwallader, K.R. Comparison of natural and roasted Turkish Tombul hazelnut (*Corylus avellana* L.) volatiles and flavor by DHA/GC/MS and descriptive sensory analysis. *J. Agric. Food Chem.* **2003**, *51*, 5067–5072. [CrossRef] [PubMed]
23. Kalogiouri, N.; Samanidou, V. Advances in the Optimization of Chromatographic Conditions for the Separation of Antioxidants in Functional Foods. *Rev. Sep. Sci.* **2019**, *1*, 17–33. [CrossRef]
24. Mandalari, G.; Tomaino, A.; Arcoraci, T.; Martorana, M.; Turco, V.L.; Cacciola, F.; Rich, G.T.; Bisignano, C.; Saija, A.; Dugo, P.; et al. Characterization of polyphenols, lipids and dietary fibre from almond skins (*Amygdalus communis* L.). *J. Food Compos. Anal.* **2010**, *23*, 166–174. [CrossRef]
25. Fuentealba, C.; Hernández, I.; Saa, S.; Toledo, L.; Burdiles, P.; Chirinos, R.; Campos, D.; Brown, P.; Pedreschi, R. Colour and in vitro quality attributes of walnuts from different growing conditions correlate with key precursors of primary and secondary metabolism. *Food Chem.* **2017**, *232*, 664–672. [CrossRef]
26. Čolić, S.D.; Akšić, M.M.F.; Lazarević, K.B.; Zec, G.N.; Gašić, U.M.; Zagorac, D.D.; Natić, M.M. Fatty acid and phenolic profiles of almond grown in Serbia. *Food Chem.* **2017**, *234*, 455–463. [CrossRef]
27. Medic, A.; Jakopic, J.; Hudina, M.; Solar, A.; Veberic, R. Identification and quantification of the major phenolic constituents in *Juglans regia* L. peeled kernels and pellicles, using HPLC–MS/MS. *Food Chem.* **2021**, *352*, 129404. [CrossRef] [PubMed]
28. Bolling, B.W.; Dolnikowski, G.; Blumberg, J.B.; Chen, C.Y.O. Polyphenol content and antioxidant activity of California almonds depend on cultivar and harvest year. *Food Chem.* **2010**, *122*, 819–825. [CrossRef]
29. de Albuquerque Mendes, M.K.; dos Santos Oliveira, C.B.; Veras, M.D.A.; Araújo, B.Q.; Dantas, C.; Chaves, M.H.; Júnior, C.A.L.; Vieira, E.C. Application of multivariate optimization for the selective extraction of phenolic compounds in cashew nuts (*Anacardium occidentale* L.). *Talanta* **2019**, *205*, 120100. [CrossRef] [PubMed]
30. Bodoira, R.; Maestri, D. Phenolic Compounds from Nuts: Extraction, Chemical Profiles, and Bioactivity. *J. Agric. Food Chem.* **2020**, *68*, 927–942. [CrossRef]
31. Wang, Y.-P.; Zou, Y.-R.; Shi, J.-T.; Shi, J. Review of the chemometrics application in oil-oil and oil-source rock correlations. *J. Nat. Gas. Geosci.* **2018**, *3*, 217–232. [CrossRef]
32. Jurado-Campos, N.; García-Nicolás, M.; Pastor-Belda, M.; Bußmann, T.; Arroyo-Manzanares, N.; Jiménez, B.; Viñas, P.; Arce, L. Exploration of the potential of different analytical techniques to authenticate organic vs. conventional olives and olive oils from two varieties using untargeted fingerprinting approaches. *Food Control* **2021**, *124*, 107828. [CrossRef]
33. Li Vigni, M.; Durante, C.; Cocchi, M. *Exploratory Data Analysis*, 1st ed.; Elsevier: Amsterdam, The Netherlands, 2013; Volume 28, ISBN 9780444595287.
34. Bolling, B.W. Almond Polyphenols: Methods of Analysis, Contribution to Food Quality, and Health Promotion. *Compr. Rev. Food Sci. Food Saf.* **2017**, *16*, 346–368. [CrossRef]
35. Keser, S.; Demir, E.; Yilmaz, O. Some bioactive compounds and antioxidant activities of the bitter almond kernel (*Prunus dulcis* var. amara). *J. Chem. Soc. Pakistan* **2015**, *36*, 922–930.
36. Yildirim, A.N.; Yildirim, F.; Şan, B.; Polat, M.; Sesli, Y. Variability of phenolic composition and tocopherol content of some commercial Almond cultivars. *J. Appl. Bot. Food Qual.* **2016**, *89*, 163–170. [CrossRef]
37. Pang, Z.; Chong, J.; Zhou, G.; De Lima Morais, D.A.; Chang, L.; Barrette, M.; Gauthier, C.; Jacques, P.É.; Li, S.; Xia, J. MetaboAnalyst 5.0: Narrowing the gap between raw spectra and functional insights. *Nucleic Acids Res.* **2021**, *49*, W388–W396. [CrossRef] [PubMed]

38. Wilks, D.S. Cluster Analysis. *Int. Geophys.* **2011**, *100*, 603–616. [CrossRef]
39. Kritikou, E.; Kalogiouri, N.P.; Kolyvira, L.; Thomaidis, N.S. Target and Suspect HRMS Metabolomics for the 13 Varieties of Olive Leaves and Drupes from Greece. *Molecules* **2020**, *25*, 4889. [CrossRef] [PubMed]
40. Martakos, I.; Kostakis, M.; Dasenaki, M.; Pentogennis, M.; Thomaidis, N.S. Simultaneous Determination of Pigments, Tocopherols, and Squalene in Greek Olive Oils: A Study of the Influence of Cultivation and Oil-Production Parameters. *Molecules* **2019**, *9*, 31. [CrossRef]
41. Samanidou, V.F.; Nikolaidou, K.I.; Papadoyannis, I.N. Development and validation of an HPLC confirmatory method for the determination of tetracycline antibiotics residues in bovine muscle according to the European Union regulation 2002/657/EC. *J. Sep. Sci.* **2005**, *28*, 2247–2258. [CrossRef]
42. Kalogiouri, N.P.; Samanidou, V.F. Liquid chromatographic methods coupled to chemometrics: A short review to present the key workflow for the investigation of wine phenolic composition as it is affected by environmental factors. *Environ. Sci. Pollut. Res.* **2020**. [CrossRef] [PubMed]
43. Jollife, I.T.; Cadima, J. Principal component analysis: A review and recent developments. *Philos. Trans. R. Soc. A Math. Phys. Eng. Sci.* **2016**, *374*, 20150202. [CrossRef] [PubMed]
44. de Lima, M.D.; Barbosa, R. Methods of authentication of food grown in organic and conventional systems using chemometrics and data mining algorithms: A review. *Food Anal. Methods* **2019**, *12*, 887–901. [CrossRef]

Article

A Rapid LC-MS/MS-PRM Assay for Serologic Quantification of Sialylated O-HPX Glycoforms in Patients with Liver Fibrosis

Aswini Panigrahi [1,2], Julius Benicky [1,2], Renhuizi Wei [1,2], Jaeil Ahn [3], Radoslav Goldman [1,2,4] and Miloslav Sanda [1,2,5,*]

1. Lombardi Comprehensive Cancer Center, Department of Oncology, Georgetown University, Washington, DC 20057, USA; ap1824@georgetown.edu (A.P.); jb2304@georgetown.edu (J.B.); rw799@georgetown.edu (R.W.); rg26@georgetown.edu (R.G.)
2. Clinical and Translational Glycoscience Research Center, Georgetown University Medical Center, Georgetown University, Washington, DC 20057, USA
3. Department of Biostatistics, Bioinformatics and Biomathematics, Georgetown University, Washington, DC 20057, USA; ja1030@georgetown.edu
4. Department of Biochemistry and Molecular & Cellular Biology, Georgetown University, Washington, DC 20057, USA
5. Max-Planck-Institut fuer Herz-und Lungenforschung, Ludwigstrasse 43, 61231 Bad Nauheim, Germany
* Correspondence: ms2465@georgetown.edu

Abstract: Development of high throughput robust methods is a prerequisite for a successful clinical use of LC-MS/MS assays. In earlier studies, we reported that nLC-MS/MS measurement of the O-glycoforms of HPX is an indicator of liver fibrosis. In this study, we show that a microflow LC-MS/MS method using a single column setup for capture of the analytes, desalting, fast gradient elution, and on-line mass spectrometry measurements, is robust, substantially faster, and even more sensitive than our nLC setup. We demonstrate applicability of the workflow on the quantification of the O-HPX glycoforms in unfractionated serum samples of control and liver disease patients. The assay requires microliter volumes of serum samples, and the platform is amenable to one hundred sample injections per day, providing a valuable tool for biomarker validation and screening studies.

Keywords: microflow LC-MS; mLC-MS/MS; liver fibrosis; hemopexin; biomarker

1. Introduction

Biomarker studies rely heavily on nano-flow liquid chromatography tandem mass spectrometry (nLC-MS/MS) for both the discovery shotgun proteomics and the targeted follow-up validation studies. In contrast to the small molecule analyte quantification, where standard HPLC flow rates for LC-MS analysis are common, the nLC-MS/MS has been favored for peptide quantification primarily because of the sensitivity of analyte detection. However, nLC-MS methods remain technically challenging, time consuming, and less robust [1], which limits their use in clinical laboratories or their applications to large sample sets.

More recently, researchers have begun to explore capillary columns with a bore wider than the conventional 75 µM ID nano-flow analytical columns [2–4]. This allows execution of the LC step of proteomic studies at a microflow rate, and at a substantially higher throughput. The increased flow rate reduces the gradient time and increases the reproducibility and robustness of the measurements [5]. However, in a conventional single spray-tip setup, the higher flow rate diminishes ionization efficiency and lowers sensitivity of detection below acceptable limits for the majority of the peptides in complex samples. This has been addressed by the development of a multi-nozzle emitter that splits the flow evenly into multiple smaller streams, which has been shown to enhance substantially the ionization efficiency [6]. In combination with advances in the sensitivity of the mass spectrometers, the microflow LC-MS/MS (mLC-MS/MS) methods reach sensitivity of detection comparable to that of nLC-MS/MS. Shotgun proteomics studies using mLC-MS/MS have reported identification of close to 10,000 proteins in cell digests, and stability and reproducibility over thousands of runs [5,7]. In these studies, the robustness of the method in high-throughput bottom-up proteomic analyses has been demonstrated using complex cell, tissue, and body fluid digests. The microflow method enabled avoidance of column overloading, resulting in good peak shapes. This, in addition to negligible carryovers, is critical for accurate quantification of compounds by the LC-MS/MS analyses. The method has been adapted for protein biomarker studies using data independent analysis (DIA), parallel reaction monitoring (PRM), and multiple reaction monitoring (MRM) [3,8–10]. However, we are not aware of any reports of the use of the mLC-MS/MS for the analysis of O-glycopeptides.

In this study, we developed a mLC-MS/MS-PRM assay for the quantification of site-specific mucin-type O-glycoforms of hemopexin, which we previously reported as a promising candidate biomarker for the serologic monitoring of liver fibrosis [11,12]. We have shown that the sialylated O-glycoforms of hemopexin (HPX) in serum of patients are associated with advancing fibrosis in hepatitis C-associated liver disease [11]. This may prove useful in the monitoring of the fibrotic liver disease, which affects a large segment of the world's population, and whose progression can be mitigated by timely lifestyle changes and interventions [13,14]. Our newly optimized method allows for capture of the analytes, desalting, and gradient elution using a one-column setup, directly in a tryptic digest of unfractionated serum, which significantly reduces the time needed for sample preparation and analysis. We used the method to quantify the HPX glycoforms in serum samples of HCV-induced liver disease, and we demonstrate that the mLC-MS/MS-PRM assay offers substantially higher throughput compared to our reported workflow [11], maintains higher sensitivity of detection, and offers a high-throughput serologic assay (100 injections/day) for an improved screening of these glycopeptide biomarker candidates.

2. Results and Discussion

Liver biopsy has been the gold standard in the diagnosis of fibrotic changes associated with chronic liver diseases, and non-invasive methods such as liver imaging, ultrasound elastography, and serologic monitoring provide additional options [13]. Serum protein biomarkers, including glycosylation pattern of liver secreted proteins, represent an attractive strategy for serologic monitoring of liver disease (reviewed in [15,16]). We have characterized O-glycoforms of HPX by mass spectrometry [11,12,17] and demonstrated that the relative abundance of the di- and mono-sialylated O-glycoforms increase significantly with the progressing fibrotic liver disease of HCV etiology [11]. Building upon our earlier studies, we aimed to develop a fast mLC-MS/MS assay to quantify the HPX glycoforms at high throughput.

2.1. Microflow LC-MS/MS for the Quantification of O-HPX

We optimized a microflow (1.5 µL/min) LC-MS/MS workflow with 5× higher throughput compared to the earlier nanoflow (0.3 µL/min) method. In a conventional metal/glass needle emitter setup this would translate to a loss of sensitivity because of the dilution of analytes. To circumvent this, we used a multi-nozzle emitter (8-nozzle, Newomics) [6], which has been reported to achieve sensitivity close to routine nLC-MS/MS applications.

The sample trapping and desalting was achieved within 2 min at a 5 µL/min flow rate using a 20 mm C18 trap column, followed by elution of the analytes at a 1.5 µL flow rate in 3 min, column washing for 2 min, followed by a 6 min equilibration step (total 13 min; for a schematic see Supplementary Figure S1). The time gap between each sample run is negligible, thus making the analysis of approximately 100 samples per day feasible. The analytes were measured by a scheduled PRM assay using an Orbitrap Fusion Lumos Mass Spectrometer (Thermo Scientific, Dreieich, Germany).

Measurement using serially diluted samples showed optimal sensitivity between 0.1 and 0.2 µg of injected serum protein sample (Figure 1). The retention time (RT) of the analytes was highly reproducible (RSD 0.20%, Figure 2) which is suitable for automated results processing. The S-HPX measurement (i.e., the ratio of disialo m/z 916.4/monosialo m/z 843.6 analyte) [11] was shown to be consistent over 50 injections (RSD 8.91%, Figure 3), demonstrating outstanding technical reproducibility of the label-free tandem mass spectrometry assay.

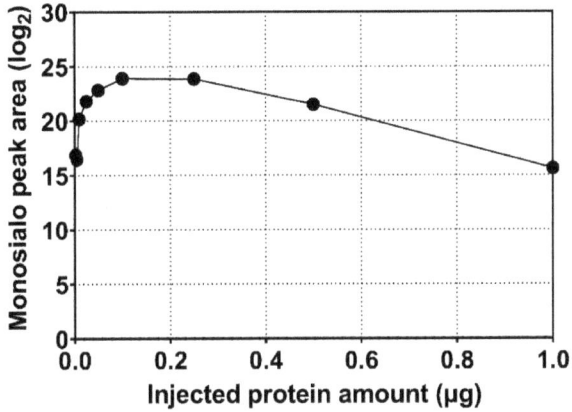

Figure 1. Peak area of tryptic monosialylated O-HPX glycopeptide in relation to the amount of serum protein analyzed by mass spectrometry.

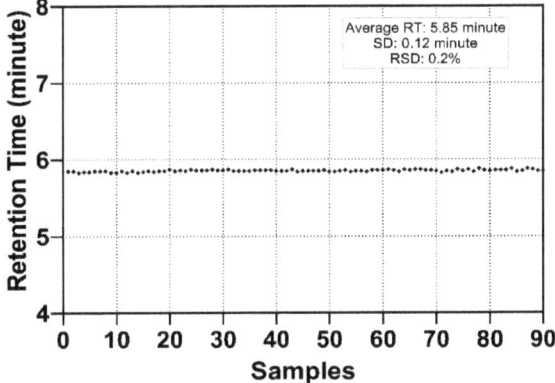

Figure 2. Retention time of a tryptic HPX O-glycopeptide on an Acclaim PepMap 100 C18 column. The consistent elution time at 5.85 ± 0.12 min demonstrates excellent reproducibility.

2.2. Application of the Micro-Flow LC-MS/MS Assay to Serum Samples of Liver Disease Patients

We reported detectability of other O-glycoforms of HPX, including the Tn-antigen, in our previous study; however, we were not able to quantify these analytes in the patient samples [11]. In our current assay, we quantify the additional analytes because of enhanced sensitivity of the current setup in spite of the introduction of faster flow rates (Supplementary Table S1). The inclusion list consisted of multiple O-glycoforms of the N-terminal HPX tryptic peptide [HexNAc (m/z 973.5), HexNAc-Hex-Neu5Ac (m/z 843.6), HexNAc-Hex-2Neu5Ac (m/z 916.4), 2HexNAc-2Hex-2Neu5Ac (m/z 1007.7), 2HexNAc-2Hex-3Neu5Ac (m/z 1080.5), 2HexNAc-2Hex-4Neu5Ac (m/z 1153.2), HexNAc-Hex (m/z 770.9)]. Their elution profile shows that the analytes elute within a short window of 5.83–5.91 min (Figure 4).

The enhanced detection of the O-HPX glycoforms in unfractionated serum samples using this microflow method may be due to the combination of sample loading capacity and excellent peak shape (Figure 5) obtained at the higher flow rate. With the assumption that minor ionization differences of the glycoforms do not affect the overall results, we calculated the ratios of multiple sialylated to respective monosialylated glycoforms. The ratios of the sialylated O-HPX analytes (S-HPX) were calculated based on the peak areas of the multiple sialylated structures to singly sialylated structures 916.4/843.6, 1080.5/1007.7, and 1153.2/1007.7 using the transitions described previously [11].

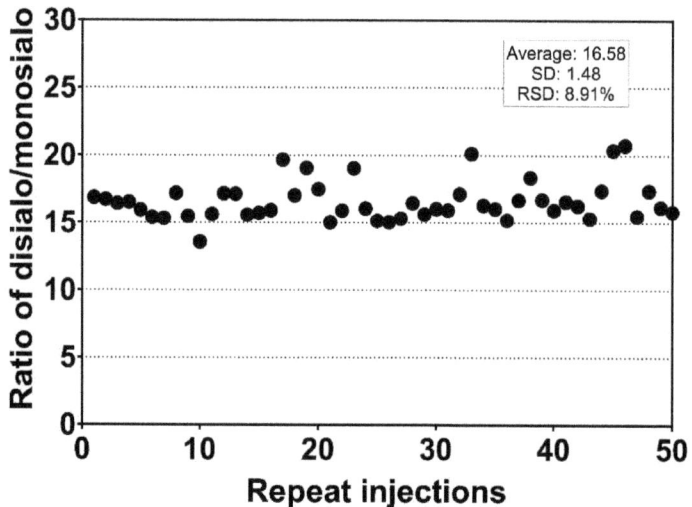

Figure 3. Repeat measurement of S-HPX from one sample by mass spectrometry. Each dot represents the ratio of the m/z 916.4 to the m/z 843.6 in one injection.

As a proof of applicability, we quantified S-HPX in serum samples of 15 HCV fibrotic and 15 HCV cirrhotic patients (HALT-C trial participants), and compared the quantities to 15 serum samples of healthy controls. The measurement was undertaken using a fixed volume of serum samples and the measure is normalized by the ratio of the glycoforms of the same protein, as described previously [11]. Statistical analyses were performed to find the association between the different analytes and the disease status. The mean ratio and standard error of 916.4/843.6 in control, fibrotic, and cirrhotic groups was 7.905 ± 0.8562, 13.69 ± 2.942, and 29.99 ± 4.950; and that of 1080.5/1007.7 was 8.802 ± 0.8, 11.65 ± 1.558, and 21.59 ± 2.587; and that of 1153.2/1007.7 was 1.07 ± 1.131, 4.261 ± 1.979, and 14.65 ± 3.49 respectively. One-way ANOVA analysis showed that the relative ratios for the three analytes, 916.4/843.6 ($p < 0.0001$), 1080.5/1007.7 ($p < 0.0001$), and 1153.2/1007.7 ($p = 0.0004$) vary significantly between the control, fibrosis, and cirrhosis groups (Figure 5). Thus, this study expands the number of meaningful analytes for the detection of liver fibrosis. It confirms the results observed in our earlier study, that the S-HPX increases progressively in fibrotic and cirrhotic participants compared to disease-free controls (Figure 5). Further studies are needed to understand the mechanism and biological processes controlling this outcome. Nevertheless, our results show that the mLC–MS/MS-PRM assay has adequate analytical performance for direct quantification of the clinically relevant S-HPX analyte in serum samples.

Overall, we demonstrate the utility of a 13 min mLC-MS/MS-PRM assay for the quantification of the S-HPX glycoforms diagnostic of liver fibrosis of HCV etiology. The assay is more sensitive compared to that of our earlier report, highly reproducible, and amenable to 100 sample injections per day. Target analyte carryover between the sample injections is negligible (results not shown). In conjunction with a simple sample preparation method without an off-line desalting step, our workflow enables analysis of at least 30 samples per day in triplicate, including necessary QC injections. These parameters would be applicable in a clinical setting. A further increase in the throughput is feasible using a wider-bore capillary column with a higher flow rate, thereby reducing the gradient run time. A multi-nozzle emitter suitable for a flow rate up to 40 µL is commercially available and would support such adjustments. Optimization of a high-flow high-sensitivity methodology would be a focus for future studies.

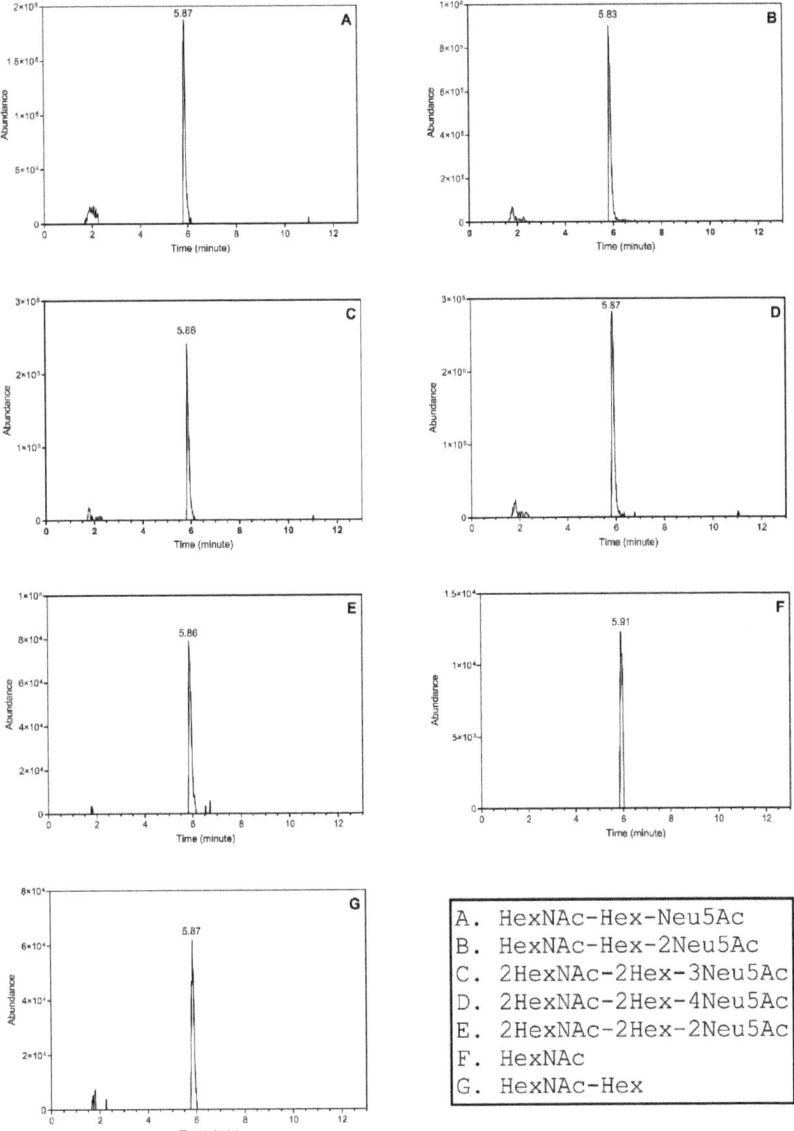

Figure 4. The extracted ion chromatograms showing the elution of the sialylated O-HPX peptides. The composition of the analytes is provided in the bottom right panel.

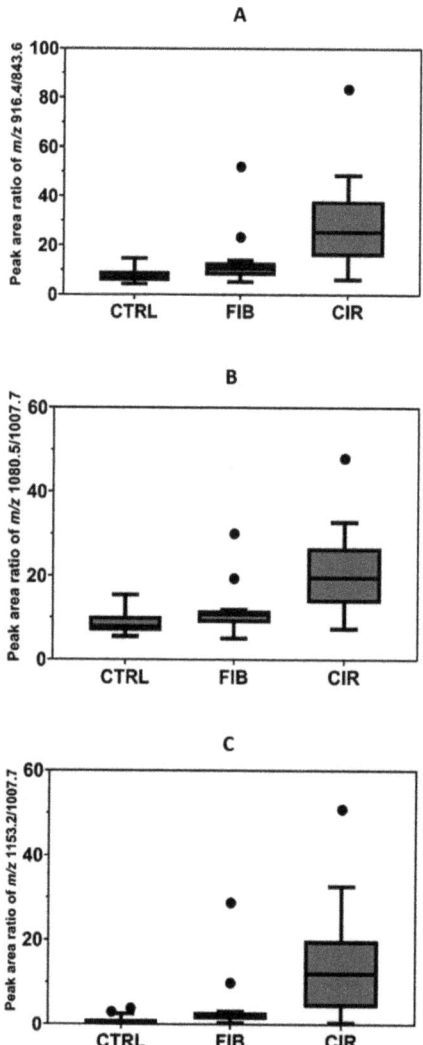

Figure 5. Quantification of S-HPX in control (CTRL, n = 15) and progressing stages of liver disease, liver fibrosis (FIB, n = 15), and cirrhosis (CIR, n = 15). S-HPX, the ratio of monosialylated glycopeptide of the same structure (disialoT/monosialoT) increases significantly ($p < 0.01$) from the control, to the fibrosis and cirrhosis groups. Ratio of (**A**) HexNAc-Hex-2Neu5Ac/HexNAc-Hex-Neu5Ac, (**B**) 2HexNAc-2Hex-3Neu5Ac/2HexNAc-2Hex-2Neu5Ac, (**C**) 2HexNAc-2Hex-4Neu5Ac/2HexNAc-2Hex-2Neu5Ac.

3. Materials and Methods

3.1. Materials

Ammonium bicarbonate, DL-dithiothreitol (DTT), iodoacetamide (IAA) (Sigma-Aldrich St. Louis, MO, USA); sequencing grade trypsin (Promega, Madison, WI, USA)). LC/MS grade Water, 0.1% formic acid in Acetonitrile, 0.1% formic acid in Water (Thermo Fisher Scientific, Waltham, MA, USA). Acclaim PepMap 100 column (Thermo Fisher Scientific, Waltham, MA, USA).

3.2. Sample Processing

Serum samples were processed by trypsin digestion, without any enrichment step, as described earlier [11]. Briefly, 2 µL of each serum sample was diluted to 140 µL with 25 mM ammonium bi-carbonate; the proteins were reduced by 5 mM DTT at 60 °C for 1 h, followed by alkylation with 15 mM iodoacetamide for 20 min at RT in the dark. Residual iodoacetamide was reduced with 5 mM DTT for 20 min at RT. The proteins (20 µL by volume from above) were digested with mass spectrometry grade trypsin (1 µg) at 37 °C O/N. Tryptic peptides were analyzed without further processing to ensure reliable quantification of the glycoforms.

3.3. Micro-Flow LC-MS/MS-PRM

LC-MS/MS analysis was performed using an Ultimate 3000 RSLCnano chromatograph and Orbitrap Fusion Lumos Mass Spectrometer platform (Thermo) with a multi-nozzle emitter (NEWOMICS, Berkeley, CA, USA) used as the microflow sprayer. Glycopeptide separation was achieved in microflow mode using an Acclaim PepMap 100 capillary column 75 µm ID × 20 mm length, packed with C18 5 µm, 300 Å (Thermo). Glycopeptides were separated as follows: starting condition flow 5 µL, 2% ACN, 0.1% formic acid; 0–1 min flow 5 µL, 2% ACN, 0.1% formic acid; 1–2 min flow 1.5 µL, 2–5% ACN, 0.1% formic acid; 2–5 min flow 1.5 µL, 5–98% ACN, 0.1% formic acid; 7–9 min flow 1.5 µL, 98% ACN, 0.1% formic acid; followed by equilibration to starting conditions for an additional 4 min (Supplementary Figure S1).

We used a Parallel Reaction Monitoring (PRM) workflow with one MS [1] full scan (400–1800 m/z, resolution 120 K, max IT 50 ms) and scheduled MS/MS fragmentation (Isolation window m/z 2.0, HCD fragmentation, resolution 30 K, scan range 200–1400, RF Lens 55%) for the analysis of the sialylated O-HPX glycopeptide TPLPPTSAHGN-VAEGETKPDPVTER (Table 1).

Table 1. Targeted PRM analysis of tryptic O-glycopeptide of HPX: analyte composition, MS data collection parameters, and transitions used for quantitation is highlighted.

Compound	m/z	z	Collision Energy (%)	Transitions Used for Quantitation
HexNAc-Hex-Neu5Ac	843.6	4	20	905.8
HexNAc-Hex-2Neu5Ac	916.4	4	20	905.8
2HexNAc-2Hex-3Neu5Ac	1080.5	4	20	905.8
2HexNAc-2Hex-4Neu5Ac	1153.2	4	20	905.8
2HexNAc-2Hex-2Neu5Ac	1007.7	4	20	905.8
HexNAc	973.5	3	20	905.8
HexNAc-Hex	770.9	4	20	905.8

3.4. Study Population

Serum samples of participants in the HALT-C trial were obtained from the central repository at the National Institute of Diabetes and Digestive and Kidney Diseases (NIDDK) as described previously [12]. In this study, O-HPX glycoforms comparison was performed in 30 participants (15 HCV fibrotic and 15 HCV cirrhotic patients) and 15 disease-free controls that donated blood samples at Georgetown University (GU) in line with approved IRB protocols. Briefly, the HALT-C trial is a prospective randomized controlled trial of 1050 patients that evaluated the effect of long-term low-dose peginterferon alpha-2a in patients who failed initial anti-HCV therapy with interferon [18]. Liver disease status of the study participants was classified based on biopsy-evaluation into groups of fibrosis (Ishak score 3–4) or cirrhosis (Ishak score 5–6). The two groups of liver disease samples, and the controls, were frequency matched on age, gender, and race (Supplementary Table S2).

3.5. Data Analysis

LC-MS/MS data were processed by Quant Browser (Thermo) with manual confirmation/integration. Peak areas were used for peptide and glycopeptide quantification and data normalization. A specific Y-ion (e.g., loss of whole glycan) was used for the quantification of the O-glycopeptides. The specific backbone fragments (y-ions) were used for the confirmation of the correct O-glycopeptides signal. The details of the MS/MS transitions used for the quantification of each glycoforms are listed in Table 1. Relative intensity of multiple sialylated analyte was calculated by normalizing its peak area to the peak area of monosialylated glycopeptide of the same structure (DisialoT/monosialoT, etc.), as described previously [11].

Statistical analysis for the HCV dataset was performed using GraphPad Prism software (v9.3.1). The ratio of three HPX-sialylated analytes 916.4, 1080.5, and 1153.2, to their respective non-sialylated forms (843.6, 1007.7, and 1007.7), was used as the quantitative measure for evaluation of the liver disease. The mean, standard error of mean, and the one-way ANOVA test was performed to determine the correlation between different analytes and disease status, and the data was visualized by nested Tukey plot.

Supplementary Materials: The following are available online at https://www.mdpi.com/article/10.3390/molecules27072213/s1, Figure S1: A schematic showing gradient conditions for LC-MS/MS analysis, Table S1: Basic information on samples analyzed in this study, participant demography and disease conditions are provided.

Author Contributions: Conceptualization, M.S.; methodology, M.S. and J.B.; validation, M.S.; formal analysis, M.S., R.W., J.A. and A.P.; investigation, M.S., J.B. and A.P.; resources, R.G.; data curation, M.S. and A.P.; writing—original draft preparation, A.P.; writing—review and editing, M.S., A.P. and R.G.; visualization, M.S. and A.P.; supervision, R.G. and M.S.; project administration, R.G. and M.S.; funding acquisition, R.G. All authors have read and agreed to the published version of the manuscript.

Funding: This work was supported in part by the National Institutes of Health (NIH grants U01CA230692 to RG and MS, R01CA238455 and R01CA135069 to RG). The content is solely the responsibility of the authors and does not necessarily represent the official views of the National Institutes of Health.

Institutional Review Board Statement: The study was conducted according to the guidelines of the Declaration of Helsinki, and approved by the Institutional Review Board of Georgetown University, IRB code: 2008-549, study: Glycans in Hepatocellular Carcinoma [12].

Informed Consent Statement: All participants provided written informed consent.

Data Availability Statement: The datasets generated during the current study are available from the corresponding author on reasonable request.

Acknowledgments: Further support was provided by the Office of The Director, National Institutes of Health under Award Number S10OD023557 supporting the operation of the Clinical and Translational Glycoscience Research Center, and Georgetown University, CCSG Grant P30 CA51008 (to Lombardi Comprehensive Cancer Center) supporting the Proteomics and Metabolomics Shared Resource.

Conflicts of Interest: The authors declare no conflict of interest. The funders had no role in the design of the study; in the collection, analyses, or interpretation of data; in the writing of the manuscript, or in the decision to publish the results.

Sample Availability: Samples of the compounds are available from the authors.

References

1. Angel, T.E.; Aryal, U.K.; Hengel, S.M.; Baker, E.S.; Kelly, R.T.; Robinson, E.W.; Smith, R.D. Mass Spectrometry-Based Proteomics: Existing Capabilities and Future Directions. *Chem. Soc. Rev.* **2012**, *41*, 3912–3928. [CrossRef] [PubMed]
2. Vowinckel, J.; Zelezniak, A.; Bruderer, R.; Mülleder, M.; Reiter, L.; Ralser, M. Cost-Effective Generation of Precise Label-Free Quantitative Proteomes in High-Throughput by MicroLC and Data-Independent Acquisition. *Sci. Rep.* **2018**, *8*, 4346. [CrossRef] [PubMed]

3. Sun, R.; Hunter, C.; Chen, C.; Ge, W.; Morrice, N.; Liang, S.; Zhu, T.; Yuan, C.; Ruan, G.; Zhang, Q.; et al. Accelerated Protein Biomarker Discovery from FFPE Tissue Samples Using Single-Shot, Short Gradient Microflow SWATH MS. *J. Proteome Res.* **2020**, *19*, 2732–2741. [CrossRef] [PubMed]
4. Bian, Y.; Zheng, R.; Bayer, F.P.; Wong, C.; Chang, Y.C.; Meng, C.; Zolg, D.P.; Reinecke, M.; Zecha, J.; Wiechmann, S.; et al. Robust, Reproducible and Quantitative Analysis of Thousands of Proteomes by Micro-Flow LC-MS/MS. *Nat. Commun.* **2020**, *11*, 157. [CrossRef] [PubMed]
5. Kuster, B.; Bian, Y.; Bayer, F.P.; Chang, Y.C.; Meng, C.; Hoefer, S.; Deng, N.; Zheng, R.; Boychenko, O. Robust Microflow LC-MS/MS for Proteome Analysis: 38 000 Runs and Counting. *Anal. Chem.* **2021**, *93*, 3686–3690. [CrossRef]
6. Chen, Y.; Mao, P.; Wang, D. Quantitation of Intact Proteins in Human Plasma Using Top-Down Parallel Reaction Monitoring-MS. *Anal. Chem.* **2018**, *90*, 10650–10653. [CrossRef] [PubMed]
7. Bian, Y.; The, M.; Giansanti, P.; Mergner, J.; Zheng, R.; Wilhelm, M.; Boychenko, A.; Kuster, B. Identification of 7000–9000 Proteins from Cell Lines and Tissues by Single-Shot Microflow LC-MS/MS. *Anal. Chem.* **2021**, *93*, 8687–8692. [CrossRef] [PubMed]
8. Bringans, S.; Ito, J.; Casey, T.; Thomas, S.; Peters, K.; Crossett, B.; Coleman, O.; Ebhardt, H.A.; Pennington, S.R.; Lipscombe, R. A Robust Multiplex Immunoaffinity Mass Spectrometry Assay (PromarkerD) for Clinical Prediction of Diabetic Kidney Disease. *Clin. Proteom.* **2020**, *17*, 37. [CrossRef] [PubMed]
9. Ni, W.; Jagust, W.; Wang, D. Multiplex Mass Spectrometry Analysis of Amyloid Proteins in Human Plasma for Alzheimer's Disease Diagnosis. *J. Proteome Res.* **2021**, *20*, 4106–4112. [CrossRef] [PubMed]
10. Ibrahim, S.; Lan, C.; Chabot, C.; Mitsa, G.; Buchanan, M.; Aguilar-Mahecha, A.; Elchebly, M.; Poetz, O.; Spatz, A.; Basik, M.; et al. Precise Quantitation of PTEN by Immuno-MRM: A Tool To Resolve the Breast Cancer Biomarker Controversy. *Anal. Chem.* **2021**, *93*, 10816–10824. [CrossRef] [PubMed]
11. Sanda, M.; Benicky, J.; Wu, J.; Wang, Y.; Makambi, K.; Ahn, J.; Smith, C.I.; Zhao, P.; Zhang, L.; Goldman, R. Increased Sialylation of Site Specific O-Glycoforms of Hemopexin in Liver Disease. *Clin. Proteom.* **2016**, *13*, 24. [CrossRef] [PubMed]
12. Benicky, J.; Sanda, M.; Pompach, P.; Wu, J.; Goldman, R. Quantification of Fucosylated Hemopexin and Complement Factor H in Plasma of Patients with Liver Disease. *Anal. Chem.* **2014**, *86*, 10716–10723. [CrossRef] [PubMed]
13. Ginès, P.; Krag, A.; Abraldes, J.G.; Solà, E.; Fabrellas, N.; Kamath, P.S. Liver Cirrhosis. *Lancet* **2021**, *398*, 1359–1376. [CrossRef]
14. El-Serag, H.B.; Rudolph, K.L. Hepatocellular Carcinoma: Epidemiology and Molecular Carcinogenesis. *Gastroenterology* **2007**, *132*, 2557–2576. [CrossRef] [PubMed]
15. Mehta, A.; Herrera, H.; Block, T. Glycosylation and Liver Cancer. *Adv. Cancer Res.* **2015**, *126*, 257–279. [CrossRef] [PubMed]
16. Zhu, J.; Warner, E.; Parikh, N.D.; Lubman, D.M. Glycoproteomic Markers of Hepatocellular Carcinoma-Mass Spectrometry Based Approaches. *Mass Spectrom. Rev.* **2019**, *38*, 265–290. [CrossRef] [PubMed]
17. Ma, J.; Sanda, M.; Wei, R.; Zhang, L.; Goldman, R. Quantitative Analysis of Core Fucosylation of Serum Proteins in Liver Diseases by LC-MS-MRM. *J. Proteom.* **2018**, *189*, 67–74. [CrossRef] [PubMed]
18. Di Bisceglie, A.M.; Shiffman, M.L.; Everson, G.T.; Lindsay, K.L.; Everhart, J.E.; Wright, E.C.; Lee, W.M.; Lok, A.S.; Bonkovsky, H.L.; Morgan, T.R.; et al. Prolonged Therapy of Advanced Chronic Hepatitis C with Low-Dose Peginterferon. *N. Engl. J. Med.* **2008**, *359*, 2429–2441. [CrossRef] [PubMed]

Article

Simultaneous Determination of Caffeine and Paracetamol in Commercial Formulations Using Greener Normal-Phase and Reversed-Phase HPTLC Methods: A Contrast of Validation Parameters

Prawez Alam [1,*], Faiyaz Shakeel [2], Abuzer Ali [3], Mohammed H. Alqarni [1], Ahmed I. Foudah [1], Tariq M. Aljarba [1], Faisal K. Alkholifi [4], Sultan Alshehri [2], Mohammed M. Ghoneim [5] and Amena Ali [6]

[1] Department of Pharmacognosy, College of Pharmacy, Prince Sattam Bin Abdulaziz University, P.O. Box 173, Al-Kharj 11942, Saudi Arabia; m.alqarni@psau.edu.sa (M.H.A.); a.foudah@psau.edu.sa (A.I.F.); t.aljarba@psau.edu.sa (T.M.A.)
[2] Department of Pharmaceutics, College of Pharmacy, King Saud University, P.O. Box 2457, Riyadh 11451, Saudi Arabia; faiyazs@fastmail.fm (F.S.); salshehri1@ksu.edu.sa (S.A.)
[3] Department of Pharmacognosy, College of Pharmacy, Taif University, P.O. Box 11099, Taif 21944, Saudi Arabia; abuali@tu.edu.sa
[4] Department of Pharmacology, College of Pharmacy, Prince Sattam Bin Abdulaziz University, P.O. Box 173, Al-Kharj 11942, Saudi Arabia; f.alkholifi@psau.edu.sa
[5] Department of Pharmacy Practice, College of Pharmacy, AlMaarefa University, P.O. Box 71666, Ad Diriyah 13713, Saudi Arabia; mghoneim@mcst.edu.sa
[6] Department of Pharmaceutical Chemistry, College of Pharmacy, Taif University, P.O. Box 11099, Taif 21944, Saudi Arabia; amrathore@tu.edu.sa
* Correspondence: p.alam@psau.edu.sa

Abstract: There has been no assessment of the greenness of the described analytical techniques for the simultaneous determination (SMD) of caffeine and paracetamol. As a result, in comparison to the greener normal-phase high-performance thin-layer chromatography (HPTLC) technique, this research was conducted to develop a rapid, sensitive, and greener reversed-phase HPTLC approach for the SMD of caffeine and paracetamol in commercial formulations. The greenness of both techniques was calculated using the AGREE method. For the SMD of caffeine and paracetamol, the greener normal-phase and reversed-phase HPTLC methods were linear in the 50–500 ng/band and 25–800 ng/band ranges, respectively. For the SMD of caffeine and paracetamol, the greener reversed-phase HPTLC approach was more sensitive, accurate, precise, and robust than the greener normal-phase HPTLC technique. For the SMD of caffeine paracetamol in commercial PANEXT and SAFEXT tablets, the greener reversed-phase HPTLC technique was superior to the greener normal-phase HPTLC approach. The AGREE scores for the greener normal-phase and reversed-phase HPTLC approaches were estimated as 0.81 and 0.83, respectively, indicated excellent greenness profiles for both analytical approaches. The greener reversed-phase HPTLC approach is judged superior to the greener normal-phase HPTLC approach based on numerous validation parameters and pharmaceutical assays.

Keywords: caffeine; greener HPTLC; paracetamol; simultaneous determination; validation

1. Introduction

Paracetamol (Figure 1A) is the commonly administered anti-inflammatory and antipyretic medicine, especially in case of pediatric and geriatric patients [1,2]. It is commercially available in a wide range of dosage forms [2]. Caffeine (Figure 1B) is a pseudo-alkaloidal drug that is commonly used in combination with paracetamol [3,4]. The combination of paracetamol and caffeine is the world's most widely used combination [4]. As a result, the qualitative and quantitative standardization of caffeine and paracetamol in commercially available formulations is necessary.

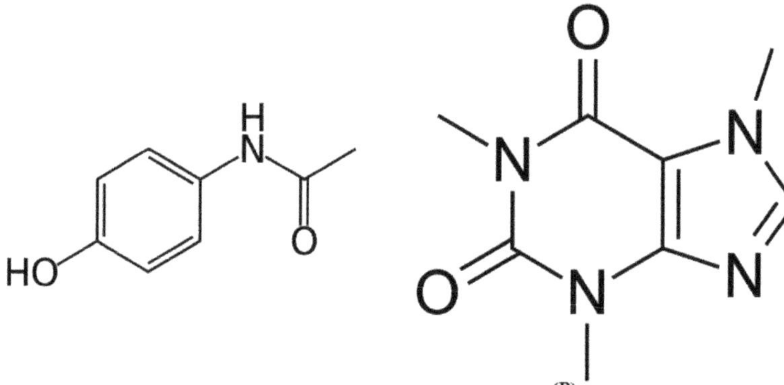

Figure 1. Chemical structures of (**A**) paracetamol and (**B**) caffeine.

An extensive literature search revealed various analytical approaches for the simultaneous determination (SMD) of caffeine and paracetamol in commercial formulations and biological fluids. For the SMD of caffeine and paracetamol in commercial formulations, different spectrometry techniques involving various chemical procedures, such as derivatization have been used [5–10]. For the SMD of caffeine and paracetamol in various commercial dosage forms, several high-performance liquid chromatography (HPLC) techniques have been used [4,11–19]. Caffeine and paracetamol were also quantified simultaneously in a human plasma sample using a HPLC method [19]. For the SMD of caffeine and paracetamol in human plasma samples, a liquid-chromatography mass-spectrometry (LC–MS) technique was also used [20]. For the SMD of caffeine and paracetamol in their pure forms and formulations, certain high-performance thin-layer chromatography (HPTLC) techniques have been used [21–23]. Various voltammetry-based approaches have also been applied for the SMD of caffeine and paracetamol in their dosage forms [24–27]. Dual-mode gradient HPLC and HPTLC methods have also been used for the SMD of caffeine and paracetamol in the presence of paracetamol impurities [28]. The electrospray laser desorption ionization mass spectrometry technique was also utilized for the SMD of caffeine and paracetamol in tablets [29]. An electrochemical cell-on-a-chip device fabricated using 3D-printing technology was also used for the SMD of caffeine and paracetamol [30]. A genetic algorithm based on wavelength selection was also applied for the SMD of caffeine and paracetamol [31]. Some other approaches, such as near-infrared spectrometry [32], flow-injection spectrometry [33], micellar liquid chromatography [34], and micellar electrokinetic capillary chromatography [35] approaches were also proposed for the SMD of caffeine and paracetamol in their dosage forms. Published reports on the SMD of caffeine and paracetamol suggested various analytical approaches for their analysis. However, the greenness scale of any of the reported analytical approach was not estimated. In addition, greener HPTLC approaches have not been utilized for the SMD of caffeine and paracetamol. For the estimation of the greenness scale, different quantitative analytical methodologies have been presented [36–40]. For the estimation of the greenness scale, only the "Analytical Greenness (AGREE)" analytical approach considers all twelve green analytical chemistry (GAC) principles [38]. As a result, the AGREE analytical methodology was applied for the estimation of greenness scale of the greener normal-phase and reversed-phase HPTLC approaches [38].

In comparison to the greener normal-phase HPTLC approach, the current study intends to establish and validate a rapid, sensitive, and greener reversed-phase HPTLC approach for the SMD of caffeine and paracetamol in commercial formulations. Following "The International Council for Harmonization (ICH)" Q2-R1 recommendations, the greener normal-phase and reversed-phase HPTLC methods for the SMD of caffeine and paracetamol were validated [41].

2. Results and Discussion

2.1. Method Development

For the development of a suitable band for the SMD of caffeine and paracetamol using the greener normal-phase HPTLC technique, various amounts of ethyl acetate (EA) and ethanol (E), including EA/E (50:50, v/v), EA/E (60:40, v/v), EA/E (70:30, v/v), EA/E (80:20, v/v), EA/E (85:15, v/v), and EA/E (90:10, v/v) were studied as the greener mobile phases.

The greener mobile phases, such as EA/E (50:50, v/v), EA/E (60:40, v/v), EA/E (70:30, v/v), EA/E (80:20, v/v), and EA/A (90:10, v/v) revealed poor chromatographic peaks of caffeine and paracetamol with high asymmetry factor (As) for caffeine (As > 1.25) and paracetamol (As > 1.30). When the greener mobile phase EA/E (85:15, v/v) was evaluated, it was discovered that this greener mobile phase provided well-resolved and intact chromatographic peaks for caffeine at a retardation factor of (R_f) = 0.40 ± 0.01 and for paracetamol of R_f = 0.59 ± 0.02 (Figure 2). Caffeine and paracetamol were also predicted to have As values of 1.06 and 1.08, respectively, which are very trustworthy. As a consequence, the EA/E (85:15, v/v) was chosen as the final mobile phase for the SMD of caffeine and paracetamol in commercial tablets utilizing the greener normal-phase HPTLC method.

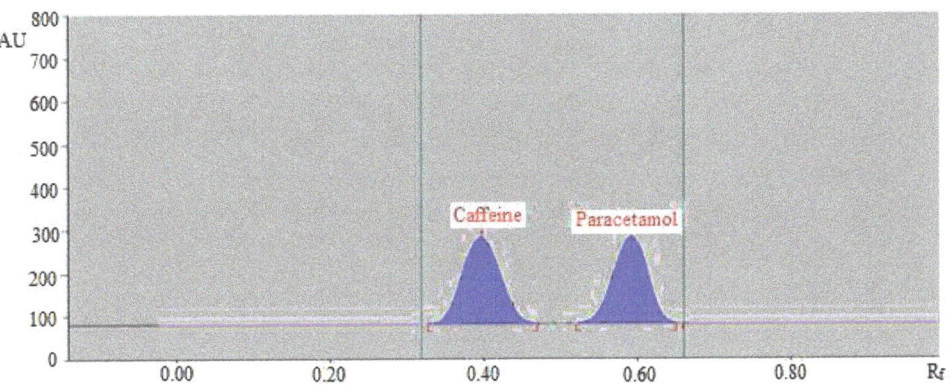

Figure 2. Normal-phase high-performance thin-layer chromatography (HPTLC) chromatogram of standard caffeine and paracetamol.

For the development of a suitable band for the SMD of caffeine and paracetamol using the greener reversed-phase HPTLC technique, various amounts of E and water (W), including E/W (50:50, v/v), E/W (60:40, v/v), E/W (70:30, v/v), E/W (80:20, v/v), and E/W (90:10, v/v) were studied as the greener mobile phases. All of the green mobile phases investigated were created under chamber saturation conditions (Figure 3).

The greener mobile phases, such as E/W (60:40, v/v), E/W (70:30, v/v), E/W (80:20, v/v), and E/W (90:10, v/v) revealed poor chromatographic peaks of caffeine and paracetamol with poor As for caffeine (As > 1.30) and paracetamol (As > 1.35). When the greener mobile phase E/W (50:50, v/v) was evaluated, it was discovered that this greener mobile phase provided well-resolved and intact chromatographic peaks of caffeine at R_f = 0.43 ± 0.01 and of paracetamol at R_f = 0.57 ± 0.02 (Figure 4). Caffeine and paracetamol were also predicted to have As values of 1.10 and 1.09, respectively, which are very trustworthy. As a consequence, the E/W (50:50, v/v) was chosen as the final mobile phase for the SMD of caffeine and paracetamol in commercial tablets utilizing the greener reversed-phase HPTLC method. The maximum response was obtained at a wavelength of 260 nm for caffeine and paracetamol when the spectral bands for caffeine and paracetamol were recorded using densitometry mode. As a result, the whole SMD of caffeine and paracetamol took place at 260 nm.

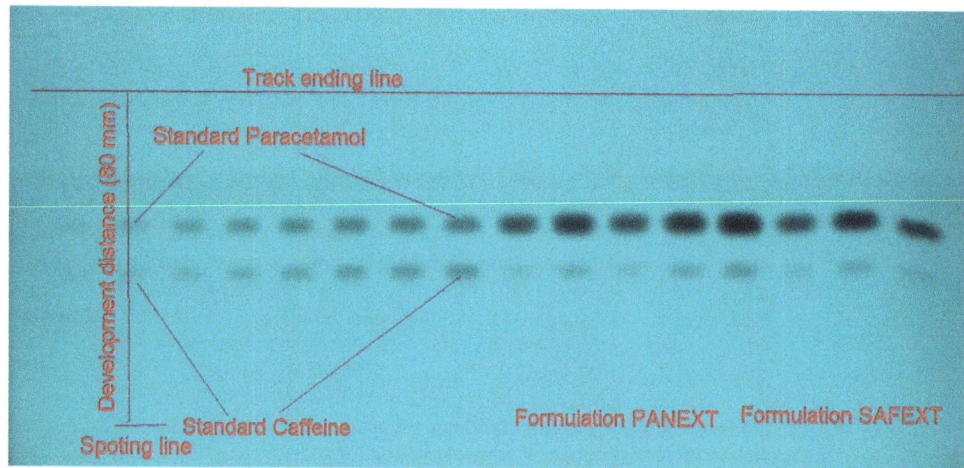

Figure 3. Developed thin-layer chromatography (TLC) plate for standard caffeine, standard paracetamol, commercial tablets PANEXT, and commercial tablets SAFEXT developed using ethanol (E)/water (W) (50:50 v/v) as the greener mobile phase for the greener reversed-phase HPTLC method.

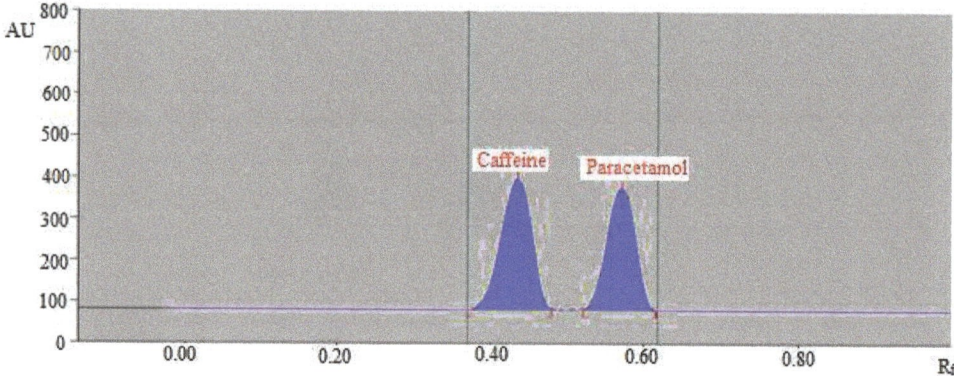

Figure 4. Reversed-phase HPTLC chromatogram of standard caffeine and paracetamol.

2.2. Method Validation

The ICH-Q2-R1 guidelines were used to estimate various parameters for the SMD of caffeine and paracetamol [41]. The results of the linear regression analysis of caffeine and paracetamol calibration curves utilizing the greener normal-phase HPTLC technique are summarized in Table 1. Caffeine and paracetamol calibration curves were linear in the 50–500 ng/band range for both drugs. Caffeine and paracetamol's determination coefficients (R^2) were found to be 0.9928 and 0.9970, respectively. Caffeine and paracetamol's regression coefficients (R) were found to be 0.9963 and 0.9984, respectively. The values of R^2 and R were highly significant for both the compounds ($p < 0.05$). These findings suggested a strong link between the concentration and measured response of caffeine and paracetamol. All these findings indicated the reliability of the greener normal-phase HPTLC approach for the SMD of caffeine and paracetamol.

Table 1. Results of the linear regression analysis for the simultaneous determination (SMD) of caffeine and paracetamol using the greener normal-phase high-performance thin-layer chromatography (HPTLC) method (mean ± SD; $n = 6$).

Parameters	Caffeine	Paracetamol
linearity range (ng/band)	50–500	50–500
R^2	0.9928	0.9970
R	0.9963	0.9984
slope ± SD	19.36 ± 0.96	18.05 ± 0.54
intercept ± SD	387.62 ± 4.38	95.66 ± 2.14
standard error of slope	0.39	0.22
standard error of intercept	1.78	0.87
95% confidence interval of slope	17.68–21.05	17.10–19.00
95% confidence interval of intercept	379.92–395.31	91.90–99.42
LOD ± SD (ng/band)	16.84 ± 0.27	17.05 ± 0.31
LOQ ± SD (ng/band)	50.52 ± 0.81	51.15 ± 0.93

R^2: determination coefficient; R: regression coefficient; LOD: limit of detection; LOQ: limit of quantification.

The resulting data for the linear regression analysis of caffeine and paracetamol calibration curves utilizing the greener reversed-phase HPTLC technique are summarized in Table 2. Caffeine and paracetamol calibration curves were linear in the 25–800 ng/band range for both drugs. Caffeine and paracetamol's R^2 were found to be 0.9976 and 0.9966, respectively. Caffeine and paracetamol's R were found to be 0.9987 and 0.9982, respectively. The values of R^2 and R were highly significant for both the compounds ($p < 0.05$). These findings again suggested a strong link between the concentration and measured response of caffeine and paracetamol. All these findings indicated the reliability of the greener reversed-phase HPTLC technique for the SMD of caffeine and paracetamol. However, the greener reversed-phase HPTLC technique was more linear than the greener normal-phase HPTLC technique.

Table 2. Results for linear regression analysis for the SMD of caffeine and paracetamol using the greener reversed-phase HPTLC method (mean ± SD; $n = 6$).

Parameters	Caffeine	Paracetamol
linearity range (ng/band)	25–800	25–800
R^2	0.9976	0.9966
R	0.9987	0.9982
slope ± SD	20.54 ± 1.05	17.12 ± 0.47
intercept ± SD	833.46 ± 7.51	696.63 ± 6.21
standard error of slope	0.42	0.19
standard error of intercept	3.06	2.53
95% confidence interval of slope	18.70–22.38	16.29–17.94
95% confidence interval of intercept	820.26–846.65	685.71–707.54
LOD ± SD (ng/band)	8.52 ± 0.12	8.71 ± 0.13
LOQ ± SD (ng/band)	25.56 ± 0.36	26.13 ± 0.39

R^2: determination coefficient; R: regression coefficient; LOD: limit of detection; LOQ: limit of quantification.

The parameters of the system appropriateness for the greener normal-phase HPTLC methodology are summarized in Table 3. For the SMD of caffeine and paracetamol, the R_f, As, and number of theoretical plates per meter (N/m) for the greener normal-phase HPTLC technique were determined to be satisfactory. The parameters of the system appropriateness for the greener reversed-phase HPTLC methodology are summarized in Table 4. For the SMD of caffeine and paracetamol, the R_f, As, and N/m for the greener reversed-phase HPTLC technique were also determined to be satisfactory.

Table 3. System suitability parameters in terms of retardation factor (R_f), asymmetry factor (As), and a number of theoretical plates per meter (N/m) of caffeine and paracetamol for the greener normal-phase HPTLC method (mean ± SD; n = 3).

Parameters	Caffeine	Paracetamol
R_f	0.40 ± 0.01	0.59 ± 0.02
As	1.06 ± 0.02	1.08 ± 0.03
N/m	5245 ± 5.81	4978 ± 5.19

Table 4. The R_f, As, and N/m values of caffeine and paracetamol for the greener reversed-phase HPTLC method (mean ± SD; n = 3).

Parameters	Caffeine	Paracetamol
R_f	0.43 ± 0.01	0.57 ± 0.02
As	1.10 ± 0.03	1.09 ± 0.02
N/m	5182 ± 5.92	5367 ± 6.32

For assessing caffeine and paracetamol, the percent of recovery was utilized to estimate the accuracy of the greener normal-phase and reversed-phase HPTLC techniques. The accuracy evaluation results for the greener normal-phase HPTLC technique are summarized in Table 5. Using the greener normal-phase HPTLC technique, the percent recoveries of caffeine and paracetamol at three separate quality control (QC) samples were expected to be 97.13–104.88 and 96.57–103.23 percent, respectively. The accuracy evaluation results for the greener reversed-phase HPTLC technique are summarized in Table 6. Using the greener reversed-phase HPTLC technique, the percent recoveries of caffeine and paracetamol at three separate QC samples were expected to be 98.84–100.62 and 98.60–101.50 percent, respectively. These results showed that both analytical techniques were accurate for the SMD of caffeine and paracetamol. For the SMD of caffeine and paracetamol, however, the greener reversed-phase HPTLC methodology was more accurate than the greener normal-phase HPTLC methodology.

Table 5. Measurement of the accuracy of caffeine and paracetamol for the greener normal-phase HPTLC method (mean ± SD; n = 6).

Conc. (ng/Band)	Conc. Found (ng/Band) ± SD	Recovery (%)	CV (%)
	Caffeine		
100	97.13 ± 2.24	97.13	2.30
300	314.65 ± 4.87	104.88	1.54
500	492.31 ± 6.13	98.46	1.24
	Paracetamol		
100	103.23 ± 3.12	103.23	3.02
300	289.71 ± 7.12	96.57	2.45
500	514.41 ± 9.12	102.88	1.77

CV: coefficient of variance.

The precision of the greener normal-phase and reversed-phase HPTLC techniques was investigated as intra/inter-assay precision and given as a percent of the coefficient of variation (CV) for the SMD of caffeine and paracetamol. Table 7 summarizes the results of intra/inter-day precisions for the SMD of caffeine and paracetamol using the greener normal-phase HPTLC technique. The percent CVs of caffeine and paracetamol for the intra-day variation were estimated as 1.30–2.39 and 1.91–3.42 percent, respectively. The percent CVs of caffeine and paracetamol for inter-day variation were estimated as 1.51–2.55 and 1.86–3.56 percent, respectively. Table 8 summarizes the results of intra/inter-day precisions for the SMD of caffeine and paracetamol using the greener reversed-phase HPTLC technique. The percent CVs of caffeine and paracetamol for the intra-day variation were estimated as 0.40–0.85 and 0.52–0.96 percent, respectively. The percent CVs of caffeine and paracetamol for inter-day variation were estimated as 0.42–0.78 and 0.55–1.03 percent, respectively. These findings indicated that both the analytical approaches were precise for the SMD of caffeine and paracetamol. However, the greener reversed-phase HPTLC methodology was more precise than the greener normal-phase HPTLC methodology for the SMD of caffeine and paracetamol.

Table 6. Measurement of the accuracy of caffeine and paracetamol for the greener reversed-phase HPTLC method (mean ± SD; n = 6).

Conc. (ng/Band)	Conc. Found (ng/Band) ± SD	Recovery (%)	CV (%)
	Caffeine		
50	50.31 ± 0.41	100.62	0.81
300	296.54 ± 1.45	98.84	0.48
800	795.61 ± 3.45	99.45	0.43
	Paracetamol		
50	49.18 ± 0.41	98.36	0.83
300	304.51 ± 1.64	101.50	0.53
800	807.54 ± 4.15	100.94	0.51

CV: coefficient of variance.

Table 7. Assessment of intra/inter-day precision of caffeine and paracetamol for the greener normal-phase HPTLC method (mean ± SD; n = 6).

Conc. (ng/Band)	Intraday Precision			Interday Precision		
	Conc. (ng/Band) ± SD	Standard Error	CV (%)	Conc. (ng/Band) ± SD	Standard Error	CV (%)
			Caffeine			
100	96.54 ± 2.31	0.94	2.39	103.65 ± 2.65	1.08	2.55
300	292.97 ± 5.02	2.04	1.71	291.98 ± 5.61	2.29	1.92
500	512.45 ± 6.68	2.72	1.30	483.27 ± 7.31	2.98	1.51
			Paracetamol			
100	96.89 ± 3.32	1.35	3.42	95.61 ± 3.41	1.39	3.56
300	313.56 ± 7.52	3.07	2.39	286.51 ± 7.59	3.09	2.64
500	486.67 ± 9.32	3.80	1.91	516.41 ± 9.61	3.92	1.86

CV: coefficient of variance.

Table 8. Assessment of intra/inter-day precision of caffeine and paracetamol for the greener reversed-phase HPTLC method (mean ± SD; n = 6).

Conc. (ng/Band)	Intraday Precision			Interday Precision		
	Conc. (ng/Band) ± SD	Standard Error	CV (%)	Conc. (ng/Band) ± SD	Standard Error	CV (%)
			Caffeine			
50	50.42 ± 0.43	0.17	0.85	49.61 ± 0.39	0.15	0.78
300	303.21 ± 1.51	0.61	0.49	297.54 ± 1.31	0.53	0.44
800	806.31 ± 3.61	1.47	0.40	797.61 ± 3.40	1.38	0.42
			Paracetamol			
50	49.54 ± 0.48	0.19	0.96	50.21 ± 0.52	0.21	1.03
300	305.61 ± 1.78	0.72	0.58	296.54 ± 1.81	0.73	0.61
800	794.65 ± 4.21	1.71	0.52	804.61 ± 4.47	1.82	0.55

CV: coefficient of variance.

By introducing slight deliberate modifications in the greener mobile phase components, the durability of the greener normal-phase and reversed-phase HPTLC techniques for the SMD of caffeine and paracetamol was examined. Table 9 summarizes the results of robustness evaluation using the greener normal-phase HPTLC approach. The percent

CVs for caffeine and paracetamol were estimated as 2.17–3.33 and 2.48–2.64 percent, respectively. Caffeine and paracetamol R_f values were also estimated to be 0.39–0.41 and 0.58–0.60, respectively.

Table 9. Results of robustness analysis of caffeine and paracetamol for the greener normal-phase HPTLC method (mean ± SD; $n = 6$).

Conc. (ng/Band)	Mobile Phase Composition (Ethyl Acetate/Ethanol)			Results		R_f
	Original	Used		(ng/Band) ± SD	% CV	
		Caffeine				
300	85:15	87:13	+2.0	294.98 ± 6.43	2.17	0.39
		85:15	0.0	302.14 ± 6.87	2.27	0.40
		83:17	−2.0	305.61 ± 7.13	2.33	0.41
		Paracetamol				
300	85:15	87:13	+2.0	287.21 ± 7.14	2.48	0.58
		85:15	0.0	291.34 ± 7.72	2.64	0.59
		83:17	−2.0	304.51 ± 8.02	2.63	0.60

CV: coefficient of variance; R_f: retardation factor.

Table 10 summarizes the results of robustness evaluation utilizing the greener reversed-phase HPTLC methodology. The percent CVs for caffeine and paracetamol were estimated as 0.91–0.94 and 0.95–1.04 percent, respectively. Caffeine and paracetamol R_f values were also estimated to be 0.42–0.44 and 0.56–0.58, respectively. These results showed that both analytical techniques were reliable for the SMD of caffeine and paracetamol. For the SMD of caffeine and paracetamol, however, the greener reversed-phase HPTLC approach was more robust than the greener normal-phase HPTLC approach.

Table 10. Results of robustness analysis of caffeine and paracetamol for the greener reversed-phase HPTLC method (mean ± SD; $n = 6$).

Conc. (ng/Band)	Mobile Phase Composition (Ethanol/Water)			Results		R_f
	Original	Used		(ng/Band) ± SD	% CV	
		Caffeine				
300	50:50	52:48	+2.0	296.31 ± 2.71	0.91	0.42
		50:50	0.0	303.54 ± 2.82	0.92	0.43
		48:52	−2.0	306.87 ± 2.91	0.94	0.44
		Paracetamol				
300	50:50	52:48	+2.0	294.87 ± 2.81	0.95	0.56
		50:50	0.0	303.21 ± 2.94	0.96	0.57
		48:52	−2.0	307.81 ± 3.21	1.04	0.58

CV: coefficient of variance; R_f: retardation factor.

The "limit of detection (LOD) and limit of quantification (LOQ)" were used to evaluate the sensitivity of the greener normal-phase and reversed-phase HPTLC methods for the SMD of caffeine and paracetamol. The predicted values of "LOD and LOQ" for caffeine and paracetamol utilizing the greener normal-phase HPTLC technique are summarized in Table 1. Using the greener normal-phase HPTLC technique, the "LOD and LOQ" for caffeine were estimated to be 16.84 ± 0.27 and 50.52 ± 0.81 ng/band, respectively. Using the greener normal-phase HPTLC technique, the "LOD and LOQ" for paracetamol were estimated to be 17.05 ± 0.31 and 51.15 ± 0.93 ng/band, respectively. The predicted values of "LOD and LOQ" for caffeine and paracetamol utilizing the greener reversed-phase HPTLC technique are summarized in Table 2. Utilizing the reversed-phase HPTLC technique, the "LOD and LOQ" for caffeine were estimated to be 8.52 ± 0.12 and 25.56 ± 0.36 ng/band, respectively. Using the greener reversed-phase HPTLC technique, the "LOD and LOQ" for paracetamol were estimated to be 8.71 ± 0.13 and 26.13 ± 0.39 ng/band, respectively. These data suggested that both analytical techniques were sensitive enough for the SMD of caffeine and paracetamol. For the SMD of caffeine and paracetamol, however, the reversed-phase HPTLC methodology was more sensitive than the normal-phase HPTLC methodology.

By comparing the R_f values and superimposed ultra-violet (UV)-absorption spectra of caffeine and paracetamol in the commercial tablets PANEXT and SAFEXT with that of standards caffeine and paracetamol, the specificity of the greener HPTLC approach for the SMD of caffeine and paracetamol was assessed. The overlaid UV spectra of standards caffeine and paracetamol, as well as caffeine and paracetamol in the commercial tablets PANEXT and SAFEXT, are shown in Figure 5.

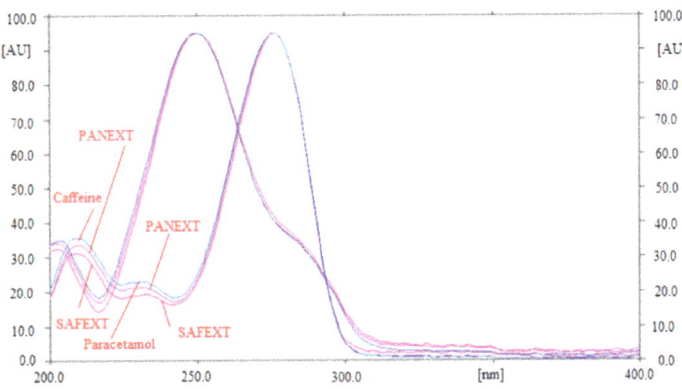

Figure 5. Superimposed ultra-violet (UV) absorption spectra of standard caffeine and paracetamol and caffeine and paracetamol in PANEXT and SAFEXT.

At a wavelength of 260 nm, the maximum densitometric responses of caffeine and paracetamol in standards and the commercial tablets PANEXT and SAFEXT were recorded. The specificity of the greener HPTLC technique for the SMD of caffeine and paracetamol was demonstrated by the identical UV spectra, R_f data, and wavelengths of caffeine and paracetamol in standards and the commercial tablets PANEXT and SAFEXT.

2.3. Application of Greener Normal-Phase and Reversed-Phase HPTLC Aapraches in the SMD of Caffeine and Paracetamol in Commercial Tablets

For the SMD of caffeine and paracetamol in their commercial formulation, the greener normal-phase and reversed-phase HPTLC techniques were used as an alternative to regular analytical approaches. The chromatograms of caffeine and paracetamol from commercial tablets were identified by comparing the TLC spots at $R_f = 0.40 \pm 0.01$ for caffeine and $R_f = 0.59 \pm 0.02$ for paracetamol in comparison with those of standards for caffeine and paracetamol using the greener normal-phase HPTLC approach. Figure 6 summarizes the recorded chromatograms of caffeine and paracetamol in the commercial tablets PANEXT (Figure 6A) and SAFEXT (Figure 6B), which showed identical peaks of caffeine and paracetamol to those of standards for caffeine and paracetamol in both the commercial tablets.

The chromatograms of caffeine and paracetamol from commercial tablets were identified by comparing their TLC spots at $R_f = 0.43 \pm 0.01$ for caffeine and $R_f = 0.57 \pm 0.02$ for paracetamol with those of standards caffeine and paracetamol using the greener reversed-phase HPTLC approach. Figure 7 summarizes the recorded chromatograms of caffeine and paracetamol in the commercial tablets PANEXT (Figure 7A) and SAFEXT (Figure 7B), which also showed identical peaks of caffeine and paracetamol to those of standards caffeine and paracetamol in both commercial tablets.

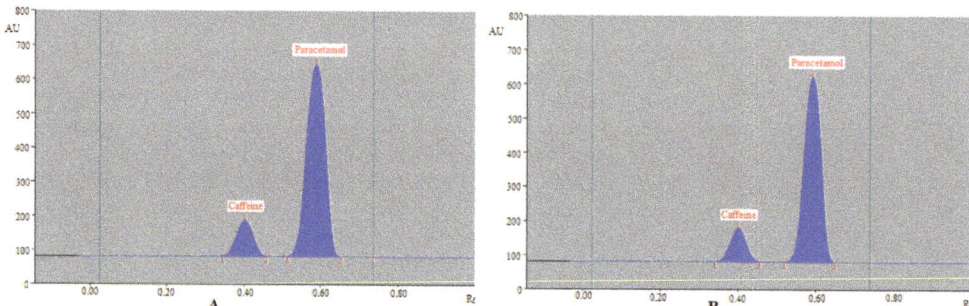

Figure 6. Normal-phase HPTLC chromatograms of caffeine and paracetamol in (**A**) commercial tablets PANEXT and (**B**) commercial tablets SAFEXT.

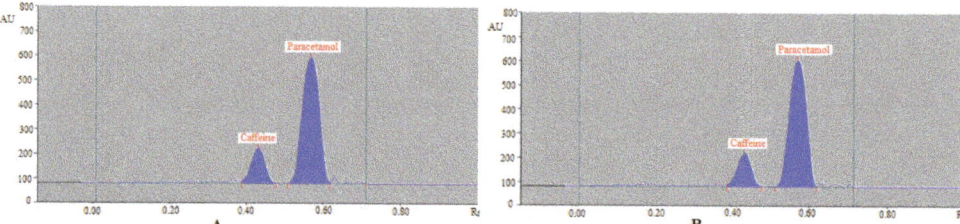

Figure 7. Reversed-phase HPTLC chromatograms of caffeine and paracetamol in (**A**) commercial tablets PANEXT and (**B**) commercial tablets SAFEXT.

Using the greener normal-phase HPTLC technique, the percent assays of caffeine in the commercial PANEXT and SAFEXT tablets were estimated to be 91.23 ± 1.14 and 92.45 ± 1.22 percent, respectively. Using the greener normal-phase HPTLC technique, the percent assays of paracetamol in commercial PANEXT and SAFEXT tablets were estimated to be 89.41 ± 1.04 and 91.13 ± 1.06 percent, respectively. Using the greener reversed-phase HPTLC technique, the percent assays of caffeine in commercial PANEXT and SAFEXT tablets were estimated to be 98.51 ± 1.42 and 101.12 ± 1.53 percent, respectively. Using the greener reversed-phase HPTLC technique, the percent assays of paracetamol in commercial PANEXT and SAFEXT tablets were estimated to be 99.42 ± 1.45 and 100.64 ± 1.49 percent, respectively. The greener normal-phase and reversed-phase HPTLC methods were shown to be suitable for the SMD of caffeine and paracetamol in commercial formulations. However, for the SMD of caffeine and paracetamol in commercial formulations, the reversed-phase HPTLC methodology was more reliable than the normal-phase HPTLC methodology.

2.4. Greenness Estimation Using AGREE

Various quantitative approaches are available for the greenness estimation of analytical approaches [36–40]. However, only AGREE applies all twelve GAC principles for greenness estimation [38]. As a result, the greenness of the greener normal-phase and reversed-phase HPTLC approaches was estimated by "AGREE: The Analytical Greenness Calculator (version 0.5, Gdansk University of Technology, Gdansk, Poland, 2020)". The typical diagram for the AGREE scale of the greener normal-phase and reversed-phase HPTLC techniques is shown in Figure 8. The AGREE scale was estimated to be 0.81 and 0.83 for the greener normal-phase and reversed-phase HPTLC methods, respectively. These findings indicated the excellent greenness nature of the greener normal-phase and reversed-phase HPTLC approaches for the SMD of caffeine and paracetamol in their commercial formulations.

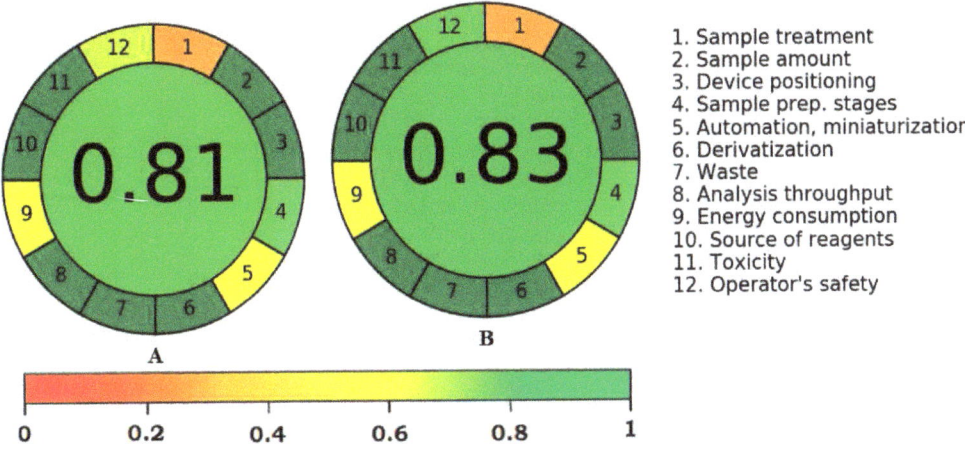

Figure 8. "Analytical GREEnness (AGREE)" scale for (**A**) greener normal-phase HPTLC and (**B**) greener reversed-phase HPTLC methods.

3. Materials and Methods

3.1. Materials

The standards of caffeine and paracetamol were provided by "Sigma Aldrich (St. Louis, MO, USA)". HPLC-grades E and EA were provided by "E-Merck (Darmstadt, Germany)". The W was obtained from the Milli-Q unit in the laboratory. The commercial tablets PANEXT and SAFEXT were obtained from the pharmacy shop in "Al-Kharj, Saudi Arabia". All other solvents utilized were of analytical grades.

3.2. Instrumentation and Analytical Procedures

The "HPTLC CAMAG TLC system (CAMAG, Muttenz, Switzerland)" was used for the SMD of caffeine and paracetamol in their standards and commercial tablets. The sample solutions were spotted as 6-mm bands utilizing a "CAMAG Automatic TLC Sampler 4 (ATS4) Sample Applicator (CAMAG, Geneva, Switzerland)". The "CAMAG microliter Syringe (Hamilton, Bonaduz, Switzerland)" was linked with sample applicator. The application rate for the SMD of caffeine and paracetamol was fixed at 150 nL/s. Under linear ascending mode, the TLC plates were developed in a "CAMAG automated developing chamber 2 (ADC2) (CAMAG, Muttenz, Switzerland)" at a distance of 80 mm. For 30 min at 22 °C, the development chamber was saturated with vapors of greener mobile phases. Caffeine and paracetamol were detected using a wavelength of 260 nm. The slit size (band length × width) and scanning rate were both set at 4 mm × 0.45 mm and 20 mm/s, respectively. Three or six replicates were used for each estimation. The software used was "WinCAT's (version 1.4.3.6336, CAMAG, Muttenz, Switzerland)".

The greener normal-phase and reversed-phase HPTLC methodologies used the same instrumentation and analytical procedures as the normal-phase and reversed-phase HPTLC approaches. The TLC plates and the greener mobile phase components were found to be the most significant differences between the two procedures. In the greener normal-phase HPTLC technique, the TLC plates were glass plates (plate size: 10 cm × 20 cm) pre-coated with normal-phase silica gel (particle size: 5 μm) 60F254S plates, but in the greener reversed-phase HPTLC approach, the TLC plates were glass plates (plate size: 10 cm × 20 cm) pre-coated with reversed-phase silica gel (particle size: 5 μm) 60F254S plates. In both cases, the polymer-binder plate was not used. In the greener normal-phase HPTLC approach, the greener mobile phase was EA/E (85:15, v/v); however, in the greener reversed-phase HPTLC approach, the greener mobile phase was E/W (50:50, v/v).

3.3. Calibration Curves and QC Sample for Caffeine and Paracetamol

Caffeine and paracetamol stock solutions were made individually by dispensing the requisite amounts of both molecules in the specified amount of respective mobile phase, resulting in a final stock solution of 100 µg/mL for both compounds. The concentrations in the 50–500 ng/band range for caffeine and paracetamol were generated using the greener normal-phase HPTLC methodology and the 25–800 ng/band range for caffeine and paracetamol using the greener reversed-phase HPTLC methodology by diluting variable volumes of caffeine or paracetamol stock solution with the respective mobile phase. For the normal-phase HPTLC methodology, 200 µL of each concentration of caffeine and paracetamol were put to normal-phase TLC plates and reversed-phase TLC plates for the reversed-phase HPTLC methodology. Using both analytical techniques, the spot area of each concentration of caffeine and paracetamol was measured. Caffeine and paracetamol calibration curves were created by graphing the concentrations of both drugs against the observed spot area in six repeats ($n = 6$). For the determination of various validation parameters, three distinct QC samples were prepared freshly.

3.4. Processing of Samples for the SMD of Caffeine and Paracetamol in Commercial Tablets

Ten commercial tablets (each containing 65 mg of caffeine and 500 mg of paracetamol) were weighed and the average weights were computed for the SMD of caffeine and paracetamol in PANEXT and SAFEXT. Each brand's tablets were coarsely crushed and powdered. A portion of each brand's powder was dissolved in 100 mL of the relevant mobile phase. For the greener normal-phase and reversed-phase HPTLC methods, 1 mL of this solution of each brand of tablet was diluted again using 10 mL of the corresponding mobile phase. The prepared solutions of PANEXT and SAFEXT commercial tablets were filtered and sonicated for around ten minutes to remove any undissolved excipients. Using the greener normal-phase and reversed-phase HPTLC methods, the generated solutions were used to determine caffeine and paracetamol in commercial tablets PANEXT and SAFEXT.

3.5. Analytical Method Validation

Utilizing the ICH-Q2-R1 recommendations, the normal-phase and reversed-phase HPTLC techniques for the SMD of caffeine and paracetamol were validated for various parameters [41]. By graphing the concentrations of caffeine and paracetamol against their measured spot area, the linearity range for caffeine and paracetamol was discovered. The normal-phase HPTLC approach's linearity for caffeine and paracetamol was evaluated in the 50–500 ng/band range ($n = 6$). For the reversed-phase HPTLC method, the linearity for caffeine and paracetamol was evaluated in the 25–800 ng/band range ($n = 6$).

The calculation of R_f, As, and N/m was used to evaluate the parameters for the system acceptability for the greener normal-phase and reversed-phase HPTLC techniques for the SMD of caffeine and paracetamol. For both analytical approaches, the R_f, As, and N/m data were computed utilizing their reported equations [39].

The percent recovery was utilized to examine the accuracy of the normal-phase and reversed-phase HPTLC methods for the SMD of caffeine and paracetamol. For caffeine and paracetamol, the accuracy of the greener normal-phase HPTLC technique was tested at three QC levels: lower QC (LQC; 100 ng/band), middle QC (MQC; 300 ng/band), and high QC (HQC; 500 ng/band). For caffeine and paracetamol, the accuracy of the greener reversed-phase HPTLC technique was tested at three QC levels: LQC (50 ng/band), MQC (300 ng/band), and HQC (800 ng/band). Using both analytical techniques, the percent of recovery for caffeine and paracetamol ($n = 6$) was assessed at each QC level.

Intra/inter-assay precision was measured for the greener normal-phase and reversed-phase HPTLC methods for caffeine and paracetamol. Quantitation of newly prepared caffeine and paracetamol solutions at LQC, MQC, and HQC on the same day for both analytical techniques ($n = 6$), was used to examine intra-assay variation for caffeine and paracetamol. Quantitation of freshly prepared solutions at LQC, MQC, and HQC on three consecutive days for both analytical techniques ($n = 6$) was used to investigate inter-assay variation for caffeine and paracetamol.

For both analytical techniques, the robustness for caffeine and paracetamol was evaluated by making some slight purposeful modification in the mobile phase composition. The greener mobile phase EA/E (85:15, v/v) for caffeine and paracetamol was altered to EA/E (87:13, v/v) and EA/E (83:17, v/v) for the greener normal-phase HPTLC technique, and the variations in chromatographic response and R_f values were recorded ($n = 6$). The greener mobile phase E/W (50:50, v/v) for caffeine and paracetamol was altered to E/W (52:48, v/v) and E/W (48:52, v/v) for the greener reversed-phase HPTLC technique, and the variations in chromatographic response and R_f values were recorded ($n = 6$).

By using a "standard deviation" technique, the sensitivity of the greener normal-phase and reversed-phase HPTLC approaches for caffeine and paracetamol was examined as "LOD and LOQ". Caffeine and paracetamol "LOD and LOQ" were computed using their published equations for both analytical procedures ($n = 6$) [41].

The R_f values and UV spectra of caffeine and paracetamol in commercial tablets PANEXT and SAFEXT were compared with those of standards caffeine and paracetamol to determine the specificity of the greener normal-phase and reversed-phase HPTLC methods for caffeine and paracetamol.

3.6. Application of Greener Normal-Phase and Reversed-Phase HPTLC Approaches in the SMD of Caffeine and Paracetamol in Commercial Tablets

For the normal-phase HPTLC technique, the obtained solutions of the commercial tablets PANEXT and SAFEXT were put on normal-phase TLC plates and on reversed-phase TLC plates for the reversed-phase HPTLC technique. For all analytical techniques, the chromatographic responses were documented using the identical experimental circumstances employed for the SMD of standards caffeine and paracetamol ($n = 3$). For both analytical procedures, the quantities of caffeine and paracetamol in commercial tablets were approximated using the calibration curves for caffeine and paracetamol.

3.7. Greenness Estimation Using AGREE

The AGREE technique [38] was utilized to assess the greenness scale for the normal-phase and reversed-phase HPTLC procedures for the SMD of caffeine and paracetamol. The AGREE scales (0.0–1.0) for the greener normal-phase and reversed-phase HPTLC approaches was estimated utilizing "AGREE: The Analytical Greenness Calculator (version 0.5, Gdansk University of Technology, Gdansk, Poland, 2020)" for both the analytical approaches.

4. Conclusions

The literature lacks greener analytical techniques for the SMD of caffeine and paracetamol. As a result, compared to the greener normal-phase HPTLC approach, this research was carried out to develop and validate the rapid, sensitive, and greener reversed-phase HPTLC approach for the SMD of caffeine and paracetamol in their commercial tablets. For the SMD of caffeine and paracetamol, the greener reversed-phase HPTLC approach is more linear, accurate, precise, robust, and sensitive than the greener normal-phase HPTLC approach. The quantities of caffeine and paracetamol in commercial tablets PANEXT and SAFEXT were found to be significantly higher using the reversed-phase HPTLC methodology compared with the normal-phase HPTLC methodology. The AGREE estimation showed the excellent green properties of both the analytical approaches. For the SMD of caffeine and paracetamol in commercial formulations, the greener reversed-phase HPTLC approach has been presented superior to the greener normal-phase HPTLC approach based on different validation criteria and pharmaceutical assays.

Author Contributions: Conceptualization, supervision—P.A. and F.S.; Methodology—A.I.F., P.A., M.H.A. and T.M.A.; Validation—A.A. (Abuzer Ali), S.A. and F.S.; Data curation—A.A. (Amena Ali), F.K.A. and M.M.G.; Funding acquisition—A.A. (Abuzer Ali); Project administration—P.A.; Software—P.A., S.A. and F.S.; Writing original draft—F.S.; Writing—review and editing—A.A. (Abuzer Ali), M.M.G. and S.A. All authors have read and agreed to the published version of the manuscript.

Funding: This research was funded by the Taif University Researchers Supporting Project (Number TURSP-2020/124), Taif University, Taif, Saudi Arabia. The APC was funded by TURSP.

Acknowledgments: Authors are thankful to the Taif University Researchers Supporting Project (Number TURSP-2020/124), Taif University, Taif, Saudi Arabia for supporting this work.

Conflicts of Interest: The authors declare no conflict of interest.

Sample Availability: Samples of the caffeine and paracetamol compounds are available from the authors.

References and Note

1. Jimenez, J.A.; Martinez, F. Thermodynamic study of the solubility of acetaminophen in propylene glycol + water cosolvent mixtures. *J. Braz. Chem. Soc.* **2006**, *17*, 125–134. [CrossRef]
2. Shakeel, F.; Alanazi, F.K.; Alsarra, I.A.; Haq, N. Solubilization behavior of paracetamol in Transcutol-water mixtures at T = (298.15 to 333.15) K. *J. Chem. Eng. Data* **2012**, *58*, 3551–3556. [CrossRef]
3. Shakeel, F.; Ramadan, W. Transdermal delivery of anticancer drug caffeine from water-in-oil nanoemulsions. *Colloids Surf. B* **2010**, *75*, 356–362. [CrossRef]
4. Rahimi, M.; Khorshidi, N.; Heydari, R. Simultaneous determination of paracetamol and caffeine in aqueous samples by ultrasound-assisted emulsification microextraction coupled with high-performance liquid chromatography. *Sep. Sci. Plus* **2020**, *3*, 561–570. [CrossRef]
5. Medina, A.R.; de Cordova, M.L.F.; Molina-Diaz, A. Simultaneous determination of paracetamol, caffeine and acetylsalicylic acid by means of FI ultraviolet pls multioptosensing device. *J. Pharm. Biomed. Anal.* **1999**, *21*, 983–992. [CrossRef]
6. Tavallali, H.; Salami, M. Simultaneous determination of caffeine and paracetamol by zero-crossing second derivative spectrophotometry in pharmaceutical preparations. *Asian J. Chem.* **2009**, *21*, 1949–1956.
7. Tavallali, H.; Sheikhaei, M. Simultaneous kinetic determination of paracetamol and caffeine by H-point standard addition method. *Afr. J. Pure Appl. Chem.* **2009**, *3*, 11–19.
8. Aktas, A.H.; Kitis, F. Spectrophotometric simultaneous determination of caffeine and paracetamol in commercial pharmaceutical by principal component regression, least squares and artificial neural networks chemometric methods. *Croat. Chem. Acta* **2014**, *87*, 69–74. [CrossRef]
9. Uddin, M.N.; Mondol, A.; Karim, M.M.; Jahan, R.A.; Rana, A.A. Chemometrics assisted spectrophotometric method for simultaneous determination of paracetamol and caffeine in pharmaceutical formulations. *Bangladesh J. Sci. Ind. Res.* **2019**, *54*, 215–222. [CrossRef]
10. Sebaiy, M.M.; Mattar, A.A. H-point assay method for simultaneous determination of paracetamol and caffeine in panadol extra dosage forms. *Can. J. Biomed. Res. Technol.* **2020**, *3*, 1–6.
11. Altun, M.L. HPLC method for the analysis of paracetamol, caffeine and dipyrone. *Turk. J. Chem.* **2002**, *26*, 521–528.
12. Issa, I.M.; Hassouna, E.M.; Zayed, A.G. Simultaneous determination of paracetamol, caffeine, domperidone, ergotamine tartrate, propyphenazole, and drotaverine HCl by high performance liquid chromatography. *J. Liq. Chromatogr. Rel. Technol.* **2012**, *35*, 2148–2161. [CrossRef]
13. Tsvetkova, B.; Kostova, B.; Pencheva, I.; Zlatkov, A.; Rachew, D.; Peikov, P. Validated method for simultaneous analysis of paracetamol and caffeine in model tablet formulation. *Int. J. Pharm. Pharm. Sci.* **2012**, *4*, 680–684.
14. Cunha, R.R.; Chaves, S.C.; Ribeiro, M.M.A.C.; Torres, L.M.F.C.; Munoz, R.A.A.; Santos, W.T.P.D.; Richter, E.M. Simultaneous determination of caffeine, paracetamol, and ibuprofen in pharmaceutical formulations by high-performance liquid chromatography with UV detection and by capillary electrophoresis with conductivity detection. *J. Sep. Sci.* **2015**, *38*, 1657–1662. [CrossRef]
15. Acheampong, A.; Gyasi, W.O.; Darko, G.; Apau, J.; Addai-Arhin, S. Validated RP-HPLC method for simultaneous determination and quantification of chlorphenaramine maleate, paracetamol and caffeine in tablet formulation. *Springer Plus* **2016**, *5*, E625. [CrossRef]
16. Narayanan, V.L.; Austin, A. Determination of acetaminophen and caffeine using reverse phase liquid (RP-LC) chromatographic technique. *Quest J. Res. Pharm. Sci.* **2016**, *3*, 5–10.
17. Aminu, N.; Chan, S.-Y.; Khan, N.H.; Farhan, A.B.; Umar, M.N.; Toh, S.-M. A simple stability-indicating HPLC method for simultaneous analysis of paracetamol and caffeine and its application to determinations in fixed-dose combination tablet dosage form. *Acta Chromatogr.* **2019**, *31*, 85–91. [CrossRef]
18. Ali, J.G.; Muhammad, I.; Hamid, S.; Muhammad, A.A.; Shoaib, H.; Tasleem, S. Simultaneous determination and quantification of paracetamol, caffeine and orphenadrine citrate using stability-indicating HPLC method in a fixed dose combination tablet dosage form. *Ann. Pharmacol. Pharm.* **2020**, *5*, E118.

19. Belal, F.; Omar, M.A.; Derayea, S.; Zayed, S.; Hammad, M.A.; Saleh, S.F. Simultaneous determination of paracetamol, caffeine and codeine in tablets and human plasma by micellar liquid chromatography. *Eur. J. Chem.* **2015**, *4*, 468–474. [CrossRef]
20. Wang, A.; Sun, J.; Feng, H.; Gao, S.; He, Z. Simultaneous determination of paracetamol and caffeine in human plasma by LC-ESI-MS. *Chromatographia* **2008**, *67*, 281–285. [CrossRef]
21. Tavallali, H.; Zareiyan, S.F.; Naghian, M. An efficient and simultaneous analysis of caffeine and paracetamol in pharmaceutical formulations using TLC with a fluorescence plate reader. *J. AOAC Int.* **2011**, *94*, 1094–1099. [CrossRef]
22. Chabukswar, A.R.; Thakur, V.G.; Dharam, D.L.; Shah, M.H.; Kuchekar, B.S.; Sharma, S.N. Development and validation of HPTLC method for simultaneous estimation of paracetamol, ibuprofen and caffeine in bulk and pharmaceutical dosage form. *Res. J. Pharm. Technol.* **2012**, *5*, 1218–1222.
23. Halka-Grysinska, A.; Slazak, P.; Zareba, G.; Markowski, W.; Klimek-Turek, A.; Dzido, T.H. Simultaneous determination of acetaminophen, propyphenazone and caffeine in cefalgin preparation by pressurized planar electrochromatography and high-performance thin-layer chromatography. *Anal. Methods* **2012**, *4*, 973–982. [CrossRef]
24. Lau, O.-H.; Luk, S.-F.; Cheung, Y.-M. Simultaneous determination of ascorbic acid, caffeine and paracetamol in drug formulations by differential-pulse voltammetry using a glassy carbon electrode. *Analyst* **1989**, *114*, 1047–1051. [CrossRef] [PubMed]
25. Yigit, A.; Yardrm, Y.; Senturk, Z. Voltametric sensor based on boron-doped diamond electrode for simultaneous determination of paracetamol, caffeine and aspirin in pharmaceutical formulations. *IEEE Sens.* **2016**, *16*, 1674–1680. [CrossRef]
26. Minh, T.T.; Phong, N.H.; Duc, H.V.; Khieu, D.Q. Microwave synthesis and voltametric simultaneous determination of paracetamol and caffeine using an MOF-199-based electrode. *J. Mater. Sci.* **2018**, *53*, 2453–2471. [CrossRef]
27. Hung, N.X.; Quang, D.A.; Toan, T.T.T.; Dung, N.N. The simultaneous determination of ascorbic acid, paracetamol, and caffeine by voltammetry method using cobalt Schiff base complex/SBA-15 modified electrode. *ECS J. Solid State Sci. Technol.* **2020**, *9*, E101004. [CrossRef]
28. Ibrahim, H.; Hamdy, A.M.; Merey, H.A.; Saad, A.S. Simultaneous determination of paracetamol, propyphenazone and caffeine in presence of paracetamol impurities using dual-mode gradient HPLC and TLC densitometry methods. *J. Chromatogr. Sci.* **2021**, *59*, 140–147. [CrossRef]
29. Meter, M.I.V.; Khan, S.M.; Taulbee-Cotton, B.V.; Dimmitt, N.H.; Hubbard, N.D.; Green, A.M.; Webster, G.K.; McVey, P.A. Diagnosis of agglomeration and crystallinity of active pharmaceutical ingredients in over the counter headache medication by electrospray desorption ionization mass spectrometry imaging. *Molecules* **2021**, *26*, 610. [CrossRef]
30. Katseli, V.; Economou, A.; Kokkinos, C. A novel all-3D-printed cell-on-a-chip device as a useful electroanalytical tool: Application to the simultaneous voltametric determination of caffeine and paracetamol. *Talanta* **2020**, *208*, 120388. [CrossRef]
31. Boltia, S.A.; Soudi, A.T.; Elzanfaly, E.S.; Zaazaa, H.E. Effect of genetic algorithm-based wavelength selection as a preprocessing tool on multivariate simultaneous determination of paracetamol, orphenadrine citrate, and caffeine in the presence of p-aminophenol impurity. *J. AOAC Int.* **2020**, *103*, 250–256. [CrossRef]
32. Muntean, D.M.; Alecu, C.; Tomuta, I. Simultaneous quantification of paracetamol and caffeine in powder blends for tableting by NIR-chemometry. *J. Spectrosc.* **2017**, *2017*, E7160675. [CrossRef]
33. Ortega-Barrales, P.; Padilla-Weigand, R.; Molina-Diaz, A. Simultaneous determination of paracetamol and caffeine by flow injection-solid phase spectrometry using C_{18} silica gel as a sensing support. *Anal. Sci.* **2002**, *18*, 1241–1246. [CrossRef]
34. Kulikov, A.U.; Verushkin, A.G. Simultaneous determination of paracetamol, caffeine, guaifenesin and preservatives in syrups by micellar LC. *Chromatographia* **2008**, *67*, 347–355. [CrossRef]
35. Emre, D.; Ozaltrn, N. simultaneous determination of paracetamol, caffeine and propyphenazone in ternary mixtures by micellar electrokinetic capillary chromatography. *J. Chromatogr. B* **2007**, *847*, 126–132. [CrossRef] [PubMed]
36. Abdelrahman, M.M.; Abdelwahab, N.S.; Hegazy, M.A.; Fares, M.Y.; El-Sayed, G.M. Determination of the abused intravenously administered madness drops (tropicamide) by liquid chromatography in rat plasma; an application to pharmacokinetic study and greenness profile assessment. *Microchem. J.* **2020**, *159*, E105582. [CrossRef]
37. Duan, X.; Liu, X.; Dong, Y.; Yang, J.; Zhang, J.; He, S.; Yang, F.; Wang, Z.; Dong, Y. A green HPLC method for determination of nine sulfonamides in milk and beef, and its greenness assessment with analytical eco-scale and greenness profile. *J. AOAC Int.* **2020**, *103*, 1181–1189. [CrossRef]
38. Pena-Pereira, F.; Wojnowski, W.; Tobiszewski, M. AGREE-Analytical GREEnness metric approach and software. *Anal. Chem.* **2020**, *92*, 10076–10082. [CrossRef] [PubMed]
39. Foudah, A.I.; Shakeel, F.; Alqarni, M.H.; Alam, P. A rapid and sensitive stability-indicating green RP-HPTLC method for the quantitation of flibanserin compared to green NP-HPTLC method: Validation studies and greenness assessment. *Microchem J.* **2021**, *164*, E105960. [CrossRef]
40. Alam, P.; Salem-Bekhit, M.M.; Al-Joufi, F.A.; Alqarni, M.H.; Shakeel, F. Quantitative analysis of cabozantinib in pharmaceutical dosage forms using green RP-HPTLC and green NP-HPTLC methods: A comparative evaluation. *Sus. Chem. Pharm.* **2021**, *21*, E100413. [CrossRef]
41. International Conference on Harmonization (ICH), Q2 (R1): Validation of Analytical Procedures–Text and Methodology, Geneva, Switzerland. 2005.

Article

Application of Skyline for Analysis of Protein–Protein Interactions In Vivo

Arman Kulyyassov

Republican State Enterprise "National Center for Biotechnology" under the Science Committee of Ministry of Education and Science of the Republic of Kazakhstan, 13/5, Kurgalzhynskoye Road, Nur-Sultan 010000, Kazakhstan; kulyyasov@biocenter.kz; Tel.: +7-7172-707534

Abstract: Quantitative and qualitative analyses of cell protein composition using liquid chromatography/tandem mass spectrometry are now standard techniques in biological and clinical research. However, the quantitative analysis of protein–protein interactions (PPIs) in cells is also important since these interactions are the bases of many processes, such as the cell cycle and signaling pathways. This paper describes the application of Skyline software for the identification and quantification of the biotinylated form of the biotin acceptor peptide (BAP) tag, which is a marker of in vivo PPIs. The tag was used in the Proximity Utilizing Biotinylation (PUB) method, which is based on the co-expression of BAP-X and BirA-Y in mammalian cells, where X or Y are interacting proteins of interest. A high level of biotinylation was detected in the model experiments where X and Y were pluripotency transcription factors Sox2 and Oct4, or heterochromatin protein HP1γ. MRM data processed by Skyline were normalized and recalculated. Ratios of biotinylation levels in experiment versus controls were 86 ± 6 (3 h biotinylation time) and 71 ± 5 (9 h biotinylation time) for BAP-Sox2 + BirA-Oct4 and 32 ± 3 (4 h biotinylation time) for BAP-HP1γ + BirA-HP1γ experiments. Skyline can also be applied for the analysis and identification of PPIs from shotgun proteomics data downloaded from publicly available datasets and repositories.

Keywords: biotin acceptor peptide (BAP); biotin ligase BirA; liquid chromatography tandem mass spectrometry (LC-MS/MS); multiple reaction monitoring (MRM); protein–protein interactions (PPIs); proximity utilizing biotinylation (PUB); proteomics

1. Introduction

Wide practical application of liquid chromatography in combination with mass spectrometry has been observed recently in proteomics [1,2] and metabolomics [3,4] as a routine method for the qualitative and quantitative analysis of biological samples. For example, when optimizing expression, performing quality control, or studying pharmacokinetics of recombinant proteins, it is crucial that the best conditions for production or analysis of the drug products are found [5,6]. Another important task is to obtain information about changes in the expression of marker proteins under different physiological conditions of the cell [7]. Examples of this include: differences in protein composition in a healthy/cancer cell or differences under the influence of external factors such as temperature, chemical agents, or radiation provide valuable information about metabolic and signaling pathways, mechanisms of stress response. In all these cases, the results are obtained as chromatograms in the multiple reaction monitoring (MRM) method, where many peptides derived from target proteins can be identified by retention time and mass spectra of fragment ions (or MS/MS spectra), and the relative amount of each peptide between samples can be determined by comparison of the peak areas [8–10].

However, information on protein composition is not sufficient to fully understand the mechanism of cell function. The quantification of protein–protein interactions (PPIs) in vivo can be a useful extension in research since more than 80% of proteins do not function separately, but rather interact and participate in the formation of stable or transient complexes [11]. These protein–protein interactions play an important role in almost all vital processes in cells, such as DNA replication, gene transcription and translation, signal transduction, cell-cycle control and proliferation, and cell–cell communication [12].

Methods based on a combination of affinity purification (AP) or tandem affinity purification (TAP) and mass spectrometry (MS/MS) have now become standard for the identification of protein partners [13–16]. However, these methods have the serious drawback of a large number of false-positive identifications [17]. In addition, the cell lysis procedure can lead to the destruction of weak protein–protein interactions, which can also lead to false-negative results. For example, the list of Oct4-interacting proteins identified by co-immunoprecipitation (Co-IP) did not include one of the most studied Oct4 partners, namely Sox2 [18].

Enzymatically catalyzed proximity labeling is an alternative to immunoprecipitation and biochemical fractionation for the proteomic analysis of macromolecular complexes and protein interaction networks [19]. In this method, ligation enzymes are expressed in cells as conjugates with proteins of interest. For example, proximity-dependent biotinylation methods are based on the use of mutant biotin ligases, BioID [20] or TurboID [21]. These BirA mutants prematurely release the highly reactive yet labile biotinoyl-AMP inside of a living cell, which readily reacts with lysine's primary amino groups of proximal proteins.

On the other hand, the proximity utilizing biotinylation (PUB) method is based on the use of humanized wild-type biotin ligase BirA fusions. The wild-type BirA uses biotin and ATP to generate biotinoyl-AMP [22,23]. Wild-type BirA holds on to the reactive biotin molecule until it is covalently attached to a very specific substrate called biotin acceptor peptide (BAP) [24]. Thus, biotinylation is a result of the direct contact of BirA and BAP parts of recombinant proteins which occurs in cases of protein–protein interaction or random collision in vivo.

Both methods are based on similar principles; this is the in vivo creation of a permanent covalent mark on one of the proteins of interest or partners interacting with them, which allows us to bypass the limitations imposed by the extraction and purification stages. Ultimately, results will be obtained with much fewer false-positive and false-negative protein identifications compared to traditional methods such as IP-MS/MS or TAP. The result is the facilitation of the bioinformatic part of data analysis.

The aim of this work was to use the Skyline program to process the results of experiments on the quantitative analysis of PPI using the proximity utilizing biotinylation (PUB) method.

2. Results and Discussion

2.1. Overview of the Method and Experimental Workflow of the PUB Protocol

The principle of the PUB method is based on using enzyme/substrate pair reactions [25–28], where two proteins to be tested for their interaction in vivo are co-expressed in mammalian cells, one as fused to the BAP, and the other fused to an enzyme BirA, which is an *Escherichia coli* protein biotin ligase [29]. When the two proteins are in proximity to each other, for example, when an interaction of X and Y occurs in vivo, a more efficient biotinylation of the BAP is to be expected (Figure 1A). The biotinylation status of the BAP fusion protein can be further monitored by Western blot, mass spectrometry, or confocal microscopy (Figure 1B,C). HEK293T, HeLa, or MRC-5 fetal lung fibroblast cell lines can be used for the transient or stable expression of recombinant proteins in the PUB method. Usually, one control experiment is performed in a parallel dish or a 6-well plate, using cells in which non-interacting proteins or other pair proteins are expressed for comparison (BirA-X and BAP-Z). Depending on the proteins of interest chosen for the experiment, biotin is added to the medium from 5 min to 9 h before harvesting the cells. The sequence of the BAP peptide was modified and compared to commonly used peptides, such as Avitag [30,31], in order to reduce the level of background biotinylation [28]. Additionally, 7His-tag was added upstream of the sequence to provide the option to purify both labeled and nonlabeled BAP fusion proteins from cell lysates.

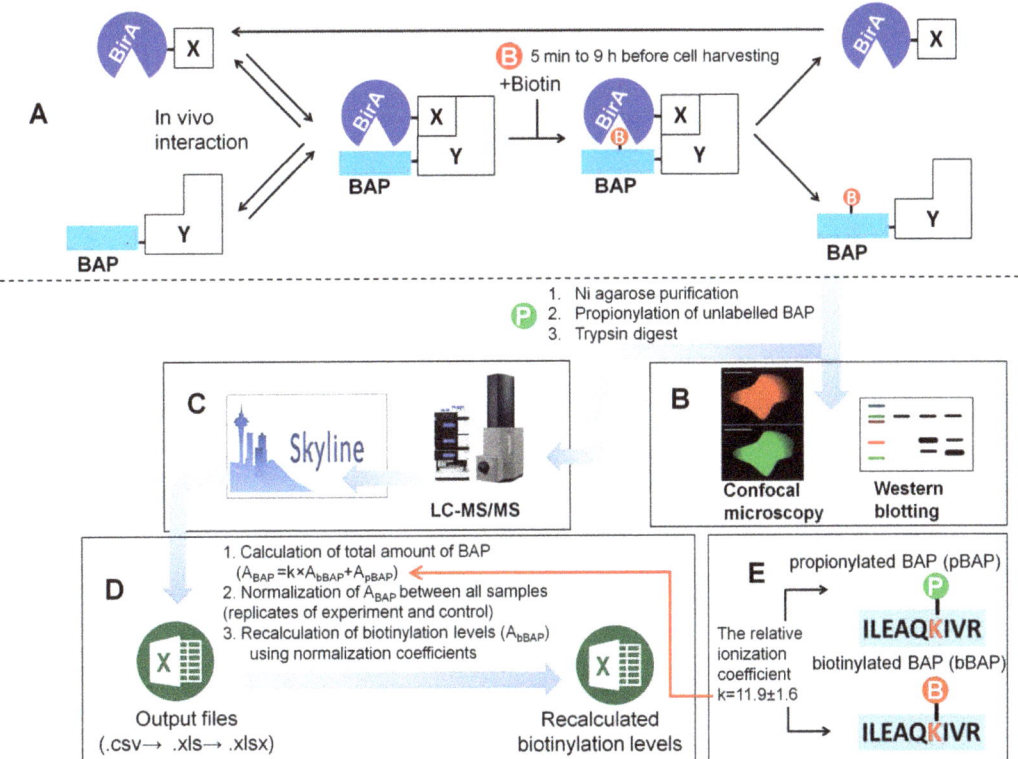

Figure 1. Principle of PUB method in living cell and workflow for quantification of PPI. In vivo interaction (**A**) of proteins X and Y results in site-specific biotinylation of the biotin acceptor peptide (BAP) by wild-type humanized biotin ligase (BirA). Biotinylated protein can be detected by WB, for example, Streptavidin-HRP, IF confocal microscopy (**B**) or LC-MS/MS (**C**). X or Y–HP1(α,β,γ), Tap54(α,β), Sox2, Oct4, or other proteins. Biotinylation levels of BAP peptides obtained after processing the results of sample analyses using Skyline (**C**) are recalculated in Microsoft Excel (**D**). P—propionylated form of BAP1070: GHHHHHHGLTR**ILEAQK(Prop)**IVRGG, B—biotinylated form of BAP1070: GHHHHHHGLTR**ILEAQK(Biot)**IVRGG, the sequence corresponding to the peptide on the chromatogram after trypsin digest is marked in bold. The relative ionization coefficient of tryptic peptides derived from propionylated and biotinylated BAP (**E**).

The experimental workflow for the LC-MS/MS analysis of samples includes additional steps, such as the purification of Ni agarose beads, propionylation, and on-gel (or on-bead) tryptic digest (Figure 1C). Propionylation was used to protect the nonbiotinylated BAP peptide from tryptic cleavage on the target lysine. This modification resulted in the production of modified and nonmodified peptides of comparable sizes, facilitating the interpretation of results. After analysis on Skyline (Figure 1D), data were exported as CSV files and processed using Microsoft Excel for the calculation of biotinylation levels (Supplementary datas S1 and S2). First, the total amount of BAP was calculated by the addition of the total area of propionylated BAP to the total area of recalculated biotinylated BAP. For the recalculation of the biotinylated BAP, the relative ionization coefficient $k = 11.9$ was used (Figure 1E), which was estimated earlier in SILAC experiments [28]. The ionization efficiency depends on the chemical structure of a molecule and would thus be different for propionyl and biotin residues. Therefore, a direct comparison between total ion chromatograms (TIC) of the biotinylated and propionylated BAP in LC-MS/MS data is not possible. After this step, the areas were normalized, and normalization coefficients of the total amounts of BAP were calculated for each sample. These normalization coefficients were then used to recalculate the biotinylation levels and for the estimation of means and standard deviations.

The total amount of BAP was calculated using the formula $A_{BAP} = k \times A_{bBAP} + A_{pBAP}$, where A_{bBAP} corresponds to the peak area of total ion chromatograms (TIC) of biotinylated, and A_{pBAP} to propionylated BAP, and k is the relative ionization coefficient between the biotinylated and propionylated BAP peptides (Figure 1D). The chromatographic elution peaks of the fragments for the four most intensive ions, y_7, y_6, y_5, and y_4, in extracted ion chromatograms (EIC) were integrated and summed to give the peak area of TIC. In the control (BAP-Y + BirA-Z) and in the experiment (BAP-Y + BirA-X), as well as in the replicates, the expression levels of recombinant BAP-Y proteins may differ. Variations in the total amount of BAP-Y can also appear during a sample preparation. Thus, direct comparison of the biotinylation levels A_{bBAP} between samples is not correct. Therefore, the total amount of BAP-Y, including its biotinylated and propionylated forms in all samples, was normalized and the normalization coefficients were determined (Supplementary data S2, Table S1).

The Skyline is an application for targeted proteomics and quantitative data analysis in the frame of the Windows operation system [32]. Its interface facilitates the improvement of mass spectrometer methods and the analysis of data from targeted MRM experiments. Skyline imports the native output files from instruments manufactured from different vendors smoothly, connecting mass spectrometer output back to the experimental design document. A rich choice of graphics displays provide powerful tools for inspecting and monitoring data integrity as data are acquired, helping instrument operators to identify problems early. It is open-source and freely available for commercial and academic use [9,33]. In addition, its output data format (csv.files) allows the performance of post-processing analysis in Microsoft Excel to recalculate biotinylation levels.

This software was successfully used to identify and quantify the target BAP peptides from all MRM data (Supplementary data S1, Figures S1 and S2). Since the amino acid sequence of the BAP1070 peptide is artificially generated and is absent in the NCBI and Swissprot databases, the sequence of this peptide was added to a client-made database, BAP1070, using the Database manager (Figure 2A). This allowed the DAT file to be generated and the spectral libraries to be created in Skyline.

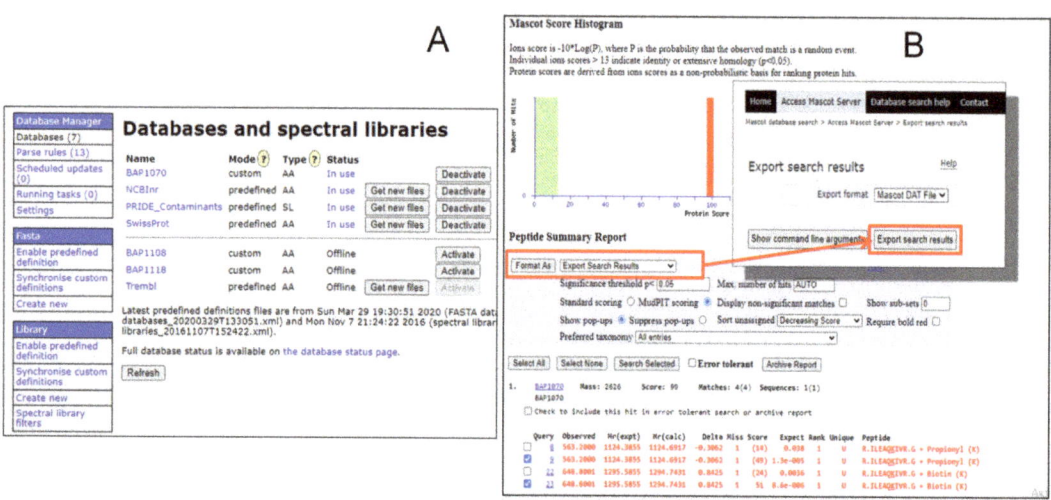

Figure 2. Creation of BAP1070 file. (**A**) Screenshots of the database manager page with BAP1070 file, (**B**) Mascot search results and exporting the DAT file.

Prior to importing raw data into Skyline, a spectral library containing the product ion spectra of the BAP target peptides was constructed using the DAT file. The spectral library consisted of MS/MS spectra of biotinylated and propionylated forms of BAP1070 peptide. A spectral library allowed for the direct comparison of BAP target peptide product ion spectra from the MRM analyses to the corresponding product ion "library match". Product ion transitions used to confirm the identity of each target peptide in the MRM analyses were automatically picked based on the four most abundant y-type product ion intensities observed in the "library match" spectrum.

2.1.1. Creation of MRM Method and LC-MS/MS Analysis of the Samples

The vendors' default method (Bruker Company) was used for the creation of the MRM method, as described earlier [34]. Precursor ions: m/z 563.2 (ILEAQK(Prop)IVR) propionylated form of BAP, and m/z 648.8 (ILEAQK(Biot)IVR) biotinylated form of BAP.

2.1.2. Creation of BAP1070 Database on Mascot Search Server Using Database Manager

Before analysis on Skyline, the raw LC-MS/MS data were processed to a special format—DAT file. First, the raw data were analyzed on Bruker DataAnalysis software to generate an MGF file. Then, the BAP1070 database was created containing a sequence of this peptide in the Database manager (Figure 2A), which is a browser-based utility for updating and configuring local copies of sequence databases. Analysis of an MGF file against the BAP1070 database in the Mascot search engine, including propionylation and biotinylation modifications, yielded a report where results could also be exported as a DAT file (Figure 2B).

2.1.3. MRM Analysis and Post-Processing of Data

All MRM data were analyzed in Skyline 19.1.0.193. A spectral library was constructed from the peptide identifications from a DAT file exported from the Mascot result page. The four product ions extracted by Skyline were determined based on the ranking of the top four most intense y-ions from the corresponding library spectrum for each peptide. Dot-product (dotp) scores were calculated based on the correlation of the measured product ion peak intensities with the peak intensities observed in the library spectrum for that same peptide [33]. Raw LC-MS/MS data and processed files were uploaded to the Panorama repository [35].

2.2. DNA Dependent Interaction of Sox2 and Oct4

After processing and recalculating Skyline results, quantitative data on biotinylation levels were obtained (Figure 3A,B). The samples from experiments with the co-expression of BAP-Sox2 and BirA-Oct4 in HEK293T cells showed a high level of biotinylation (86 ± 6 and 71 ± 5 for different biotinylation times). This is due to the presence of DNA binding domains, HMG present in Sox2 and POU in Oct4, which recognize *Utf1* or other motifs [36] and result in close contact between target BAP and biotin ligase BirA. Contrary to Sox2, GFP lacks DNA binding domains, and in a control experiment with the coexpression of BAP-GFP and BirA-Oct4, very low biotinylation levels were observed (sample 0–9 on Figure 3A). Recombinant proteins BAP-GFP and BAP-Sox2 from HEK293T cell nuclear lysates were purified on Ni sepharose beads and propionylated before trypsin digest, as described earlier [34].

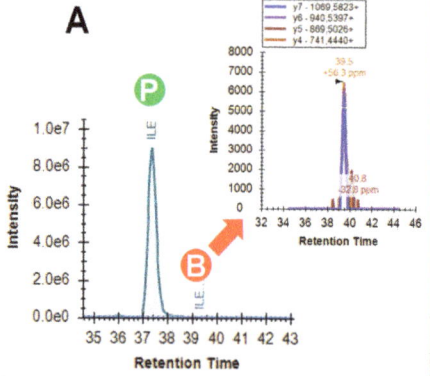
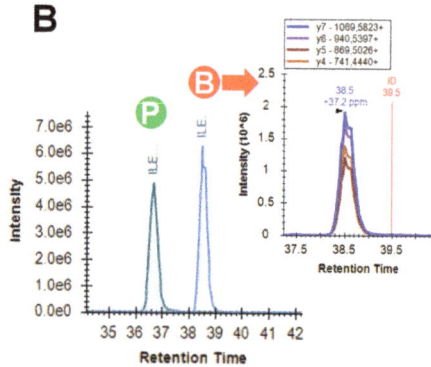
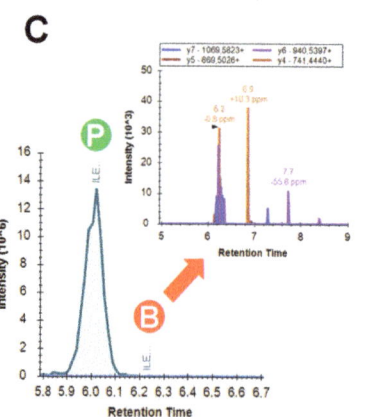
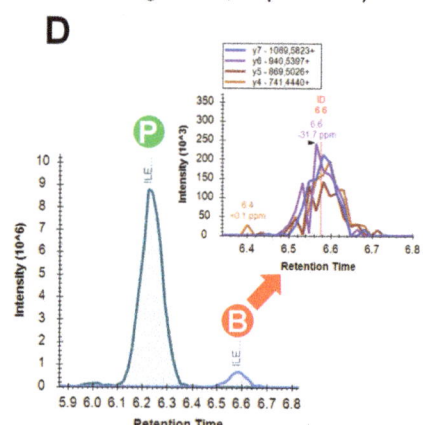

Figure 3. Skyline graphical representation of chromatograms from PUB experiments. Total ion chromatograms of propionylated (P in green circle) and biotinylated (B in red circle) BAP peptides for two examples of experimental PPIs (**B**,**D**) obtained using different instrument platforms (**A–B**, **C–D**). Four most intense fragment ions, y_7, y_6, y_5 and y_4, were chosen for area calculation of biotinylated BAP peptide in extracted ion chromatograms. Left side are controls BAP-GFP + BirA-Oct4 (**A**) and BAP-HP1γ + BirA-Tap54α (**C**) and right side of the figure represents experiments with interacting proteins—BAP-Sox2 + BirA-Oct4 (**B**) and BAP-HP1γ + BirA-HP1γ (**D**). The average ratios of biotinylation levels were obtained from three experiments after recalculation and normalization.

2.3. Protein Oligomerization HP1γ-HP1γ

Heterochromatin protein HP1γ was chosen as another example of protein–protein interactions. The proteins of this family contain a chromo shadow domain (CSD), which allows them to form dimers and oligomers [37,38]. The formation of these oligomeric structures is critical for the organization of heterochromatin in the cell nucleus [39]. In the experiment, BAP-HP1γ and BirA-HP1γ protein pairs were expressed. Another protein, TAP54α, participates in the formation of hexamers with ATPase activity and is a component of histone acetyltransferase complexes [40,41]. Since Tap54α is not a protein that interacts with HP1γ, we chose a model where other protein pairs, BAP-HP1γ and BirA-Tap54α, were expressed in a separate dish as a control. The difference in the biotinylation level of the experiment (BAP-HP1γ + BirA-HP1γ) versus control (BAP-HP1γ + BirA-Tap54α) after processing the raw mass spectrometer data with Skyline was also significant and was 32 ± 3 (Figure 3C,D). The raw data were obtained on an Agilent nanoHPLC-Chip-3D6340 Iontrap instrument and converted to mzML format [42] using ProteoWizard software [43,44].

2.4. Analysis of Shotgun Proteomics Data (Mutant Biotin Ligase BioID Application)

The Skyline program was mainly developed for the analysis of MRM results in targeted proteomics [32]. However, this application can also be used to analyze the results of Shotgun proteomics where an instrument is operated in data-dependent acquisition (DDA) mode. For validation, the raw data from the results of the BioID mutant biotin ligase experiments, published recently by Go et al. [45] and publicly available in the Massive repository, were downloaded. Go et al. identified 35,902 interactions with 4424 unique high-confidence proximity interactors for 192 BioID fused bait proteins from different cellular compartments. The MGF file was used to obtain the DAT file, as described earlier in Section 2.1.2, which was used in Skyline to build the spectral library. Since these results were obtained on a different instrument platform (Eksigent NanoLC-Ultra 2D plus HPLC system-Orbitrap Elite) and under different modes of operation, the parameters in the Peptide setting and Transition settings tabs were changed, as described in the experimental part from paper [45]. An example of interacting (or proximal) proteins of mitochondrial Pyruvate Dehydrogenase E1 Subunit Alpha 1 (PDHA1) fused with BioID is shown in Figure 4.

2.5. Perspectives of Enzymatically Catalyzed Labeling for Biological Research

Methods based on the use of wild-type biotin ligase BirA and their mutant versions, BioID or TurboID, have a similar principle: the creation of a permanent covalent label on the partner protein in vivo. This facilitates subsequent steps in protocols and especially easy and efficient purification of biotinylated proteins from cell lysates using commercially available reagents and kits.

In addition, these methods can be complementary to each other. For example, while the use of BioID or TurboID allows the identification of proximal or partner proteins in cells, the PUB method can be used to quantitatively compare the identified protein–protein proximity.

The MRM method, where the output results are presented as coupled data of chromatographic parameters and mass spectra for each peptide, is now widely used to study the mechanisms of external influences (temperature, radiation, or chemical reagents) on the expression of various proteins in a cell. By analogy, the PUB method can be used to study the influence of various external factors on protein–protein interactions in a living cell, which can extend a given research area and provide additional information about cell organization and function.

Thus, the use of modern bioinformatics programs such as Skyline in combination with PUB, BioID, and TurboID methods will facilitate the analyses of large amounts of data to solve various problems of cell and molecular biology.

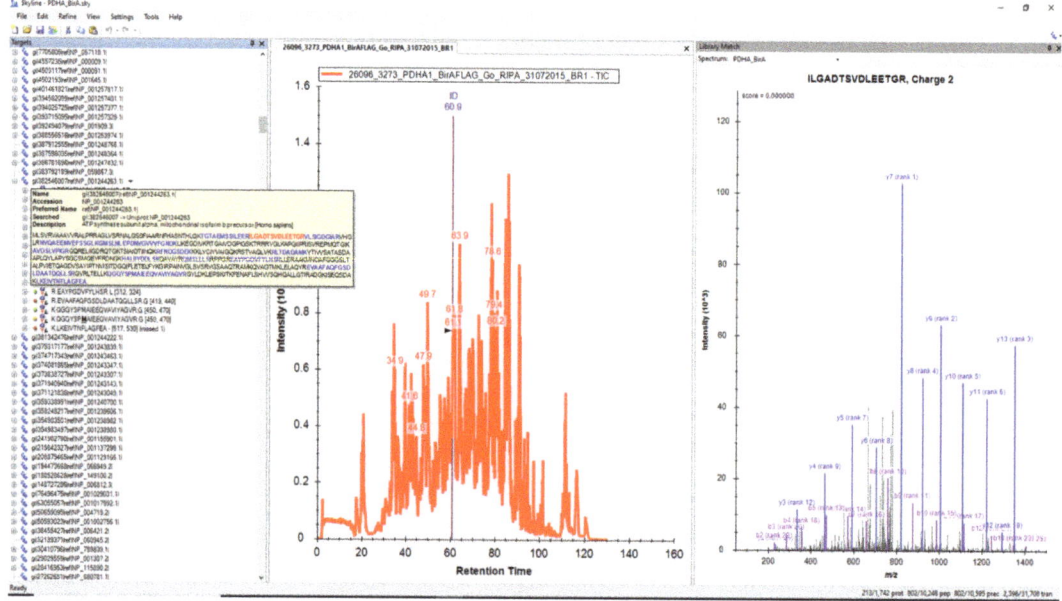

Figure 4. Results of analysis of data from paper Go et al. [45] using Skyline program. All peptides are grouped into lists from interacting or proximal proteins on the left side, the total ion chromatogram (TIC) in the center and MS/MS spectra are shown in the right part of this figure. Example of ATP synthase (inset on left side) as a protein, proximal to PDHA1.

3. Materials and Methods

Cell culture, transient transfection, and sample preparation steps were described earlier [28,34].

The peptide mixtures were analyzed using two LC-MS/MS systems:

1. Nanoflow HPLC system (Thermo Dionex Ultimate 3000, ThermoScientific) with Acclaim PepMap100 C18 pre-column, 5 mm × 300 μm; 5 μm particles (Thermo Scientific, #160454) and Acclaim Pep-Map RSLC column 15 cm × 75μm, 2 μm particles (Thermo Scientific, #164534) coupled via CaptiveSpray to the QTOF Impact II mass spectrometer (Bruker). Raw LC-MS/MS data were interpreted with the Bruker Compass DataAnalysis (version 4.3) software. The separation gradient was 48 min from 2% to 50% acetonitrile. Flow rates—300 nL/min.

2. Nano-HPLC (Agilent Technologies 1200) was coupled to an ion-trap mass spectrometer (Bruker 6300 series) equipped with a nanoelectrospray source via protein HPLC Chip (Agilent Technologies, G4240-62001) with 40 nL trap 75 um × 43 mm 5 um 300SB-C18-ZX and analytical column packed with ZORBAX 300SB-C18, 5 μm particle size. The separation gradient was 7 min from 5% to 90% acetonitrile. Flow rates—300 nL/min.

The LC-MS/MS instruments were set to monitor transitions of biotinylated (m/z 648.8, collision energy 33.0 eV) and propionylated (m/z 563.2, collision energy 27.0 eV) forms of BAP peptide in samples.

3.1. Data Preparation and Creation of MGF File Using DataAnalysis

For preliminary analysis of data and generation of peak lists, DataAnalysis (DA 4.1) software was used. Retention-time information was changed to seconds.

The MGF file was generated from raw data by clicking the following tabs on the menu: Find/Compounds MS(n)→Deconvolute/Mass spectra → File/Export/Compounds.

Subsequent database searches were performed using Mascot search engine. Then, the results were imported as the DAT file which were used to build a spectral library in Skyline.

3.2. Creation of BAP1070 Fasta Database on Mascot Search Server

In MS Notepad text editor, the aminoacid sequence of BAP1070 was pasted with the description line as follows:

>BAP1070
GHHHHHHHGLTRILEAQKIVRGG

This file was saved as BAP1070_fasta.txt

On mascot server http://mascot-server/mascot/index.html (Configuration last updated Thu Apr 15 10:36:27 2021), the BAP1070 database was created using the following steps: Home subpage → MascotUtilities → ConfigurationEditor → Database Manager → Fasta → Create new. Configuration details for BAP1070: Database name—BAP1070, Database type—aminoacid, Accession parse rule— > [^] *\(.*\), Description parse rule > [^] *\(.*\), Taxonomy source—none, Sequence report source—FASTA file, Full-text report source—None, Number of threads—automatic, Use memory mapping?—Yes, Lock to memory?—No.

Analysis of data, including Building a Spectral Library in Skyline, Configuring Transition Settings, Populating the Skyline Peptide Tree, Importing Raw Data into Skyline and Subsequent Filtering, and data processing and calculation of biotinylation levels are described in Supplementary data S1, Figures S1 and S2.

4. Conclusions

In this study, the Skyline program was used for the first time to analyze results obtained by using a proximity utilizing biotinylation method based on expression in mammalian target cells BAP-X and wild-type BirA-Y protein conjugates (first example: X-Sox2, Y-Oct4, versus control X-GFP, Y-Oct4 and second: example X, Y-HP1γ versus control X-HP1γ, Y- Tap54α). Peak areas of biotinylated BAP were used for the estimation of PPI, while peak areas of propionylated BAP on MRM chromatograms were used for the recalculation and normalization of data between different samples. This program allowed for fast processing of raw data, the calculation of peak areas, and provided the output file in CSV format, which is convenient for subsequent analysis on Microsoft Excel.

Skyline was also used to analyze data on protein–protein interactions and proximities obtained by using mutant biotin ligase BioID [45]. These raw data were downloaded from the MassIVE Repository database and were sourced from another LC-MS/MS instrument platform, demonstrating that the Skyline program is not "instrument or vendor-oriented".

Overall, the Skyline program offers an advantage in that it provides a good graphic representation of data and reduces analysis time. This protocol could be applicable, not only to BAP, but also to other synthetic peptides which are absent in NCBI or SwissProt databases.

Supplementary Materials: The following are available online, Figure S1: Extracted ion chromatograms (EICs) of the top four ranked y-ions, Figure S2: Relative quantification diagrams for biotinylated and propionylated forms of BAP after processing on Skyline Table S1: Calculation of corrected biotinylation levels.

Funding: This research was funded by a grant from the Ministry of Education and Science of the Republic of Kazakhstan AP09259838 "Application of new proteomics methods in studying the mechanism of action of pluripotency transcription factors expressed in mammalian cell lines" for 2021-2023 (State registration number 0121PK00163).

Institutional Review Board Statement: Not applicable.

Informed Consent Statement: Not applicable.

Data Availability Statement: The MS proteomics data have been deposited into the ProteomeXchange Consortium via the PRIDE partner repository, with the data set identifier PXD015756.

Acknowledgments: A.K. is greatly thankful to Yerlan Ramankulov (National Center for Biotechnology, Kazakhstan) for funding acquisition, Ruslan Kalendar (Department of Agricultural Sciences, University of Helsinki, FI-00014 Helsinki, Finland,) for help in reading and editing this manuscript, and Madina Zhunusova (National Center for Biotechnology, Kazakhstan) for assistance with HEK293T cell culture.

Conflicts of Interest: The author declares no conflict of interest.

Sample Availability: The vector plasmids, pcDNA3-BAP-Sox2 and pOz-humBirA-GFP, for transient transfection in cells are available from Addgene (Addgene ID 133281 and 133283, respectively).

References

1. Aebersold, R.; Mann, M. Mass-spectrometric exploration of proteome structure and function. *Nature* **2016**, *537*, 347–355. [CrossRef]
2. Sobsey, C.A.; Ibrahim, S.; Richard, V.R.; Gaspar, V.; Mitsa, G.; Lacasse, V.; Zahedi, R.P.; Batist, G.; Borchers, C.H. Targeted and Untargeted Proteomics Approaches in Biomarker Development. *Proteomics* **2020**, *20*, e1900029. [CrossRef]
3. Domenick, T.M.; Gill, E.L.; Vedam-Mai, V.; Yost, R.A. Mass Spectrometry-Based Cellular Metabolomics: Current Approaches, Applications, and Future Directions. *Anal. Chem.* **2021**, *93*, 546–566. [CrossRef] [PubMed]
4. Gonzalez-Riano, C.; Dudzik, D.; Garcia, A.; Gil-de-la-Fuente, A.; Gradillas, A.; Godzien, J.; Lopez-Gonzalvez, A.; Rey-Stolle, F.; Rojo, D.; Ruperez, F.J.; et al. Recent Developments along the Analytical Process for Metabolomics Workflows. *Anal. Chem.* **2020**, *92*, 203–226. [CrossRef]
5. Ezan, E.; Dubois, M.; Becher, F. Bioanalysis of recombinant proteins and antibodies by mass spectrometry. *Analyst* **2009**, *134*, 825–834. [CrossRef] [PubMed]
6. Hickey, J.M.; Sahni, N.; Toth, R.T., IV; Kumru, O.S.; Joshi, S.B.; Middaugh, C.R.; Volkin, D.B. Challenges and opportunities of using liquid chromatography and mass spectrometry methods to develop complex vaccine antigens as pharmaceutical dosage forms. *J. Chromatogr. B Analyt. Technol. Biomed. Life Sci.* **2016**, *1032*, 23–38. [CrossRef]
7. Ebhardt, H.A.; Root, A.; Sander, C.; Aebersold, R. Applications of targeted proteomics in systems biology and translational medicine. *Proteomics* **2015**, *15*, 3193–3208. [CrossRef] [PubMed]
8. Schubert, O.T.; Rost, H.L.; Collins, B.C.; Rosenberger, G.; Aebersold, R. Quantitative proteomics: Challenges and opportunities in basic and applied research. *Nat. Protoc.* **2017**, *12*, 1289–1294. [CrossRef]
9. Carr, S.A.; Abbatiello, S.E.; Ackermann, B.L.; Borchers, C.; Domon, B.; Deutsch, E.W.; Grant, R.P.; Hoofnagle, A.N.; Huttenhain, R.; Koomen, J.M.; et al. Targeted peptide measurements in biology and medicine: Best practices for mass spectrometry-based assay development using a fit-for-purpose approach. *Mol. Cell Proteom.* **2014**, *13*, 907–917. [CrossRef]
10. Kulyyassov, A.; Fresnais, M.; Longuespee, R. Targeted liquid chromatography-tandem mass spectrometry analysis of proteins: Basic principles, applications, and perspectives. *Proteomics* **2021**, e2100153. [CrossRef]
11. Berggard, T.; Linse, S.; James, P. Methods for the detection and analysis of protein-protein interactions. *Proteomics* **2007**, *7*, 2833–2842. [CrossRef]
12. Braun, P.; Gingras, A.C. History of protein-protein interactions: From egg-white to complex networks. *Proteomics* **2012**, *12*, 1478–1498. [CrossRef]
13. Fang, X.; Yoon, J.G.; Li, L.; Tsai, Y.S.; Zheng, S.; Hood, L.; Goodlett, D.R.; Foltz, G.; Lin, B. Landscape of the SOX2 protein-protein interactome. *Proteomics* **2011**, *11*, 921–934. [CrossRef]
14. Huang, X.; Wang, J. The extended pluripotency protein interactome and its links to reprogramming. *Curr. Opin. Genet. Dev.* **2014**, *28*, 16–24. [CrossRef]
15. Gao, Z.; Cox, J.L.; Gilmore, J.M.; Ormsbee, B.D.; Mallanna, S.K.; Washburn, M.P.; Rizzino, A. Determination of Protein Interactome of Transcription Factor Sox2 in Embryonic Stem Cells Engineered for Inducible Expression of Four Reprogramming Factors. *J. Biol. Chem.* **2012**, *287*, 11384–11397. [CrossRef]
16. Ng, P.M.; Lufkin, T. Embryonic stem cells: Protein interaction networks. *Biomol. Concepts* **2011**, *2*, 13–25. [PubMed]
17. Titeca, K.; Lemmens, I.; Tavernier, J.; Eyckerman, S. Discovering cellular protein-protein interactions: Technological strategies and opportunities. *Mass Spectrom. Rev.* **2019**, *38*, 79–111. [CrossRef] [PubMed]
18. Wang, J.; Rao, S.; Chu, J.; Shen, X.; Levasseur, D.N.; Theunissen, T.W.; Orkin, S.H. A protein interaction network for pluripotency of embryonic stem cells. *Nature* **2006**, *444*, 364–368. [CrossRef] [PubMed]
19. Xu, Y.; Fan, X.; Hu, Y. In vivo interactome profiling by enzyme-catalyzed proximity labeling. *Cell Biosci.* **2021**, *11*, 27. [CrossRef]
20. Roux, K.J.; Kim, D.I.; Raida, M.; Burke, B. A promiscuous biotin ligase fusion protein identifies proximal and interacting proteins in mammalian cells. *J. Cell Biol.* **2012**, *196*, 801–810. [CrossRef]
21. Branon, T.C.; Bosch, J.A.; Sanchez, A.D.; Udeshi, N.D.; Svinkina, T.; Carr, S.A.; Feldman, J.L.; Perrimon, N.; Ting, A.Y. Efficient proximity labeling in living cells and organisms with TurboID. *Nat. Biotechnol.* **2018**, *36*, 880–887. [CrossRef]
22. Kwon, K.; Streaker, E.D.; Beckett, D. Binding specificity and the ligand dissociation process in the E. coli biotin holoenzyme synthetase. *Protein Sci.* **2002**, *11*, 558–570. [CrossRef]
23. Kwon, K.; Beckett, D. Function of a conserved sequence motif in biotin holoenzyme synthetases. *Protein Sci.* **2000**, *9*, 1530–1539. [CrossRef] [PubMed]

24. Beckett, D.; Kovaleva, E.; Schatz, P.J. A minimal peptide substrate in biotin holoenzyme synthetase-catalyzed biotinylation. *Protein Sci.* **1999**, *8*, 921–929. [CrossRef] [PubMed]
25. Chen, I.; Howarth, M.; Lin, W.; Ting, A.Y. Site-specific labeling of cell surface proteins with biophysical probes using biotin ligase. *Nat. Methods* **2005**, *2*, 99–104. [CrossRef]
26. Fernandez-Suarez, M.; Chen, T.S.; Ting, A.Y. Protein-protein interaction detection in vitro and in cells by proximity biotinylation. *J. Am. Chem. Soc.* **2008**, *130*, 9251–9253. [CrossRef] [PubMed]
27. Slavoff, S.A.; Liu, D.S.; Cohen, J.D.; Ting, A.Y. Imaging protein-protein interactions inside living cells via interaction-dependent fluorophore ligation. *J. Am. Chem. Soc.* **2011**, *133*, 19769–19776. [CrossRef] [PubMed]
28. Kulyyassov, A.; Shoaib, M.; Pichugin, A.; Kannouche, P.; Ramanculov, E.; Lipinski, M.; Ogryzko, V. PUB-MS: A Mass Spectrometry-based Method to Monitor Protein-Protein Proximity in vivo. *J. Proteome Res.* **2011**, *10*, 4416–4427. [CrossRef]
29. Chapman-Smith, A.; Cronan, J.E. The enzymatic biotinylation of proteins: A post-translational modification of exceptional specificity. *Trends Biochem. Sci.* **1999**, *24*, 359–363. [CrossRef]
30. Tenzer, S.; Moro, A.; Kuharev, J.; Francis, A.C.; Vidalino, L.; Provenzani, A.; Macchi, P. Proteome-wide characterization of the RNA-binding protein RALY-interactome using the in vivo-biotinylation-pulldown-quant (iBioPQ) approach. *J. Proteome Res.* **2013**, *12*, 2869–2884. [CrossRef]
31. Lectez, B.; Migotti, R.; Lee, S.Y.; Ramirez, J.; Beraza, N.; Mansfield, B.; Sutherland, J.D.; Martinez-Chantar, M.L.; Dittmar, G.; Mayor, U. Ubiquitin profiling in liver using a transgenic mouse with biotinylated ubiquitin. *J. Proteome Res.* **2014**, *13*, 3016–3026. [CrossRef]
32. MacLean, B.; Tomazela, D.M.; Shulman, N.; Chambers, M.; Finney, G.L.; Frewen, B.; Kern, R.; Tabb, D.L.; Liebler, D.C.; MacCoss, M.J. Skyline: An open source document editor for creating and analyzing targeted proteomics experiments. *Bioinformatics* **2010**, *26*, 966–968. [CrossRef]
33. Pino, L.K.; Searle, B.C.; Bollinger, J.G.; Nunn, B.; MacLean, B.; MacCoss, M.J. The Skyline ecosystem: Informatics for quantitative mass spectrometry proteomics. *Mass Spectrom. Rev.* **2020**, *39*, 229–244. [CrossRef] [PubMed]
34. Kulyyassov, A.; Ogryzko, V. In Vivo Quantitative Estimation of DNA-Dependent Interaction of Sox2 and Oct4 Using BirA-Catalyzed Site-Specific Biotinylation. *Biomolecules* **2020**, *10*, 142. [CrossRef]
35. Sharma, V.; Eckels, J.; Schilling, B.; Ludwig, C.; Jaffe, J.D.; MacCoss, M.J.; MacLean, B. Panorama Public: A Public Repository for Quantitative Data Sets Processed in Skyline. *Mol. Cell Proteom.* **2018**, *17*, 1239–1244. [CrossRef] [PubMed]
36. Kulyyassov, A.; Kalendar, R. In Silico Estimation of the Abundance and Phylogenetic Significance of the Composite Oct4-Sox2 Binding Motifs within a Wide Range of Species. *Data* **2020**, *5*, 111. [CrossRef]
37. Eissenberg, J.C.; Elgin, S.C. The HP1 protein family: Getting a grip on chromatin. *Curr. Opin. Genet. Dev.* **2000**, *10*, 204–210. [CrossRef]
38. Lomberk, G.; Wallrath, L.; Urrutia, R. The Heterochromatin Protein 1 family. *Genome Biol.* **2006**, *7*, 228. [CrossRef]
39. Sanulli, S.; Trnka, M.J.; Dharmarajan, V.; Tibble, R.W.; Pascal, B.D.; Burlingame, A.L.; Griffin, P.R.; Gross, J.D.; Narlikar, G.J. HP1 reshapes nucleosome core to promote phase separation of heterochromatin. *Nature* **2019**, *575*, 390–394. [CrossRef]
40. Puri, T.; Wendler, P.; Sigala, B.; Saibil, H.; Tsaneva, I.R. Dodecameric structure and ATPase activity of the human TIP48/TIP49 complex. *J. Mol. Biol.* **2007**, *366*, 179–192. [CrossRef]
41. Ikura, T.; Ogryzko, V.V.; Grigoriev, M.; Groisman, R.; Wang, J.; Horikoshi, M.; Scully, R.; Qin, J.; Nakatani, Y. Involvement of the TIP60 histone acetylase complex in DNA repair and apoptosis. *Cell* **2000**, *102*, 463–473. [CrossRef]
42. Martens, L.; Chambers, M.; Sturm, M.; Kessner, D.; Levander, F.; Shofstahl, J.; Tang, W.H.; Rompp, A.; Neumann, S.; Pizarro, A.D.; et al. mzML—A community standard for mass spectrometry data. *Mol. Cell Proteom.* **2011**, *10*. [CrossRef] [PubMed]
43. Kessner, D.; Chambers, M.; Burke, R.; Agus, D.; Mallick, P. ProteoWizard: Open source software for rapid proteomics tools development. *Bioinformatics* **2008**, *24*, 2534–2536. [CrossRef] [PubMed]
44. Adusumilli, R.; Mallick, P. Data Conversion with ProteoWizard msConvert. *Methods Mol. Biol.* **2017**, *1550*, 339–368. [PubMed]
45. Go, C.D.; Knight, J.D.R.; Rajasekharan, A.; Rathod, B.; Hesketh, G.G.; Abe, K.T.; Youn, J.Y.; Samavarchi-Tehrani, P.; Zhang, H.; Zhu, L.Y.; et al. A proximity-dependent biotinylation map of a human cell. *Nature* **2021**, *595*, 120–124. [CrossRef] [PubMed]

Article

Development of HPLC Method for Catechins and Related Compounds Determination and Standardization in Miang (Traditional Lanna Fermented Tea Leaf in Northern Thailand)

Sunanta Wangkarn [1,2,3], Kate Grudpan [1,2,3,4], Chartchai Khanongnuch [3,4,5], Thanawat Pattananandecha [2,3,4], Sutasinee Apichai [2,3,4] and Chalermpong Saenjum [2,3,4,6,*]

1. Department of Chemistry, Faculty of Science, Chiang Mai University, Chiang Mai 50200, Thailand; sunanta.w@cmu.ac.th (S.W.); kgrudpan@gmail.com (K.G.)
2. Center of Excellence for Innovation in Analytical Science and Technology (I-ANALY-S-T), Chiang Mai University, Chiang Mai 50200, Thailand; thanawat.pdecha@gmail.com (T.P.); sutasinee.apichai@gmail.com (S.A.)
3. Cluster of Excellence on Biodiversity-based Economic and Society (B.BES-CMU), Chiang Mai University, Chiang Mai 50200, Thailand; ck_biot@yahoo.com
4. Research Center for Multidisciplinary Approaches to Miang, Chiang Mai University, Chiang Mai 50200, Thailand
5. Division of Biotechnology, Faculty of Agro-Industry, Chiang Mai University, Chiang Mai 50200, Thailand
6. Department of Pharmaceutical Sciences, Faculty of Pharmacy, Chiang Mai University, Chiang Mai 50200, Thailand
* Correspondence: chalermpong.s@cmu.ac.th; Tel.: +66-89-9504227

Abstract: High performance liquid chromatography (HPLC) for catechins and related compounds in *Miang* (traditional Lanna fermented tea leaf) was developed to overcome the matrices during the fermentation process. We investigated a variety of columns and elution conditions to determine seven catechins, namely (+)-catechin, (−)-gallocatechin, (−)-epigallocatechin, (−)-epicatechin, (−)-epigallocatechin gallate, (−)-gallocatechin gallate, (−)-epicatechin gallate, as well as gallic acid and caffeine, resulting in the development of reproducible systems for analyses that overcome sample matrices. Among the three reversed-phase columns, column C (deactivated, with extra dense bonding, double endcapped monomeric C18, high-purity silica at 3.0 mm × 250 mm and a 5 μm particle size) significantly improved the separation between *Miang* catechins in the presence of acid in the mobile phase within a shorter analysis time. The validation method showed effective linearity, precision, accuracy, and limits of detection and quantitation. The validated system was adequate for the qualitative and quantitative measurement of seven active catechins, including gallic acid and caffeine in *Miang*, during the fermentation process and standardization of *Miang* extracts. The latter contain catechins and related compounds that are further developed into natural active pharmaceutical ingredients (natural APIs) for cosmeceutical and nutraceutical products.

Keywords: HPLC; method validation; *Miang*; catechins; caffeine; gallic acid

1. Introduction

Tea (*Camellia sinensis*, family Theaceae) is the most frequently consumed beverage worldwide and a rich natural source of polyphenols, flavonoids, and alkaloids. Numerous studies have identified the characteristic constituents of tea leaves into two main groups: catechins and alkaloids. The active catechins are (+)-catechin (C), (−)-epicatechin (EC), (−)-gallocatechin (GC), (−)-epigallocatechin (EGC), (−)-catechin gallate (CG), (−)-gallocatechin gallate (GCG), (−)-epicatechin gallate (ECG), and (−)-epigallocatechin gallate (EGCG), whereas the major active alkaloid is caffeine (Caf.). EGCG is the major component of unfermented green tea, accounting for approximately 10–50% of tea catechins overall [1]. In fermented green tea, the major component is GC, with less EGCG content [2]. One of the most intriguing properties of tea catechins is protection against cancer, diabetes, hypertension, dyslipidemia, and cardiovascular diseases [3–6]. Several studies have reported that the antioxidant activity is higher in green tea due to higher amounts of EGCG and EGC [7–9]. In addition, tea contains caffeine, which stimulates the central nervous system and induces short-term increases in blood pressure [10].

In Northern Thailand, two varieties of tea are cultivated, namely Assam tea (*Camellia sinensis* var. *assamica*) and Chinese tea (*Camellia sinensis* var. *sinensis*). Assam tea leaves are larger than those of the Chinese variety. In 2007, 77% of fresh tea leaves produced in Thailand were processed into dried tea, and 23% produced *Miang* (traditional Lanna fermented tea leaf) [11]. In Northern Thailand, every generation has inherited *Miang* production, which is known as a typical fermented tea produced from the Assam variety. After harvesting fresh tea leaves, the tea leaves are steamed, bunched, and fermented via endo-oxidation from 13 days to 4 months [12]. The color of *Miang* ranges from yellow-green to dark green. According to local wisdom, traditional *Miang* production is categorized into two processes: the filamentous fungi growth-based process or two-step fermentation process and the non-filamentous fungi-based fermentation process [13]. *Miang*'s taste ranges from tart to sour. As a caffeine source, *Miang* is often eaten as a snack during the workday to increase alertness. As a unique product exclusive to the northern provinces of Thailand, it is also used in local ceremonial events, i.e., Chiang Rai, Chiang Mai, Nan, Lampang, Phare, Phayao, and Mae Hong Son. However, some are exported to Laos, Myanmar, and Southern China [13,14]. *Miang* contains high amounts of bioactive compounds, including EC, C, GC, EGCG, ECG, and EGC [15]. During the fermentation period, the phytochemicals and nutritional compounds of steamed tea leaves are used by enzymes derived from various micro-organisms linked to catalytic biotransformation processes that produce metabolites, including polyphenolic compounds, organic acids, amino acids, and health-related bioactive metabolites [16–18]. *Miang* plays a key role as a natural anti-oxidative agent in the body; therefore, consuming *Miang* has many health benefits. Owing to increasing interest in the quality of *Miang* products, there is a strong demand for efficient quality control measures to ensure the proper content of active catechins and related compounds.

Extensive studies on determining catechins and caffeine in several types of tea have been reported using high-performance liquid chromatography (HPLC) and capillary electrophoresis (CE) [19–27]. Although these methods demonstrate the separation and detection of tea catechins, they have notable limitations regarding sample matrices and their complexity. For example, Kanpiengjai et al. [14] and Chaikaew et al. [28] reported that tannin-tolerant lactic acid bacteria and tannin-tolerant yeasts produce health-benefiting compounds, including phenolic compounds, organic acids, and volatile acids during the fermentation process. These compounds become the interfere matrix. Therefore, a new method must be developed to overcome them. Furthermore, few data have been reported on determining individual catechin and caffeine in *Miang*. Individual catechin amounts in *Miang* tea were reported by Sirisa-Ard et al. [15]. The amount of catechin and catechin derivatives was analyzed by HPLC equipped with a UV detector for wavelengths between 280 and 210 nm. A reversed-phase C18 column (4.6 mm × 250 mm; Waters, Ireland) with a column temperature between 25 and 30 °C was used. The linear gradient of elution was followed by 0–100% of mobile phase A (86% v/v phosphoric acid (0.2% v/v) in 12% acetonitrile and 1.5% v/v tetrahydrofuran) for 30 min and gradually increased mobile phase B (73.5% v/v phosphoric acid (0.2% v/v) in 25% acetonitrile and 1.5% v/v tetrahydrofuran) from 0–100% for 10 min and holding for 20 min with a flow rate of 1 mL/min. The results showed that GC, EGC, C, EC, EGCG, and ECG amounts in *Miang* tea (B02D) were 9.65, 0.84, 16.13, 61.60, 6.46, and 1.93 mg/g in dry samples, respectively. The number of active ingredients in *Miang* was also determined using the HPLC method [29,30]. However, long-term analysis was required in the previous reports because the optimum condition lacked investigation. There have been no reports of using HPLC to separate and detect individual catechins, GA, and Caf.to simultaneously overcome the matrix interference of the compounds during Assam tea fermentation (*Miang*). Small amounts of catechin

(1.34–8.71 mg/g) were found in all *Miang* samples, whereas relatively high EGCG contents (range from 18.50 to 37.24 mg/g) varied among treatments with total phenolic compounds at around 26.24–48.76 mg/g [26].

The total phenolic content, flavonoid content, proanthocyanidin content, and antioxidant activities were reported for three different maturities of *Camellia sinensis* var. *assamica* leaves from Northern Thailand using various extracting solvents. The results revealed that the highest yields were for shoot tea with hot DI water extraction [27]. Sampanvejsobha et al. reported that the amount of total catechins (0.767–3.543% dry weight), caffeine (0.747–1.428% dry weight), tannins (0.963–1.831% dry weight), theanine (1.993–3.686% dry weight), and other ions in astringent *Miang* collected from markets in Chiang Mai, Chiang Rai, and the Phare provinces [12]. Huang et al. [31] studied the microbial transformation of traditional pickled tea fermented under anaerobic conditions. Based on the analysis of changes in the chemical components and sensory quality of pickled tea, properly controlling the fermentation time is a key step for obtaining the desired quality. After 7 days of submerged fermentation, the pickled tea improved sensory quality, and its taste was less bitter and astringent [31].

In this study, we developed an HPLC system for catechins and related compounds -determination and standardization in *Miang* extracts. The study involved comparing three HPLC columns for separating catechin, their derivatives, GA, and Caf. in *Miang*. To obtain accurate data and an efficient HPLC routine method, our research concerning *Miang* analyzed the linearity, accuracy, precision, limit of detection (LOD), and limit of quantitation (LOQ) of the validated HPLC method.

2. Results and Discussion

2.1. Comparative Separation of Columns

The comparative separation of columns was performed according to the various matrices produced during the biotransformation of *Miang*. The study began with an attempt to reproduce several separations of catechins, GA, and Caf. using three HPLC columns for determining and standardizing the amount of compounds of interest in *Miang* samples. Chromatographic conditions using three columns of C18 were optimized for specificity, resolution, and analysis time at room temperature. Methanol or acetonitrile mixed with either phosphoric acid or acetic acid were studied for their use in mobile phases. The conditions of each column were modified from previous works. Oboh et al. used column A to separate catechin, GA, and Caf., but must be modified to separate the individual catechins [32]. Under various gradient elution systems, the initial effort was performed on column A (deactivated, non-endcapped monomeric C18 column and silica purity were not provided). As shown in Figure 1, six compounds of interest were separated completely within 29 min, and the gradient elution shown in Table 1 GCG, EC, and Caf. were coeluted. Additionally, ECG showed significant peak tailing; according to a previous study, this was likely caused by unfavorable interaction of the basic compound with accessible acidic silanols [23]. An occurring peak fronting in some compounds including GA, GC, EGC, and C may cause by the concentrations or volume injected were overloading of column.

Column B (deactivated, endcapped monomeric C18, high-purity silica) was tested using methanol in the presence and absence of ethyl acetate mixed with either phosphoric acid or acetic acid as mobile phases modified from a previous report [33]. The separation was performed with ethyl acetate added to the mobile phase gave sharper peak shapes (C, EGCG, and GCG) than those with ethyl acetate absent [34–36]. Furthermore, acetic acid has the same effect on the separation, but is not as effective as phosphoric acid. The amount of phosphoric acid in the mobile phase was used in the range of 0.05–0.10% to improve the peak shapes of GA, EGCG, and GCG. Therefore, the separation of seven catechins (GC, EGC, C, EC, EGCG, GCG, and ECG), GA, and Caf. was achieved within 84 min under suitable isocratic condition, as shown in Figure 2 and Table 1.

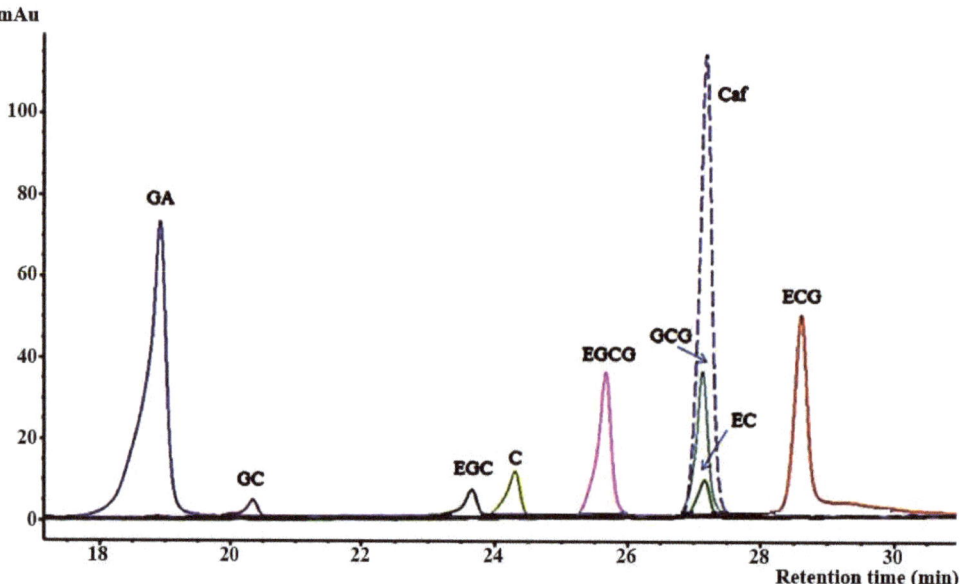

Figure 1. Chromatographic separation of seven catechins containing gallic acid and caffeine in a standard mixture using column A under a suitable gradient elution (see Table 1).

Table 1. The HPLC conditions of three columns used for separating catechins and related compounds.

Column		Condition	
(1) Column A Deactivated, non-endcapped monomeric C18 4.6 mm × 250 mm, 5 µm particle size with no silica purity provided	Mobile phase Gradient elution	MeOH: 0.05% HOAc in H_2O Time (min) % of MeOH 0 3 1 3 21 50 26 55 40 95	
	Wavelength Flow rate	270 nm 0.5 mL/min	
(2) Column B Deactivated, endcapped monomeric C18, High-purity silica, 3.0 mm × 250 mm, 5 µm particle size and a 2 µm filter attached to both ends of column	Mobile phase Isocratic elution	A: 1% ethyl acetate in MeOH B: 0.1% H_3PO_4 in H_2O A:B = 15:85 (v/v)	
	Wavelength Flow rate	270 nm 0.45 mL/min	
(3) Column C Deactivated, extra dense bonding, double endcapped monomeric C18, high-purity silica, 3.0 mm × 250 mm, 5 µm particle size, and 10% carbon loading	Mobile phase Gradient elution	A: (90:10 MeOH-ACN) + 0.1% HOAc B: 0.1% HOAc in H_2O Time (min) %A 0 10 2 10 25 21 28 25 30 100 35 10 40 10	
	Wavelength Flow rate	210 nm 1.0 mL/min	

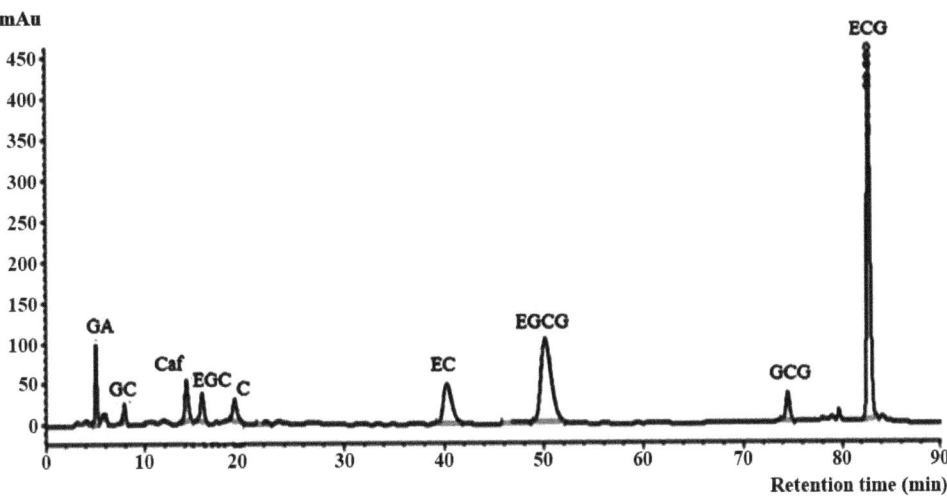

Figure 2. Chromatographic separation of seven catechins containing gallic acid and caffeine in a standard mixture using column B under a suitable isocratic elution (see Table 1).

In most published studies, the mobile phase containing water, acids (trifluoroacetic acid, phosphoric acid, and acetic acid), and either methanol or acetonitrile has been used for catechin analysis of green tea and dried tea leaves [7,22,23,34,35]. Column C and the mobile phase, which comprised a mixture of an eluent A (acetonitrile + 0.1% acetic acid) and B (0.1% acetic acid in water), were employed under various gradient elution systems to shorten the analysis time and improve separation between the seven catechins, GA, and Caf. Among nine compounds of interest, the separation between catechin and caffeine was poor (Rs < 1.0) under the gradient elution system illustrated in Table 1. Adding the volume fraction of methanol into the eluent A (MeOH-ACN mixture) was varied from 0% to 100% in 10% increments. The volume ratio 90:10 MeOH-ACN for eluent A was the only combination that resulted in the separation of all nine compounds within 30 min, as shown in Figure 3 and Table 1. According to the principle of separation in reversed-phase chromatography, the obtained separate order was GA, GC, EGC, C, Caf., EC, EGCG, GCG, and ECG. Regarding the log P-value, logarithms of the partition coefficient are used between solute concentrations in immiscible binary phase solvents, namely water and octanol, which measure the lipophilicity or hydrophobicity of each compound. The hydrophobic compounds, observed from the high log P-values, were distributed into then stationary phase and eluted slowly. By contrast, the hydrophilic compounds, observed from the low values of log P, distributed efficiently into the mobile phase, resulting in quick processing. The log P-values of GA, GC, EGC, C, Caf., EC, EGCG, GCG, and ECG were 1.13, 1.49, 1.49, 1.80, −0.55, 1.80, 3.08, 3.08, and 3.88, respectively, as shown in Table 2, which correspond to the sequence in this study [37]. Moreover, the functional and size of the molecules shown in Figure 4 are involved causing the separation sequence as illustrated in Figure 3. A resolution >1.0 was achieved for all neighboring peaks; this is considered acceptable for analytical purposes as it indicates a 98% separation between two neighboring peaks. Although the baseline shifted due to the change in the mobile phase, it did not affect the detection of individual peaks. The detection wavelength was selected at 210 nm because nearly all compounds of interest exhibited maximum absorbance compared with 230 and 270 nm [7,22,23].

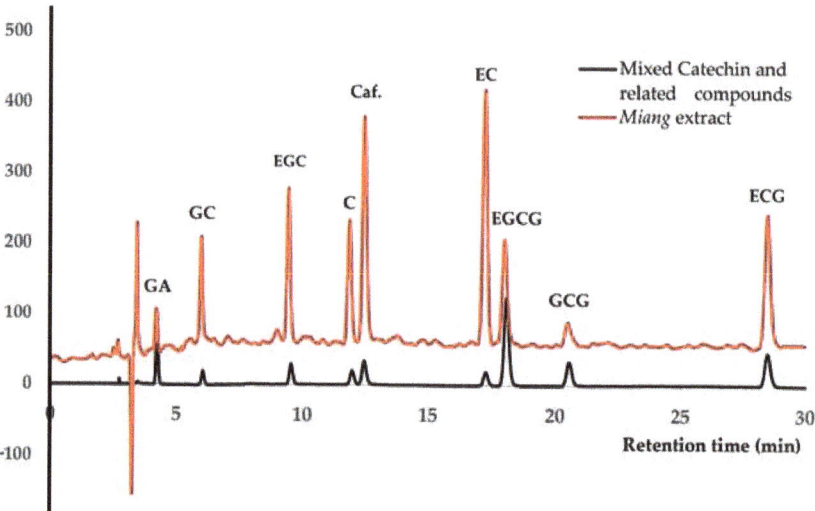

Figure 3. Chromatographic separation of seven catechins containing gallic acid and caffeine in a standard mixture using column C under a suitable gradient elution (see Table 1) comparable to the *Miang* extract.

Table 2. Performance characteristics of column C during method validation and evaluation.

Compounds	Molecular Weight	Log P	Precision (% RSD)				Linear Range (mg/L)	Correlation Coefficient	LOD (mg/L)	LOQ (mg/L)
			Retention Time		Peak Area					
			Intraday	Interday	Intraday	Interday				
GA	170.12	0.70	1.02	1.25	0.69	0.87	2–20	0.9993	0.58	2.01
GC	306.27	1.49	0.89	1.07	0.77	0.92	2–15	0.9998	0.52	1.77
EGC	306.27	1.49	0.86	1.26	0.86	1.06	2–15	0.9994	0.64	2.09
C	290.27	1.80	1.05	1.28	0.82	0.98	2–20	0.9997	0.49	1.59
Caf.	194.19	−0.55	0.49	0.97	0.49	0.88	2–20	0.9998	0.23	0.67
EC	290.27	1.80	0.77	0.93	0.87	1.33	2–20	0.9995	0.68	2.18
EGCG	458.37	3.08	1.02	1.13	0.79	1.07	2–15	0.9994	0.58	1.86
GCG	458.37	3.08	0.65	0.88	0.95	1.17	2–20	0.9997	0.62	2.13
ECG	442.37	3.88	0.58	0.70	0.48	1.29	2–20	0.9997	0.33	1.07

Among the three columns, column B and column C are used in high-purity and inert silica support, further deactivating the C18 chains through endcapping. Our results demonstrated that both columns separated all nine compounds under suitable conditions. In addition to double end-capping via the extra dense bonding of column C, these columns provided a higher quality chromatography than the tested columns.

2.2. Method Validation

As illustrated in Figures 2 and 3, both columns separated all nine compounds successfully. As shown in Table 2, some analytical parameters such as linearity, limit of detection (LOD), limit of quantitation (LOQ), and precision were examined to evaluate the method's performance. The calibration curves of columns C and B were linear in the ranges between 2–20 and 5–35 mg/L, showing correlation coefficients (R^2) of more than 0.9993 and 0.9970 for each compound, respectively. LOD and LOQ were determined as 3 and 10 standard deviations from the blank signal ($n = 7$), respectively. The values of LOD and LOQ ranged between 0.23–0.68 and 0.67–2.18 mg/L using column C, between 0.10–2.45 mg/L and 0.23–8.19 mg/L using column B, indicating this method's sufficient sensitivity.

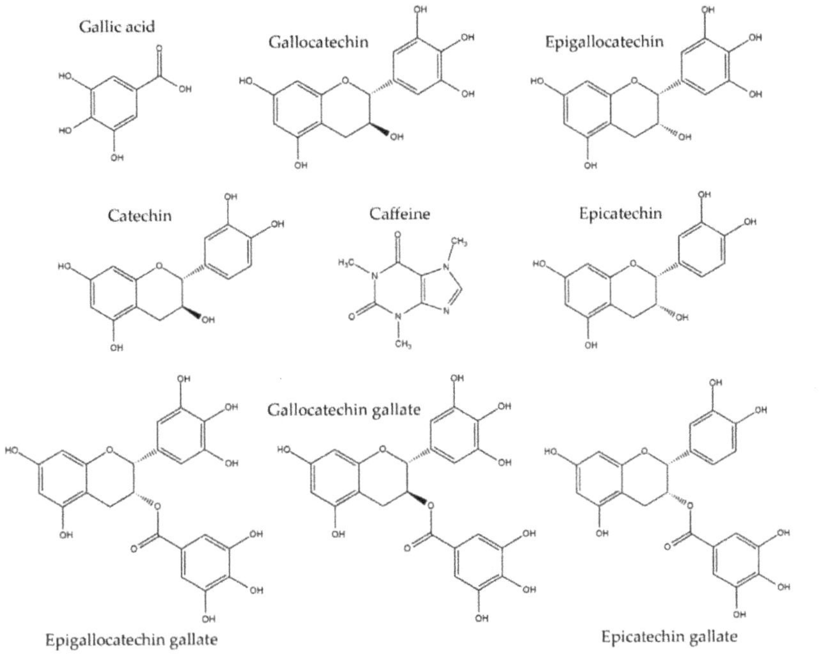

Figure 4. Structure of gallic acid, caffeine, catechin and derivatives.

For column C, intraday and interday precisions of retention time expressed as RSD were less than 2%, whereas precisions using column B were less than 3.0%. Considering the peak area, the RSD values of both precisions for all nine compounds using column C were less than 2%. The RSD values of intraday precision for column B were less than 2%, whereas interday precision was less than 4%

As shown in Table 2, the performance of column C for the separation of all target compounds was superior to that of column B. The method's accuracy was determined by investigating recovery studies of nine compounds using column C. Assays were performed on three *Miang* extracts in three replicates at each concentration. The recovery of spiked *Miang* extract, in terms of method accuracy, was within the range of 85–106%, and RSD values were less than 8%, as shown in Table 3. The result was satisfactory for the intended purpose and adequate for routine analysis.

Table 3. Mean recoveries of catechins and related compounds from the *Miang* extracts.

Compounds	*Miang* Extract-1		*Miang* Extract-2		*Miang* Extract-3	
	Recovery (%)	% RSD	Recovery (%)	% RSD	Recovery (%)	% RSD
GA	87–106	7	90–101	6	85–101	6
GC	85–101	6	85–98	4	85–97	5
EGC	89–101	5	85–97	7	88–102	6
C	88–102	5	93–102	5	95–105	5
Caf.	90–102	5	92–102	6	92–102	5
EC	86–96	4	89–98	5	91–102	6
EGCG	87–101	5	87–97	5	85–101	5
GCG	88–102	5	85–98	6	88–102	6
ECG	90–101	6	88–101	5	86–95	4

Mean value of three replicates for two concentrations (5LOQ and 10LOQ; $n = 6$).

2.3. Quantitative Analysis in Miang Extracts

In *Miang* extract, sample matrices may cause a bias by increasing or decreasing the peak signal attributed to the measurement. Various extracts of *Miang* samples were analyzed by the validated method with column C to confirm the method's suitability in determining and standardizing catechins and related compounds. The resulting chromatogram in Figure 3 was compared to a mixed standard. The contents of individual compounds in *Miang* extracts are shown in Table 4. A total of 75% ethanolic solvent exhibited the highest extractability of total catechins (at 60 °C for C, 70 °C for EGC and EGCG, and 80 °C for EC, GC, GCG, and ECG whereas 50% ethanolic solvent showed the highest extractability for GA at 80 °C, and Caf. at 60 °C. Note that the individual catechins, GA, and Caf. were stable at the obtained optimal extraction temperature below 90 °C, as confirmed from the results of previous studies [38]. Previous extraction kinetic studies found that some of the compound contents, including EGC, EC, and Caf., decrease when the extraction temperature increases to 90 °C. The results correspond with those of Liang et al. [39], who reported that 75% ethanol is the highest extractability of total catechins for fresh tea leaves. The individual catechin contents obtained from this optimal extraction, including EGC, C, ECG and EGCG, were higher than previous studies. Nonetheless, only GC and ECG showed lower extractability [15].

Table 4. Amount of catechins and related compounds in *Miang* extracts.

Compounds	-	S1T1	S1T2	S1T3	S2T1	S2T2	S2T3	S3T1	S3T2	S3T3
GA	Mean	0.64	1.58	0.86	1.70	1.96	0.80	2.85	1.40	3.23
	%RSD	1.52	3.78	4.58	3.33	3.59	3.79	2.20	4.08	2.89
GC	Mean	3.52	3.47	4.23	2.61	1.64	1.35	3.62	1.98	1.18
	%RSD	2.21	1.75	2.15	3.67	3.40	1.52	2.82	4.29	4.58
EGC	Mean	8.90	19.41	9.51	4.75	1.04	1.76	4.71	4.07	4.41
	%RSD	3.38	1.09	1.47	1.49	4.24	3.45	1.95	1.48	4.35
C	Mean	45.05	18.19	16.75	8.16	5.44	6.21	8.49	8.16	9.46
	%RSD	1.10	2.04	3.58	2.63	4.25	3.76	2.64	2.63	2.93
Caf.	Mean	13.68	23.89	5.69	35.63	25.84	26.20	41.82	40.81	30.69
	%RSD	3.53	1.76	3.76	1.20	2.21	1.19	1.27	1.85	1.59
EC	Mean	2.91	13.04	15.62	1.14	0.82	3.23	7.78	3.55	10.48
	%RSD	4.67	4.61	4.41	4.14	3.08	3.96	3.96	3.93	4.79
EGCG	Mean	4.36	12.89	10.69	8.76	3.58	4.92	5.86	11.22	7.26
	%RSD	4.53	2.47	2.96	2.55	2.25	2.80	3.42	2.22	2.85
GCG	Mean	1.07	1.03	2.80	1.08	0.84	0.99	2.03	1.34	2.10
	%RSD	5.17	2.33	4.93	5.09	4.29	2.22	3.66	4.25	3.72
ECG	Mean	0.72	ND	1.62	ND	ND	0.39	0.52	0.81	0.86
	%RSD	2.76	-	5.12	-	-	3.74	3.83	4.44	4.28

Mean = average amount of each compound in mg/g of the *Miang* extract; mean value of three replicates and three injections for each replicate. ND = not detected (below LOD value). Extraction solvents 75% ethanol (S1), deionized water (S2), and 50% ethanol (S3) at 60, 70, and 80 °C (T1, T2, and T3), respectively).

3. Materials and Methods

3.1. Chemicals and Reagents

Standards for catechins and related compounds were used in this work. Catechins and other related compounds were purchased from Sigma-Aldrich (St Louis, Missouri, MO, USA): (+)-catechin (C), (−)-gallocatechin (GC), (−)-epigallocatechin (EGC), (−)-epicatechin (EC), (−)-epigallocatechin gallate (EGCG), (−)-gallocatechin gallate (GCG), (−)-epicatechin gallate (ECG), gallic acid (GA), and caffeine (Caf.). HPLC-grade acetonitrile and methanol, including ethyl acetate, were supplied by Merck (Darmstadt, Germany). Analytical-grade acetic acid (Sigma-Aldrich) and orthophosphoric acid (BDH, Poole, U.K.) were also purchased. HPLC-grade water (18 MΩ) was prepared using a Millipore Milli-Q purification system (Millipore Corp. Bedford, MA, USA) and used to prepare all solutions.

3.2. Instrumentation

An HP 1200 series liquid chromatography system (Agilent Technologies, Santa Clara, CA, USA) comprising a vacuum degasser, quaternary pump, auto-sampler, thermostated column compartment, and diode array detector was used. The three reversed-phase LC columns used were column A (4.6 mm × 250 mm, 5 µm particle size; Vertical Chromatography Co., Ltd., Nonthaburi, Thailand), column B (3.0 mm × 250 mm, 5 µm particle size, Wako Pure Chemical Industries, Ltd., Japan), and column C (3.0 mm × 150 mm, 5 µm particle size, Agilent Technologies, Santa Clara, CA, USA) and all columns were equipped with a specific C18 guard column. Isocratic and gradient elution systems were developed using different mobile phases to separate seven tea catechins, GA, and Caf. at flow rates of 0.45, 0.50, and 1.0 mL/min. The detection of analytes was performed by UV detection at 210 and 270 nm.

3.3. Sample Extraction

The *Miang* samples were produced by a non-filamentous fungi-based fermentation process and collected from Chiang Dao district, Chiang Mai, Thailand, in October 2018. Prior to the HPLC analysis, they were extracted by three different solvents including 75% ethanol (S1), deionized water (S2), and 50% ethanol (S3) at 60, 70, and 80 °C (T1, T2, and T3), respectively) for 1 h. The duration of each extraction was chosen according to previous studies [38,39] that reported the highest efficiency of all compound extractions at 40 and 10 min and stability at 80 and 65 min by water and ethanol solvent. Then, the extracted solution was evaporated under reduced pressure and dried with a vacuum dryer. Subsequently, the *Miang* extracts dissolved and were filtered through a 0.45 µm membrane filter and 10–80 µL of extracts were analyzed directly by HPLC under suitable conditions. Each extract was analyzed for individual catechins, GA, and Caf. content in three replicates.

3.4. Method Validation

Our method was validated according to EURACHEM guidelines [40]. At the concentration corresponding to the middle of the calibration range; the standard mixture of nine compounds was injected with ten replicates for the suitability of the system's test measurements. The intraday precision (repeatability) and interday precision (within laboratory reproducibility, measurements were performed on three different days) were monitored. We determined the following validation parameters: range, linearity, limit of detection (LOD) and quantitation (LOQ), and accuracy. Linearity was assessed using mixed standard solutions at five concentration levels of each compound. The selected *Miang* extracts were spiked with a mixed standard solution at 5 and 10 times the LOQ used for determining the method accuracy, or relative spiked recovery.

4. Conclusions

We found that the monomeric C18 column is preferable to non-endcapped and non-deactivated columns due to their complexity and sample matrices; the qualitative and quantitative analysis of catechins and related compounds in *Miang* samples were successful using endcapped and deactivated columns. Moreover, the presence of acid in the mobile phase is essential for complete separation, especially for GA and GC. The mobile phase was column-dependent in the presence and absence of ethyl acetate for catechin analysis. The proposed HPLC method using column C (3.0 mm × 150 mm, 5 µm particle size) allowed for an accurate quantitation of catechins, GA, and Caf. in *Miang* extracts without interference from other components and performed a single separation in 30 min. The method we developed provides a shorter analysis time compared with previous methods, and effectively overcomes the interference of other metric compounds in *Miang*. Therefore, our method serves as an important reference for the quality control and standardization of *Miang* production, especially since the amounts of active compounds in *Miang* are prone to variation from environmental factors and manufacturing conditions.

Author Contributions: S.W., K.G. and C.S. designed the study; S.W., C.K., T.P., S.A. and C.S. collected and extracted the plant material; S.W. and C.S. performed the experiments; S.W., K.G., T.P. and C.S. wrote, reviewed, and edited the manuscript. All authors have read and agreed to the published version of the manuscript.

Funding: This research was funded by Agricultural Research Development Agency (ARDA), Thailand and National Research Council of Thailand (NRCT), Grant No. PRP5805021460 and the APC was funded by Cluster of Excellence on Biodiversity-based Economic and Society (B.BES-CMU), Chiang Mai University, Thailand.

Institutional Review Board Statement: Not applicable.

Informed Consent Statement: Not applicable.

Data Availability Statement: The original contributions to this study are included in this article. The data presented in this study are available upon request from the corresponding author.

Acknowledgments: The authors are grateful for the financial support from the Agricultural Research Development Agency (ARDA), Thailand; National Research Council of Thailand (NRCT), and the Cluster of Excellence on Biodiversity-based Economic and Society (B.BES-CMU), Chiang Mai University, Thailand. This study was also supported by the Distinguished Research Professor Award in "Green Innovation in Chemical Analysis with Local Wisdom" (grant No. DPG6080002; K. Grudpan), Thailand Research Fund (TRF), Thailand and postdoctoral fellowship granted by Chiang Mai University, Chiang Mai, Thailand (T.P. and S.A.).

Conflicts of Interest: The authors declare no conflict of interest.

Sample Availability: Not available.

References

1. Liang, Y.; Ma, W.; Lu, J.; Wu, Y. Comparison of chemical compositions of *Ilex latifolia* Thumb and *Camellia sinensis* L. *Food Chem.* **2001**, *75*, 339–343. [CrossRef]
2. Liang, Y.; Lu, J.; Zhang, L.; Wu, S.; Wu, Y. Estimation of black tea quality by analysis of chemical composition and colour difference of tea infusions. *Food Chem.* **2003**, *80*, 283–290. [CrossRef]
3. Da Silva Pinto, M. Tea: A new perspective on health benefits. *Food Res. Int.* **2013**, *53*, 558–567. [CrossRef]
4. Sharangi, A.B. Medicinal and therapeutic potentialities of tea (*Camellia sinensis* L.)—A review. *Food Res. Int.* **2009**, *42*, 529–535. [CrossRef]
5. Yang, C.S.; Lambert, J.D.; Ju, J.; Lu, G.; Sang, S. Tea and cancer prevention: Molecular mechanisms and human relevance. *Toxicol. Appl. Pharmacol.* **2007**, *224*, 265–273. [CrossRef]
6. Zielinski, A.A.F.; Haminiuk, C.W.I.; Alberti, A.; Nogueira, A.; Demiate, I.M.; Granato, D. A comparative study of the phenolic compounds and the in vitro antioxidant activity of different Brazilian teas using multivariate statistical techniques. *Food Res. Int.* **2014**, *60*, 246–254. [CrossRef]
7. Bronner, W.E.; Beecher, G.R. Method for determining the content of catechins in tea infusions by high-performance liquid chromatography. *J. Chromatogr. A.* **1998**, *805*, 137–142. [CrossRef]
8. Burana-osot, J.; Yanpaisan, W. Catechins and caffeine contents of green tea commercialized in Thailand. *J. Pharm. Biomed. Sci.* **2012**, *22*, 1–7.
9. Toschi, T.G.; Bordoni, A.; Hrelia, S.; Bendini, A.; Lercker, G.; Biagi, P.L. The protective role of different green tea extracts after oxidative damage is related to their catechin composition. *J. Agric. Food Chem.* **2000**, *48*, 3973–3978. [CrossRef]
10. Ashihara, H.; Crozier, A. Caffeine: A well-known but little mentioned compound in plant science. *Trends Plant Sci.* **2001**, *6*, 407–413. [CrossRef]
11. Sampanvejsobha, S.; Theppakorn, T.; Winyayong, P.; Eungwanichayapant, P. *A study on the current status of tea in Thailand*; Thailand Reserch Fund: Bangkok, Thailand, 2008.
12. Sampanvejsobha, S.; Laohakunjit, N.; Sumonpun, P. *A Study on the Current Status of Tea in Thailand*; Thailand Reserch Fund: Bangkok, Thailand, 2012.
13. Khanongnuch, C.; Unban, K.; Kanpiengjai, A.; Saenjum, C. Recent research advances and ethno-botanical history of *miang*, a traditional fermented tea (*Camellia sinensis* varassamica) of northern Thailand. *J. Ethn. Foods.* **2017**, *4*, 135–144. [CrossRef]
14. Kanpiengjai, A.; Chui-Chai, N.; Chaikaew, S.; Khanongnuch, C. Distribution of tannin-'tolerant yeasts isolated from *Miang*, a traditional fermented tea leaf (*Camellia sinensis* var. *assamica*) in northern Thailand. *Int. J. Food Microbiol.* **2016**, *238*, 121–131. [CrossRef]
15. Sirisa-Ard, P.; Peerakam, N.; Sutheeponhwiroj, S.; Shimamura, T.; Kiatkarun, S. Biological evaluation and application of fermented *Miang* (*Camellia sinensis* var. *assamica* (J.W.Mast.) Kitam.) for tea production. *J. Food Nutr. Res.* **2017**, *5*, 48–53.

16. Unban, K.; Khatthongngam, N.; Pattananandecha, T.; Saenjum, C.; Shetty, K.; Khanongnuch, C. Microbial community dynamics during the non-filamentous fungi growth-based fermentation process of *Miang*, a traditional fermented tea of north Thailand and their product characterizations. *Front. Microbiol.* **2020**, *11*, 1515. [CrossRef]
17. Zhu, M.Z.; Li, N.; Zhou, F.; Ouyang, J.; Lu, D.M.; Xu, W.; Li, J.; Lin, H.Y.; Zhang, Z.; Xiao, J.B.; et al. Microbial bioconversion of the chemical components in dark tea. *Food Chem.* **2019**, *312*, 126043. [CrossRef]
18. Kodchasee, P.; Nain, K.; Abdullahi, A.D.; Unban, K.; Saenjum, C.; Shetty, K.; Khanongnuch, C. Microbial dynamics-links properties and functional metabolites during *Miang* fermentation using the filamentous fungi growth-based process. *Food Biosci.* **2021**, *41*, 100998. [CrossRef]
19. Yang, X.R.; Ye, C.X.; Xu, J.K.; Jiang, Y.M. Simultaneous analysis of purine alkaloids and catechins in *Camellia sinensis*, *Camellia ptilophylla* and *Camellia assamica* var. kucha by HPLC. *Food Chem.* **2007**, *100*, 1132–1136. [CrossRef]
20. Zuo, Y.; Chen, H.; Deng, Y. Simultaneous determination of catechins, caffeine and gallic acids in green, Oolong, black and pu-erh teas using HPLC with a photodiode array detector. *Talanta* **2002**, *57*, 307–316. [CrossRef]
21. Lee, B.L.; Ong, C.N. Comparative analysis of tea catechins and theaflavins by high-performance liquid chromatography and capillary electrophoresis. *J. Chromatogr. A* **2000**, *881*, 439–447. [CrossRef]
22. Wang, H.; Helliwell, K.; You, X. Isocratic elution system for the determination of catechins, caffeine and gallic acid in green tea using HPLC. *Food Chem.* **2000**, *68*, 115–121. [CrossRef]
23. Dalluge, J.J.; Nelson, B.C.; Brown Thomas, J.; Sander, L.C. Selection of column and gradient elution system for the separation of catechins in green tea using high-performance liquid chromatography. *J. Chromatogr. A* **1998**, *793*, 265–274. [CrossRef]
24. Hadad, G.M.; Salam, R.A.; Soliman, R.M.; Mesbah, M.K. Rapid and simultaneous determination of antioxidant markers and caffeine in commercial teas and dietary supplements by HPLC-DAD. *Talanta* **2012**, *101*, 38–44. [CrossRef] [PubMed]
25. Nováková, L.; Spácil, Z.; Seifrtová, M.; Opletal, L.; Solich, P. Rapid qualitative and quantitative ultra-high performance liquid chromatography method for simultaneous analysis of twenty nine common phenolic compounds of various structures. *Talanta* **2010**, *80*, 1970–1979. [CrossRef] [PubMed]
26. Mirasoli, M.; Gotti, R.; Di Fusco, M.; Leoni, A.; Colliva, C.; Roda, A. Electronic nose and chiral-capillary electrophoresis in evaluation of the quality changes in commercial green tea leaves during a long-term storage. *Talanta* **2014**, *129*, 32–38. [CrossRef] [PubMed]
27. El-Hady, D.A.; El-Maali, N.A. Determination of catechin isomers in human plasma subsequent to green tea ingestion using chiral capillary electrophoresis with a high-sensitivity cell. *Talanta* **2008**, *76*, 138–145. [CrossRef] [PubMed]
28. Chaikaew, S.; Baipong, S.; Sone, T.; Kanpiengjai, A.; Chui-Chai, N.; Asano, K.; Khanongnuch, C. Diversity of lactic acid bacteria from *Miang*, a traditional fermented tea leaf in northern Thailand and their tannin-tolerant ability in tea extract. *J. Microbiol.* **2017**, *55*, 720–729. [CrossRef] [PubMed]
29. Phromrukachat, S.; Tiengburanatum, N.; Meechui, J. Assessment of active ingredients in pickled tea. *AJOFAI* **2010**, *3*, 312–318.
30. Dorkbuakaew, N.; Ruengnet, P.; Pradmeeteekul, P.; Nimkamnerd, J.; Nantitanon, W.; Thitipramote, N. Bioactive compounds and antioxidant activities of *Camellia sinensis* var. *assamica* in different leave maturity from northern Thailand. *Int. Food Res. J.* **2016**, *23*, 2291–2295.
31. Huang, Y.; Liu, C.; Xiao, X. Quality characteristics of a pickled tea processed by submerged fermentation. *Int. J. Food Prop.* **2016**, *19*, 1194–1206. [CrossRef]
32. Obon, G.; Adewuni, T.M.; Ademiluyi, A.O.; Olasehinde, T.A.; Ademosun, A.O. Phenolic constituents and inhibitory effects of *Hibiscus sabdariffa* L. (Sorrel) calyx on cholinergic monoaminergic, and purinergic enzyme activities. *J. Diet. Suppl.* **2018**, *15*, 910–922.
33. Acar, E.T.; Celep, M.E.; Charehsaz, M.; Akyüz, G.S.; Yeşlada, E. Development and validation of a high-performance liquid chomatography-diode-array detection method for the determination of eight phenolic constituents in extracts of deferent wine species. *Turk. J. Pharm. Sci.* **2018**, *15*, 22–28.
34. Goto, T.; Yoshida, Y.; Kiso, M.; Nagashima, H. Simultaneous analysis of individual catechins and caffeine in green tea. *J. Chromatogr. A* **1996**, *749*, 295–299. [CrossRef]
35. Nishitani, E.; Sagesaka, Y.M. Simultaneous determination of catechins, caffeine and other phenolic compounds in tea using new HPLC method. *J. Food Compost. Anal.* **2004**, *17*, 675–685. [CrossRef]
36. Saito, S.T.; Welzel, A.; Suyenaga, E.S.; Bueno, F. A method for fast determination of epigallocatechin gallate (EGCG), epicatechin (EC), catechin (C) and caffeine (CAF) in green tea using HPLC. *Food Sci. Technol.* **2006**, *26*, 394–400. [CrossRef]
37. Marczyń, Z.; Skibska, B.; Nowak, S.; Jambor, J.; Zgoda, M. Actual solubility ($S_{|real.|}$), level of hydrophilic-lipophilic balance ($HLBR_{equ.}$, HLB_D, HLB_G) and partition coefficient (log P) of phytochemicals contained in *Ext. Camellia sinensis* L. *aqu. siccum* in the light of general Hildebrand-Scatchard-Fedors theory of solubility. *Herba Pol.* **2018**, *64*, 47–59.
38. Ziaedini, A.; Jafai, A.; Zakeri, A. Extraction of antioxidants and caffeine from green tea (*Camelia sinensis*) leaves: Kinetics and modeling. *Food Sci. Technol. Int.* **2010**, *16*, 505–510. [CrossRef]
39. Liang, H.; Liang, Y.; Dong, J.; Lu, J. Tea extraction methods in relation to control of epimerization of tea catechins. *J. Sci. Food Agric.* **2007**, *87*, 1748–1752. [CrossRef]
40. Magnusson, B.; Örnemark, U. *Eurachem Guide: The Fitness for Purpose of Analytical Methods—A Laboratory Guide to Method Validation and Related Topics*, 2nd ed.; Eurachem: Teddington, UK, 2014; ISBN 978-91-87461-59-0.

Article

The Evaluation of Multiple Linear Regression–Based Limited Sampling Strategies for Mycophenolic Acid in Children with Nephrotic Syndrome

Joanna Sobiak [1,*], Matylda Resztak [1], Maria Chrzanowska [1], Jacek Zachwieja [2] and Danuta Ostalska-Nowicka [2]

1. Department of Physical Pharmacy and Pharmacokinetics, Poznan University of Medical Sciences, 60-781 Poznań, Poland; mresztak@ump.edu.pl (M.R.); mchrzan@ump.edu.pl (M.C.)
2. Department of Pediatric Nephrology and Hypertension, Poznan University of Medical Sciences, 60-572 Poznań, Poland; zachwiej@mp.pl (J.Z.); dostalska@ump.edu.pl (D.O.-N.)
* Correspondence: jsobiak@ump.edu.pl

Abstract: We evaluated mycophenolic acid (MPA) limited sampling strategies (LSSs) established using multiple linear regression (MLR) in children with nephrotic syndrome treated with mycophenolate mofetil (MMF). MLR-LSS is an easy-to-determine approach of therapeutic drug monitoring (TDM). We assessed the practicability of different LSSs for the estimation of MPA exposure as well as the optimal time points for MPA TDM. The literature search returned 29 studies dated 1998–2020. We applied 53 LSSs ($n = 48$ for MPA, $n = 5$ for free MPA [fMPA]) to predict the area under the time-concentration curve (AUC_{pred}) in 24 children with nephrotic syndrome, for whom we previously determined MPA and fMPA concentrations, and compare the results with the determined AUC (AUC_{total}). Nine equations met the requirements for bias and precision ±15%. The MPA AUC in children with nephrotic syndrome was predicted the best by four time-point LSSs developed for renal transplant recipients. Out of five LSSs evaluated for fMPA, none fulfilled the ±15% criteria for bias and precision probably due to very high percentage of bound MPA (99.64%). MPA LSS for children with nephrotic syndrome should include blood samples collected 1 h, 2 h and near the second MPA maximum concentration. MPA concentrations determined with the high performance liquid chromatography after multiplying by 1.175 may be used in LSSs based on MPA concentrations determined with the immunoassay technique. MPA LSS may facilitate TDM in the case of MMF, however, more studies on fMPA LSS are required for children with nephrotic syndrome.

Keywords: mycophenolate mofetil; mycophenolic acid; pediatric patients; limited sampling strategy; multiple linear regression; therapeutic drug monitoring

1. Introduction

Mycophenolate mofetil (MMF) is an immunosuppressive drug administered in the prophylaxis against acute rejection after solid organ transplantation as well as in autoimmune diseases [1], nephrotic syndrome [2,3], and atopic dermatitis [4]. The MMF active moiety, mycophenolic acid (MPA), is characterized by complex and variable pharmacokinetics and high serum albumin binding (97–99%) [1,5]. MPA pharmacokinetics in renal transplant recipients are widely described in the literature [1,6–10], however, although the pharmacokinetics are assumed to be different, there are few studies concerning children with nephrotic syndrome treated with MMF [11–14]. In our previous study [11], we observed that the target values of the pharmacokinetic parameters, such as the concentration before the next dose (C_0) and the area under the concentration—time curve from 0 to 12 h (AUC_{total}), in children with nephrotic syndrome treated with MMF should be higher than those recommended after renal transplantation [1]. Similar observations were described by other authors [12,15].

MPA therapeutic drug monitoring (TDM) is frequently recommended, mainly to avoid underexposure [1,16]. TDM was shown to be favorable not only in renal transplant

recipients [6], but also in patients with lupus nephritis [17] and steroid-dependent nephrotic syndrome [12,13]. One method of TDM is the limited sampling strategy (LSS), which allows us to predict AUC_{total} on the basis of only few blood samples [6] instead of the time-consuming, expensive, and uncomfortable to patients method of collecting 8 to 15 blood samples over 12 h for a full pharmacokinetic profile [18]. LSS may be calculated using the Bayesian approach or multiple linear regression (MLR) analysis, which uses an equation derived from stepwise regression analysis based on concentrations measured at pre-defined times after dosing [16,19]. MLR is easier to use than Bayesian analysis, although one important limitation of the MLR approach is the reliance of the equations on the accuracy of the exact times of blood sample collection [7,16]. MLR LSSs have been proposed for MPA in many groups of patients [8,9,20,21]. Whereas many authors emphasize that each LSS should be applied to the same group of patients as it was established [22], Ting et al. [20] observed that the application of LSSs established for lung transplant recipients to the heart transplant population yielded satisfactory prediction results, Gellermann et al. [15] applied the LSSs established for children after renal transplantation and adult heart transplant recipients to evaluate AUC in children with nephrotic syndrome, and Katsuno et al. [17] used the LSS established for renal transplant recipients to predict AUC in patients with lupus nephritis. Additionally, Tong et al. [23] applied the LSS established with the high performance liquid chromatography (HPLC) method to evaluate the AUC for patients for whom the enzyme multiplied immunoassay technique (EMIT) was used for MPA determination, while Neuberger et al. [24] applied an MPA LSS established after the administration of another MPA formulation, enteric-coated mycophenolic sodium (EC-MPS), in MMF treated patients.

Due to the small number of studies on MPA pharmacokinetics in children with nephrotic syndrome, in this study we evaluated MLR-based LSSs found in the literature in children with nephrotic syndrome treated with MMF. The evaluation aimed to assess the practicability of different LSSs for the estimation of MPA exposure as well as to find the optimal time points for MPA TDM.

2. Results

2.1. MPA and fMPA Pharmacokinetics

The MPA and free MPA (fMPA) concentrations versus time in 24 children with nephrotic syndrome treated with MMF are presented in Figure 1. The results of MPA and fMPA maximum concentration (C_{max}), time to reach C_{max} (t_{max}), and AUC_{total} values are presented in Table 1. MPA C_0 was above 2.0 μg/mL and above 3.0 μg/mL in 67% ($n = 16$) and 42% ($n = 10$) of children, respectively. MPA C_{max} was observed 1 h after MMF administration in 79% of children. Out of 24 children, 63% ($n = 15$) had MPA AUC_{total} within the 30–60 μg·h/mL range. For 21% ($n = 5$) of children, MPA AUC_{total} was above 60 μg·h/mL. Mean MPA binding to plasma protein was 99.65%, with only 0.35% of fMPA.

Table 1. Plasma concentrations and exposure of MPA and fMPA in children with nephrotic syndrome.

	Parameter	Mean ± SD	Range
MPA	C_{max} (μg/mL)	18.20 ± 9.34	4.96–44.22
	t_{max} (h)	1 ± 1	1–3
	AUC_{total} (μg·h/mL)	53.14 ± 17.77	22.27–94.54
fMPA	C_{max} (μg/mL)	0.0660 ± 0.0081	0.1605–0.0409
	AUC_{total} (μg·h/mL)	0.1837 ± 0.0867	0.0551–0.3806

MPA, mycophenolic acid; fMPA, free mycophenolic acid; AUC_{total}, area under the time–concentration curve from 0 to 12 h; SD, standard deviation.

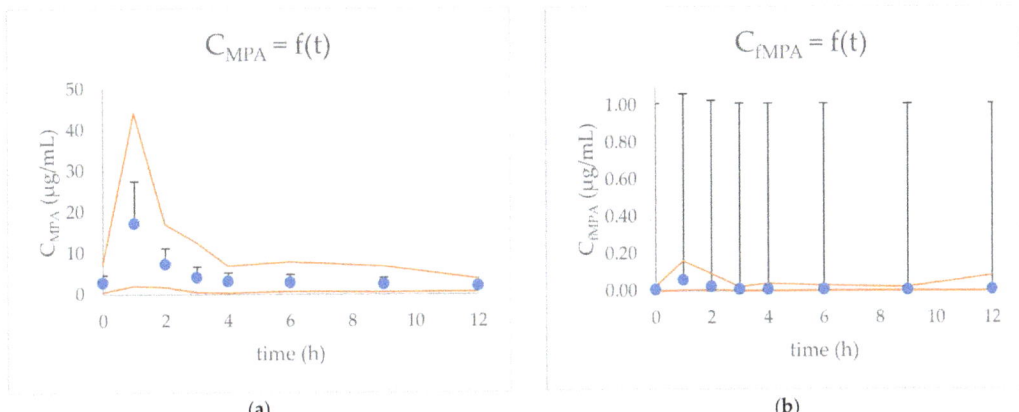

Figure 1. The concentration (+SD) versus time graphs for: (**a**) MPA and (**b**) fMPA for 24 children included in the study. Orange curves indicate the maximum and minimum concentrations at each time-point.

2.2. The Evaluation of MLR LSSs in Children with Nephrotic Syndrome

The search of the literature returned 29 studies meeting the requirements concerning MLR LSSs for MPA and fMPA, dated 1998–2020. We applied 48 MPA LSSs [8,9,14,21,22,25–48] and five fMPA LSSs [35,36,42] found in the literature to calculate the predicted area under the (0–12 h) time–concentration curve (AUC_{pred}) in children with nephrotic syndrome treated with MMF, and compared the results with AUC_{total}. In the majority of studies, calcineurin inhibitors (CsA or tacrolimus (Tac)) were co-administered with MMF. In two studies, only MMF was administered and in one other study, only 8% of patients received CsA concomitantly. The majority of studies concerned patients after solid organ transplantation. We found seven studies including pediatric patients after renal transplantation ($n = 4$), with nephrotic syndrome ($n = 2$), and with lupus erythematosus ($n = 1$). In order to better describe the results, we divided the LSSs according to the methods of MPA determination and subdivided according to the indications for MMF treatment (Tables 2 and 3). The LSSs for fMPA are presented separately (Table 4).

Table 2. Predictive performance of MLR-based HPLC–MPA LSSs available in the literature for estimation of MPA AUC_{pred} in children with nephrotic syndrome treated with MMF.

No	Equation	Indication for MMF Treatment	Drugs Co-Administered	Reference	%MPE (95% CI)	%MAE (95% CI)	r^2	% of AUC_{pred} within ±15% of AUC_{total}
1	$AUC_{pred} = 9.328 + 1.311 \times C_1 + 1.455 \times C_2 + 2.901 \times C_4$	adult renal Tx	Tac	[43]	−0.55 (−6.89–5.78)	11.68 (7.83–15.53)	0.807	67
2	$AUC_{pred} = 15.94 + 1.77 \times C_2 + 2.34 \times C_4 + 4.76 \times C_9$	adult renal Tx	Tac, steroids	[41]	−5.08 (−13.08–2.92)	15.74 (10.97–20.51)	0.619	50
3	$AUC_{pred} = 20.38 + 0.26 \times C_0 + 2.06 \times C_2 + 3.82 \times C_4$	adult renal Tx	Tac, steroids	[41]	−4.19 (−13.17–4.78)	17.23 (11.87–22.58)	0.465	46
4	$AUC_{pred} = 9.02 + 3.77 \times C_0 + 1.33 \times C_1 + 1.68 \times C_3 + 2.96 \times C_6$	adult renal Tx	CsA, steroids	[37]	12.91 (5.53–20.30)	18.00 (12.97–23.03)	0.773	54
5	$AUC_{pred} = 6.02 + 5.61 \times C_0 + 1.28 \times C_1 + 0.9 \times C_2 + 2.54 \times C_4$	adult renal Tx	CsA, steroids	[48]	13.43 (1.59–25.28)	18.35 (9.55–27.14)	0.723	50
6	$AUC_{pred} = 3.504 + 1.098 \times C_1 + 0.670 \times C_2 + 5.659 \times C_4$	adult renal Tx	CsA, steroids	[36]	−14.12 (−21.65–[−6.58])	19.95 (15.50–24.40)	0.684	33
7	$AUC_{pred} = 15.19 + 6.92 \times C_0 + 1.08 \times C_1 + 0.72 \times C_2$	adult renal Tx	CsA, steroids	[48]	16.28 (4.44–28.13)	24.56 (15.76–33.36)	0.527	42
8	$AUC_{pred} = -0.247 + 11.73 \times C_6 + 2.92 \times C_2$	adult renal Tx	CsA, steroids	[39]	3.04 (−11.63–17.71)	26.45 (17.13–35.76)	0.487	46
9	$AUC_{pred} = 9.57 \times C_6 + 27.238$	adult renal Tx	no data	[38]	9.88 (−4.88–24.63)	28.31 (18.99–37.62)	0.265	29
10	$AUC_{pred} = 10.403 + 0.841 \times C_2 + 1.105 \times C_3 + 0.447 \times C_4$	adult renal Tx	CsA, steroids	[36]	−54.94 (−59.83–[−50.05])	54.94 (50.05–59.83)	0.372	0
11	$AUC_{pred} = 10.229 + 0.925 \times C_1 + 1.750 \times C_2 + 4.586 \times C_6$	adult liver Tx	Tac, steroids	[29]	0.49 (−6.08–7.05)	12.57 (8.85–16.28)	0.823	63
12	$AUC_{pred} = 17.930 + 1.992 \times C_2 + 4.136 \times C_6$	adult liver Tx	Tac, steroids	[29]	−12.17 (−20.08–[−4.25])	18.22 (12.89–23.54)	0.565	50
13	$AUC_{pred} = 1.783 + 1.248 \times C_1 + 0.888 \times C_2 + 8.027 \times C_4$	adult islet Tx	Tac	[22]	4.18 (−6.31–14.68)	17.47 (9.94–24.99)	0.648	50
14	$AUC_{pred} = 2.778 + 1.413 \times C_1 + 0.963 \times C_3 + 7.511 \times C_4$	adult islet Tx	Tac	[22]	4.04 (−6.34–14.41)	17.93 (10.80–25.06)	0.619	50

Table 2. Cont.

No	Equation	Indication for MMF Treatment	Drugs Co-Administered	Reference	%MPE (95% CI)	%MAE (95% CI)	r^2	% of AUC_{pred} within ±15% of AUC_{total}
15	$AUC_{pred} = 1.547 + 1.417 \times C_1 + 9.448 \times C_4$	adult islet Tx	Tac	[22]	5.48 (−7.15–18.10)	21.29 (12.31–30.28)	0.557	50
16	$AUC_{pred} = 1.410 − 0.259 \times C_0 + 1.443 \times C_1 + 9.622 \times C_4$	adult islet Tx	Tac	[22]	5.60 (−4.78–15.97)	21.86 (14.73–28.99)	0.551	50
17	$logAUC_{pred} = 1.024 + 0.192 \times logC_0 + 0.213 \times logC_1 + 0.355 \times logC_2$	adult lung Tx	CsA, steroids	[44]	−14.11 (−20.76–[−7.45])	17.79 (13.05–22.53)	0.718	42
18	$logAUC_{pred} = 1.14 + 0.241 \times logC_0 + 0.406 \times logC_2$	adult lung Tx	CsA, steroids	[44]	−25.96 (−34.21–[−17.72])	28.88 (22.70–35.07)	0.427	21
19	$AUC_{pred} = 4.43 + 2.76 \times C_0 + 0.51 \times C_1 + 1.97 \times C_2 + 4.27 \times C_6$	adult HSCT	CsA	[42]	−8.34 (−15.19–[−1.50])	15.79 (12.13–19.45)	0.708	54
20	$AUC_{pred} = 1.2039 \times AUC_{1-4} + 8.9727$	adult HSCT	CsA	[34]	−31.85 (−35.91–[−27.80])	31.85 (27.80–35.91)	0.841	4
21	$AUC_{pred} = 0.10 + 11.15 \times C_0 + 0.42 \times C_1 + 2.80 \times C_2$	adult heart Tx	CsA, steroids	[45]	15.24 (−1.66–32.14)	31.94 (20.15–43.72)	0.366	33
22	$AUC_{pred} = −0.51 + 11.47 \times C_0 + 3.24 \times C_2$	adult heart Tx	CsA, steroids	[45]	8.19 (−10.63–27.02)	35.54 (24.06–47.02)	0.264	25
23	$AUC_{pred} = 13.81 + 0.68 \times C_1 + 1.08 \times C_2 + 2.21 \times C_3 + 4.62 \times C_0$	children systemic lupus erythematosus	none	[21]	9.82 (1.38–18.25)	16.26 (9.95–22.57)	0.738	50

AUC_{pred}, predicted area under the time(0–12 h)–concentration curve; AUC_{total}, determined area under the concentration—time curve from 0 to 12 h; CI, confidence interval; CsA, cyclosporine; HPLC, high performance liquid chromatography; HSCT, hematopoietic stem cell transplantation; LSSs, limited sampling strategies; MMF, mycophenolate mofetil; MPA, mycophenolic acid; %MAE, percentage of mean absolute relative prediction error; %MPE, mean relative prediction error; Tac, tacrolimus; Tx, transplantation.

Table 3. The predictive performance of MLR-based EMIT/PETINIA-MPA LSSs available in the literature for estimation of MPA AUC_{pred} in children with nephrotic syndrome treated with MMF.

No	Equation	Indication for MMF Treatment	Drugs Co-Administered	Reference	%MPE (95% CI)	%MAE (95% CI)	r^2	% of AUC_{pred} within ±15% of AUC_{total}
1	$AUC_{pred} = 10.6 + 1.1 \times C_1 + 1.1 \times C_2 + 2.0 \times C_4 + 3.9 \times C_6$	adult renal Tx	Tac, steroids	[30][1]	2.90 (−2.92–8.73)	11.56 (8.30–14.82)	0.860	67
2	$AUC_{pred} = 7.4 + 2.3 \times C_0 + 1.2 \times C_1 + 2.3 \times C_3 + 4.4 \times C_6$	adult renal Tx	Tac, steroids	[30][1]	7.32 (1.50–13.14)	12.21 (8.20–16.22)	0.829	71
3	$AUC_{pred} = 3.8 + 3.5 \times C_0 + 1.2 \times C_1 + 1.9 \times C_3 + 5.4 \times C_6$	adult renal Tx	Tac, steroids	[30][1]	9.85 (2.24–17.47)	15.90 (10.51–21.28)	0.742	63
4	$AUC_{pred} = 4.42 + 1.74 \times C_1 + 2.99 \times C_4 + 5.43 \times C_9$	adult renal Tx	CsA	[40]	8.16 (0.88–15.43)	15.92 (11.67–20.18)	0.826	58
5	$AUC_{pred} = 17.3 + 4.4 \times C_0 + 1.1 \times C_1 + 2.9 \times C_4$	adult renal Tx	Tac, steroids	[27]	9.13 (0.09–18.17)	18.63 (12.91–24.35)	0.638	50
6	$AUC_{pred} = 23.37 + 4.21 \times C_0 + 3.60 \times C_4$	adult renal Tx	Tac	[47]	−12.35 (−22.77–[−1.92])	21.82 (14.85–28.78)	0.198	46
7	$AUC_{pred} = 4.38 + 2.14 \times C_1 + 7.19 \times C_9$	adult renal Tx	CsA	[40]	11.62 (0.49–22.75)	22.50 (15.10–29.91)	0.722	42
8	$AUC_{pred} = 20.30 + 5.80 \times C_0 + 3.06 \times C_4$	adult renal Tx	Tac	[47]	−12.12 (−25.13–0.88)	23.57 (18.18–28.96)	0.160	42
9	$AUC_{pred} = 8.149 + 1.442 \times C_2 + 1.056 \times C_4 + 7.133 \times C_6$	adult renal Tx	Tac, steroids	[26]	−20.52 (−29.19–[−11.85])	25.56 (19.90–31.21)	0.501	25
10	$AUC_{pred} = 22.93 + 4.63 \times C_0 + 5.60 \times C_6$	adult renal Tx	Tac	[47]	−1.86 (−14.87–11.14)	27.50 (22.11–32.89)	0.208	17
11	$AUC_{pred} = 14.9 + 1.3 \times C_1 + 3 \times C_4 + 3.7 \times C_6$	adult renal Tx	Tac, steroids	[27]	96.25 (71.31–121.19)	98.30 (74.90–121.71)	0.549	4
12	$AUC_{pred} = 5.92 + 1.10 \times C_1 + 1.01 \times C_2 + 1.77 \times C_4 + 4.80 \times C_6$	adult liver Tx	Tac, steroids	[28]	−3.29 (−9.47–2.88)	11.84 (8.09–15.59)	0.829	67
13	$AUC_{pred} = 8.144 + 2.880 \times C_3$	adult liver Tx	Tac, steroids	[31]	−62.44 (−68.53–[−56.35])	62.44 (56.35–68.53)	0.134	0
14	$AUC_{pred} = 8.22 + 3.16 \times C_0 + 0.99 \times C_1 + 1.33 \times C_2 + 4.18 \times C_4$	children renal Tx	CsA	[32]	7.93 (1.47–14.39)	12.58 (7.68–17.48)	0.799	67
15	$AUC_{pred} = 8.217 + 3.163 \times C_0 + 0.994 \times C_1 + 1.334 \times C_2 + 4.183 \times C_4$	children renal Tx	CsA	[8]	8.14 (1.68–14.61)	12.65 (7.71–17.58)	0.799	67

Table 3. Cont.

No	Equation	Indication for MMF Treatment	Drugs Co-Administered	Reference	%MPE (95% CI)	%MAE (95% CI)	r^2	% of AUC$_{pred}$ within ±15% of AUC$_{total}$
16	AUC$_{pred}$ = 7.73 + 0.94 × C$_1$ + 2.55 × C$_2$ + 5.48 × C$_6$	children renal Tx	CsA	[32]	8.94 (2.19–15.68)	14.67 (10.17–19.18)	0.829	58
17	AUC$_{pred}$ = 10.75 + 0.98 × C$_1$ + 2.38 × C$_2$ + 4.86 × C$_6$	children renal Tx	CsA	[33]	10.08 (3.46–16.66)	14.76 (10.10–19.42)	0.842	50
18	AUC$_{pred}$ = 12.62 + 7.78 × C$_0$ + 0.9 × C$_1$ + 1.3 × C$_2$	children renal Tx	CsA	[9]	13.81 (2.00–25.62)	23.20 (14.55–31.85)	0.515	50
19	AUC$_{pred}$ = 13.73 + 9.024 × C$_0$ + 1.779 × C$_2$	children renal Tx	CsA	[9]	0.31 (−14.71–15.34)	28.79 (20.33–37.25)	0.203	21
20	AUC$_{pred}$ = 15.1 + 9.68 × C$_0$ + 1.28 × C$_1$	children renal Tx	CsA	[9]	23.57 (8.22–38.91)	33.21 (21.65–44.77)	0.374	29
21	AUC$_{pred}$ = 12.3 + 4.7 × C$_0$ + 1.2 × C$_1$ + 2.7 × C$_3$ + 1.8 × C$_6$	adult autoimmune disease	CsA	[46]	18.85 (11.45–26.25)	20.15 (13.42–26.88)	0.811	50
22	AUC$_{pred}$ = 17.5 + 7.1 × C$_0$ + 1.0 × C$_1$ + 2.6 × C$_3$	adult autoimmune disease	CsA	[46]	24.84 (13.36–36.02)	27.45 (17.47–37.43)	0.607	33
23	AUC$_{pred}$ = 38.3 + 11.7 × C$_0$	adult autoimmune disease	CsA	[46]	35.64 (13.52–57.76)	47.39 (29.84–64.94)	0.051	21
24	AUC$_{pred}$ = 21.971 + 2.6059 × C$_2$	children INS	CsA	[14] [1]	−24.57 (−32.54–[−16.59])	26.14 (19.16–33.12)	0.455	33
25	AUC$_{pred}$ = 8.7 + 4.63 × C$_0$ + 1.90 × C$_1$ + 1.52 × C$_2$	children NS	none	[25]	24.21 (14.28–34.13)	29.03 (21.90–36.15)	0.718	17

AUC$_{pred}$, predicted area under the (0–12 h) time–concentration curve; AUC$_{total}$, determined area under the concentration—time curve from 0 to 12 h; CI, confidence interval; CsA, cyclosporine; EMIT, enzyme multiplied immunoassay technique; INS, idiopathic nephrotic syndrome; LSSs, limited sampling strategies; MMF, mycophenolate mofetil; MLR, multiple linear regression; MPA, mycophenolic acid; %MAE, percentage of mean absolute relative prediction error; %MPE, mean relative prediction error; NS, nephrotic syndrome; PETINIA, particle enhanced turbidimetric inhibition immunoassay; Tac, tacrolimus; Tx, transplantation. [1] MPA determined with particle enhanced turbidimetric inhibition immunoassay (PETINIA).

Table 4. The predictive performance of MLR-based HPLC-fMPA LSSs available in the literature for the estimation of fMPA AUC_{pred} in children with nephrotic syndrome treated with MMF.

No	Equation	Indication for MMF Treatment	Drugs Co-Administered	Reference	%MPE (95% CI)	%MAE (95% CI)	r^2	% of AUC_{pred} within ±15% of AUC_{total}
1	fMPA AUC_{pred} = 34.2 + 1.12 × C_1 + 1.29 × C_2 + 2.28 × C_4 + 3.95 × C_6	liver Tx	Tac, steroids	[35]	13.68 (6.44–20.91)	18.53 (13.71–23.35)	0.871	38
2	fMPA AUC_{pred} = 63.92 + 2.01 × C_0 + 0.67 × C_1 + 2.05 × C_2 + 4.26 × C_6	HSCT	CsA	[42]	−14.45 (−23.61–[−5.28])	22.17 (16.56–27.77)	0.725	33
3	fMPA AUC_{pred} = 136.826 + 0.76 × C_1 + 0.84 × C_2 + 3.914 × C_4	renal Tx	CsA, steroids	[36]	52.65 (29.91–75.39)	54.69 (32.86–76.52)	0.768	21
4	fMPA AUC_{pred} = 178.167 + 0.954 × C_2 + 4.001 × C_4	renal Tx	CsA, steroids	[36]	59.46 (28.68–90.25)	63.35 (34.04–92.65)	0.564	43
5	fMPA AUC_{pred} = 180.543 + 0.956 × C_2 − 0.223 × C_3 + 4.342 × C_4	renal Tx	CsA, steroids	[36]	61.48 (30.43–92.54)	64.84 (35.08–94.60)	0.560	25

AUC_{pred}, predicted area under the (0–12 h) time–concentration curve; AUC_{total}, determined area under the concentration—time curve from 0 to 12 h; CI, confidence interval; CsA, cyclosporine; fMPA, free mycophenolic acid; LSSs, limited sampling strategies; HSCT, hematopoietic stem cell transplantation; MMF, mycophenolate mofetil; MLR, multiple linear regression; %MAE, percentage of mean absolute relative prediction error; %MPE, mean relative prediction error; Tac, tacrolimus; Tx, transplantation.

The predictive performances for the estimation of MPA AUC_{pred} using the 23 MPA MLR LSSs available in the literature in which MPA was determined based on HPLC method are presented in Table 2. Only two out of 23 equations (9%) met the requirements of ±15% for %MPE and 15% for %MAE. If the acceptable %MPE and %MAE were extended to ±20%, 13 equations (57%) would fulfill the criteria. For two of the 23 LSSs (9%), AUC_{pred} was within ±15% of AUC_{total} for more than 60% of children, concomitantly with r^2 above 0.800. These LSSs included C_1-C_2-C_4 and C_1-C_2-C_6, both of which were established for Tac co-administration. High r^2 was found in the Gota et al. [34] equation, concomitantly with low predictive performance. A number of 11 LSSs (48%) gave an AUC_{pred} within ±15% of the AUC_{total} for less than 50% of children.

The predictive performances of 25 MPA MLR LSSs in which MPA was determined based on EMIT or particle enhanced turbidimetric inhibition immunoassay (PETINIA) are presented in Table 3. Seven of 25 LSSs (28%) met the requirements of ±15% for %MPE and 15% for %MAE. If the acceptable %MPE and %MAE were extended to ±20%, ten equations (40%) would fulfill the criteria. For three of 25 LSSs (12%), the AUC_{pred} was within ±15% of the AUC_{total} for more than 60% of children, concomitantly with r^2 above 0.800. These LSSs included C_1-C_2-C_4-C_6 (two LSSs) and C_0-C_1-C_3-C_6, all of which were established for Tac co-administration. In 13 of 25 LSSs (52%), the AUC_{pred} was within ±15% of the AUC_{total} in less than 50% of children.

We found five MLR LSSs for fMPA in three studies which we applied to calculate the fMPA AUC_{pred} for children with nephrotic syndrome. The predictive performance of the fMPA MLR LSSs is presented in Table 4. In all three studies, MPA was determined with the HPLC method. None of the equations fulfilled the criteria for %MPE and %MAE. There was one four time point equation (C_1-C_2-C_4-C_6), which was established for patients after liver transplantation and co-treated with Tac, which met the requirements of ±20% for %MPE and %MAE, and demonstrated an r^2 above 0.800.

2.3. Comparison of the Best Matched MLR LSSs

Nine LSSs with %MPE and %MAE ±15%, and $r^2 \geq 0.799$ were considered the best. These equations were established for adult renal transplant recipients (n = 3), adult liver transplant recipients (n = 2), and pediatric renal transplant recipients (n = 4). For these equations, the graphs describing the correlations between the AUC_{total} and the AUC_{pred} were drawn (Figure 2), and Bland–Altman (Figure 3) tests were performed. For the majority of equations, the Bland–Altman test showed only one or two values exceeding the fixed range of the mean ± 1.96 SD, which confirmed the agreement between the AUC_{total} and the AUC_{pred}.

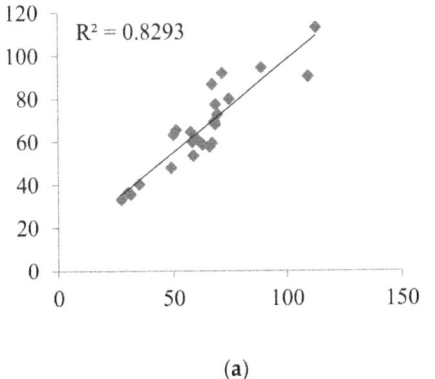

(a)

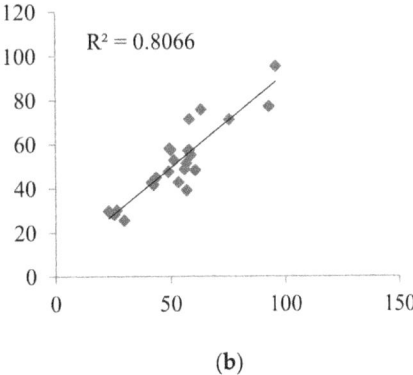

(b)

Figure 2. Cont.

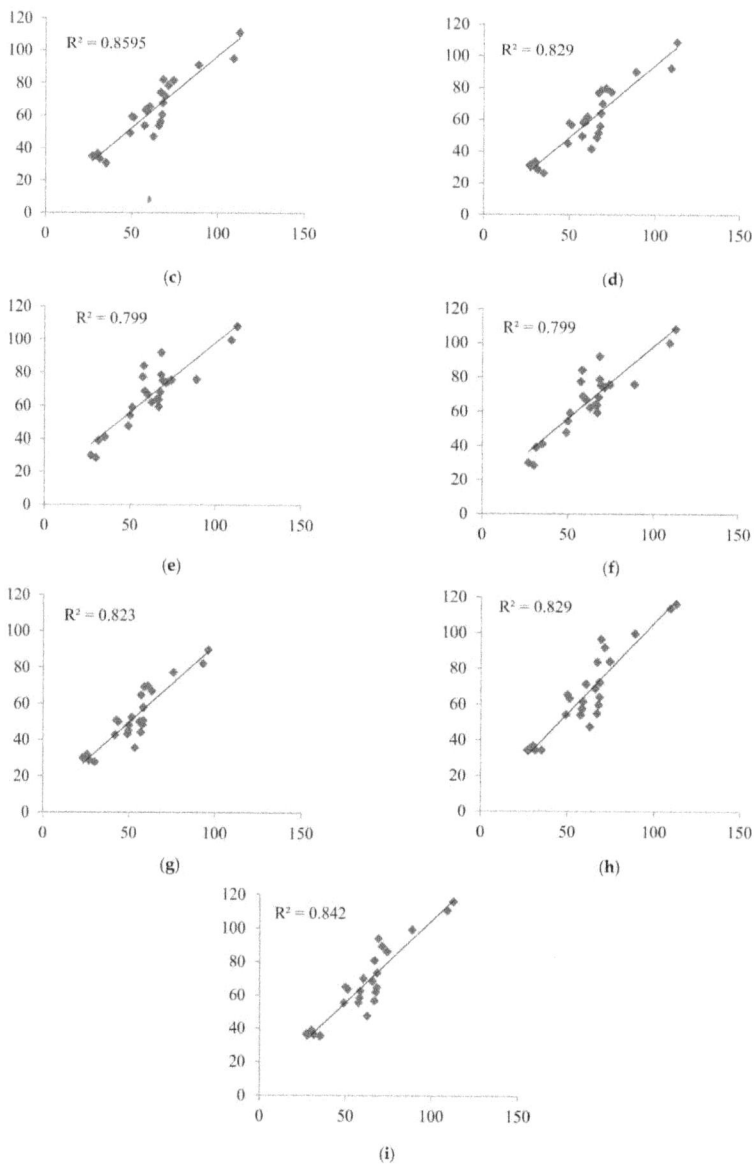

Figure 2. Correlations between the MPA AUC_{total} and the MPA AUC_{pred} calculated for children with nephrotic syndrome using MLR LSS equations found in the literature that fulfilled the criteria for %MPE and %MAE ±15%; (**a**) $AUC_{pred} = 7.4 + 2.3 \times C_0 + 1.2 \times C_1 + 2.3 \times C_3 + 4.4 \times C_6$ [30]; (**b**) $AUC_{pred} = 9.328 + 1.311 \times C_1 + 1.455 \times C_2 + 2.901 \times C_4$ [43]; (**c**) $AUC_{pred} = 10.6 + 1.1 \times C_1 + 1.1 \times C_2 + 2.0 \times C_4 + 3.9 \times C_6$ [30]; (**d**) $AUC_{pred} = 5.92 + 1.10 \times C_1 + 1.01 \times C_2 + 1.77 \times C_4 + 4.80 \times C_6$ [28]; (**e**) $AUC_{pred} = 8.22 + 3.16 \times C_0 + 0.99 \times C_1 + 1.33 \times C_2 + 4.18 \times C_4$ [32]; (**f**) $AUC_{pred} = 8.217 + 3.163 \times C_0 + 0.994 \times C_1 + 1.334 \times C_2 + 4.183 \times C_4$ [8]; (**g**) $AUC_{pred} = 10.229 + 0.925 \times C_1 + 1.750 \times C_2 + 4.586 \times C_6$ [29]; (**h**) $AUC_{pred} = 7.73 + 0.94 \times C_1 + 2.55 \times C_2 + 5.48 \times C_6$ [32]; (**i**) $AUC_{pred} = 10.75 + 0.98 \times C_1 + 2.38 \times C_2 + 4.86 \times C_6$ [33].

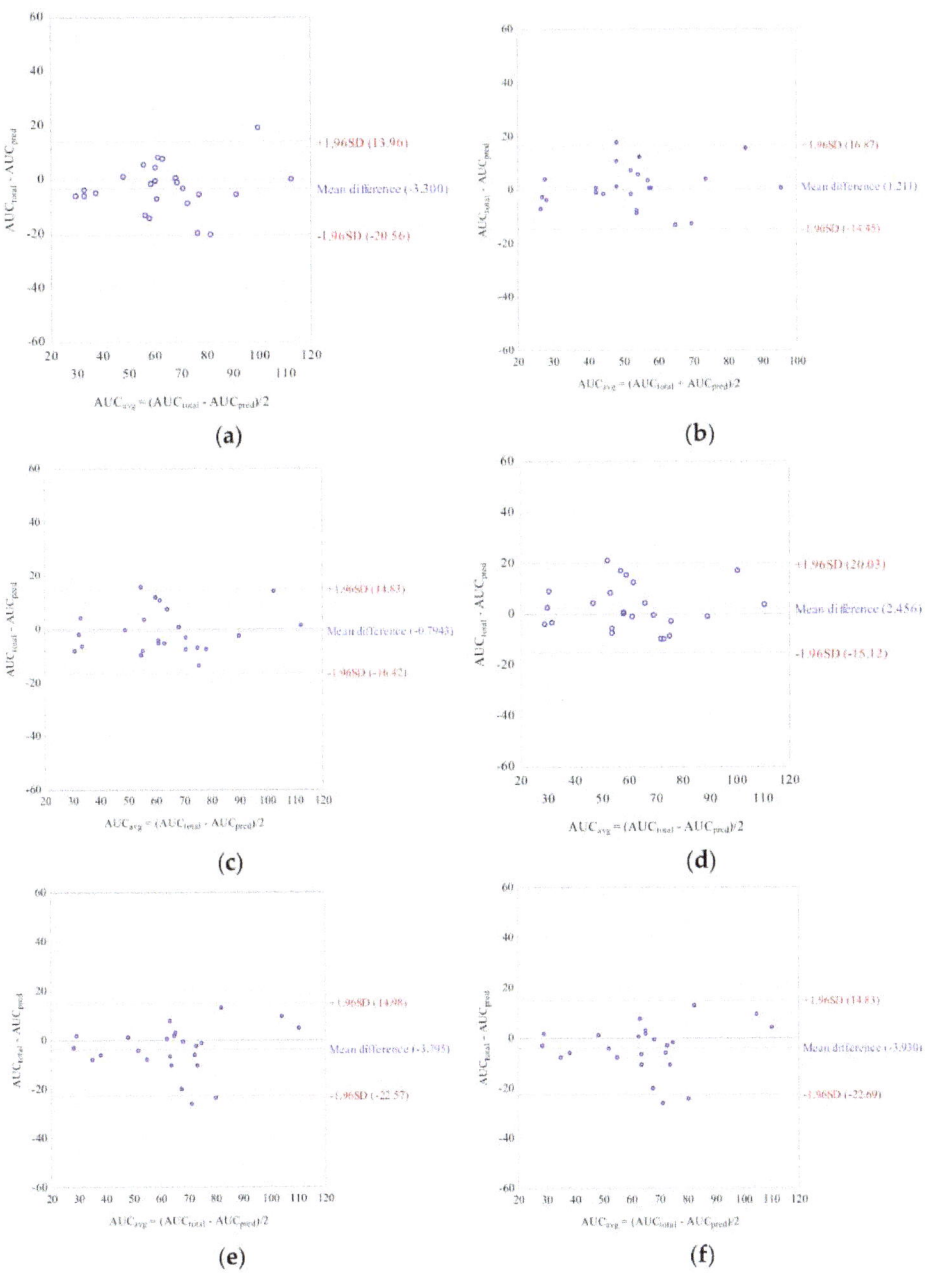

Figure 3. *Cont.*

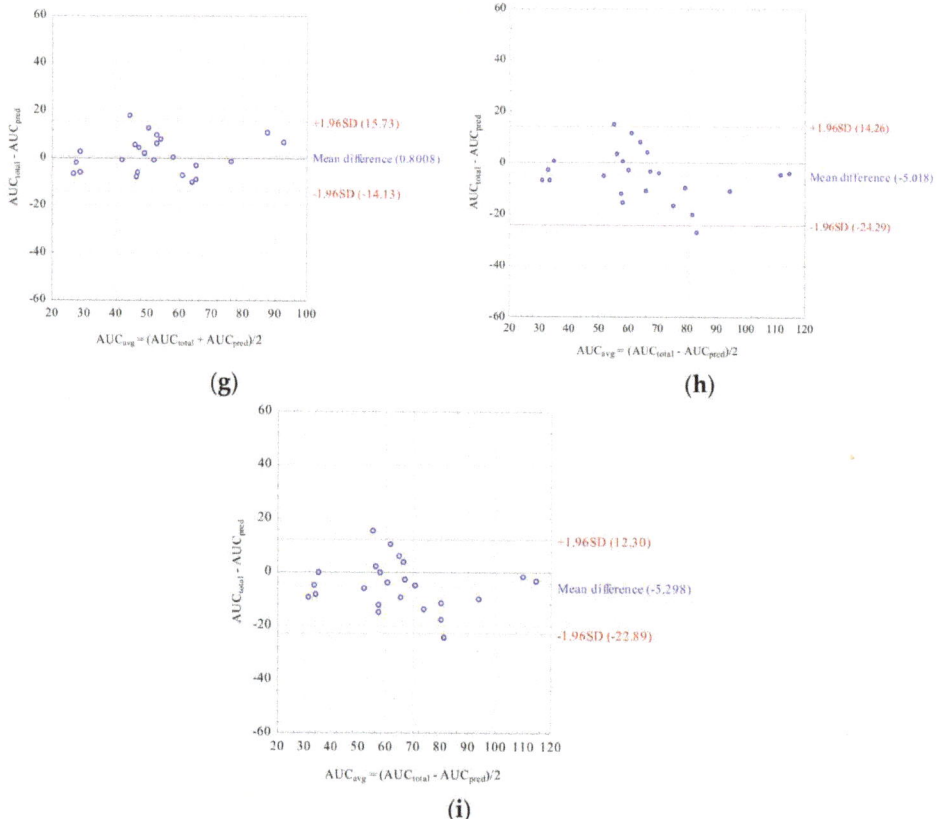

Figure 3. Bland–Altman analyses testing agreement between the MPA AUC$_{total}$ and the MPA AUC$_{pred}$ calculated for children with nephrotic syndrome using the MLR LSS equations found in the literature that fulfilled the criteria for %MPE and %MAE ± 15%; (**a**) AUC$_{pred}$ = 7.4 + 2.3 × C$_0$ + 1.2 × C$_1$ + 2.3 × C$_3$ + 4.4 × C$_6$ [30]; (**b**) AUC$_{pred}$ = 9.328 + 1.311 × C$_1$ + 1.455 × C$_2$ + 2.901 × C$_4$ [43]; (**c**) AUC$_{pred}$ = 10.6 + 1.1 × C$_1$ + 1.1 × C$_2$ + 2.0 × C$_4$ + 3.9 × C$_6$ [30]; (**d**) AUC$_{pred}$ = 5.92 + 1.10 × C$_1$ + 1.01 × C$_2$ + 1.77 × C$_4$ + 4.80 × C$_6$ [28]; (**e**) AUC$_{pred}$ = 8.22 + 3.16 × C$_0$ + 0.99 × C$_1$ + 1.33 × C$_2$ + 4.18 × C$_4$ [32]; (**f**) AUC$_{pred}$ = 8.217 + 3.163 × C$_0$ + 0.994 × C$_1$ + 1.334 × C$_2$ + 4.183 × C$_4$ [8]; (**g**) AUC$_{pred}$ = 10.229 + 0.925 × C$_1$ + 1.750 × C$_2$ + 4.586 × C$_6$ [29]; (**h**) AUC$_{pred}$ = 7.73 + 0.94 × C$_1$ + 2.55 × C$_2$ + 5.48 × C$_6$ [32]; (**i**) AUC$_{pred}$ = 10.75 + 0.98 × C$_1$ + 2.38 × C$_2$ + 4.86 × C$_6$ [33].

3. Discussion

Estimating LSS is the approach of TDM applied for many drugs, e.g., MPA, levofloxacin, and etoposide [49–51]. We recently established and compared LSS for MPA in children with nephrotic syndrome using two different approaches [52]. In the present study, we used the MPA LSSs found in the literature in the attempt to assess their practicability for the estimation of MPA exposure and to find the optimal time points for MPA TDM in children with nephrotic syndrome. We verified the LSSs established for different indications, as in the literature we found studies in which LSS developed for one population was used to evaluate LSS in other population [20,24].

The novelty of our study is that we converted MPA concentrations determined with HPLC to evaluate the MPA LSSs established for EMIT or PETINIA. As MPA concentrations are 15–20% higher when established with EMIT or PETINIA due to MPA cross reaction with the MPA metabolite acyl-glucuronide [16,53], we multiplied the HPLC determined concentration by 1.175. Tong et al. [23] used MPA LSSs established for adult heart transplant recipients with the HPLC method to predict the AUC in children with nephrotic syndrome for whom MPA concentrations were determined with EMIT without any adjustment. Our results of predictive performance for both HPLC and EMIT/PETINIA did not differ significantly, and therefore we concluded that this approach may enable using LSSs established with EMIT or PETINIA to predict the MPA AUC based on HPLC-determined concentrations.

Nine MPA LSSs fulfilled the criteria of the best predictive performance. Because MMF is mainly administered as an acute rejection prophylaxis after renal transplantation and most of the studies concerned adults, five out of nine the best MLR LSSs were established for adults [28–30,43]. Four LSSs considered as the best were established for pediatric patients [8,32,33]. Among these four LSSs, although two equations were very similar, they were published in two different articles, and we therefore evaluated both of them [8,32]. Seven of nine LSSs included renal transplant recipients, both adult ($n = 3$) [30,43] and pediatric ($n = 4$) [8,32,33]. Two of nine the best LSSs included liver transplant recipients [28,29]. Surprisingly, the LSSs established for children with nephrotic syndrome [14,25] or lupus erythematosus [21] performed poorly as they did not fulfill the criteria: the values of r^2 were below 0.800, and $\leq 50\%$ of the AUC_{pred} values were within $\pm 15\%$ of the AUC_{total}. These poor results may be explained by one time point equation in the Hibino et al. study [14] and the relatively high intercept.

In our opinion, in the case of MPA, accurate and precise LSSs should consist of at least three time points. Among the best LSSs, four and five LSSs included four and three time points, respectively. The predictive performance for one and two time point LSSs were unsatisfactory. If the criteria were extended to $\pm 20\%$ for %MPE and %MAE, only one two-time-point equation would have fulfilled the criteria. However, the percentage of AUC_{pred} within $\pm 15\%$ of AUC_{total} was rather poor for this equation (50%). Moreover, equations with only one time-point performed poorly with respect to the percentage of the AUC_{pred} within $\pm 15\%$ of the AUC_{total} ($\leq 33\%$). Interestingly, for one LSS, which included AUC_{1-4} instead of concentration at defined time points [34], r^2 was >0.800, while the predictive performance and the percentage of the AUC_{pred} within $\pm 15\%$ of the AUC_{total} were unsatisfactory. Moreover, the LSSs which included logarithmic concentrations did not perform well [44].

The inclusion of particular time points may be of significant importance as they reflect MPA pharmacokinetics. In our study, eight of the nine (89%) best-matched equations included C_1 and C_2, and six equations included C_6. Those three time points coincide with the MPA C_{max} (1–2 h after dosing) and the second maximum concentration (C_{max2}; 6–12 h after MMF administration) [10]. This evidence suggests that the MPA C_{max} and C_{max2} influence its AUC the most, and the blood samples should be collected at least in three time points near C_{max} and C_{max2} to precisely predict the AUC. According to the literature, for children with nephrotic syndrome C_2 or time points up to 2 h after MMF administration should be included in the MPA LSS [14,25]. The inclusion of C_6 makes using LSS cumbersome. However, according to the literature, better predictive performance was observed for LSSs which included time points in the latter half of the dosing interval [16]. Out of the nine best matched equations, only 3 (33%) included C_0. This observation is in accordance with the literature data, as MPA C_0 correlates poorly with AUC_{total} [6].

We evaluated the MLR LSSs found in the literature regardless the drugs co-administered with MMF. Five of nine the best LSSs were established for MMF- and Tac-treated patients. According to the literature, Tac does not influence MPA clearance [3], and in patients with autoimmune disease MPA clearance is likely to be in close agreement with estimates from renal allograft recipients co-treated with Tac [54]. On the other hand, MPA concentrations are lower if co-administered with CsA [10]. CsA inhibits MPA enterohepatic recirculation, causing a decrease in MPA exposition, and therefore blood sampling does not require including time-points around the MPA C_{max2} when MMF is co-administered with CsA [16]. Among the LSSs applied in this study, only in three studies with MLR LSSs [21,25,46] did the patients not receive concomitant medications (in one study only 8% of patients received CsA [46]). Surprisingly, in our study, for these LSSs the predictive performance fell beyond ±15% range. The equation from the Prabha et al. [21] study would have fulfilled the extended criteria (±20%). One equation, which included C_6, from the de Winter et al. study [46], was characterized by the r^2 being >0.800, however, it did not fulfill even the extended criteria, therefore, we confirmed that choosing model equations based only on their r^2 values may be misleading [55].

Out of five LSSs developed for fMPA [35,36,42], none fulfilled the criteria when used to evaluate the fMPA AUC_{pred} in children with nephrotic syndrome. One equation would fulfill the criteria extended to ±20%, but the percentage of the AUC_{pred} within ±15% of the AUC_{total} for this formula was poor (38%). The obtained results may indicate differences in MPA protein binding in children with nephrotic syndrome. According to the literature, MPA is bound to plasma proteins in 97% to 99% [29,56]. In our previous study [11], similarly as in this study, the median fMPA fraction was 0.36%, which gives very high percentage of bound MPA (99.64%).

The limitation of our study is the fact that we were unable to apply the LSSs with time points 0.5, 0.75, or 1.5 h after MMF administration as blood sampling was not so frequent in the children included in the study.

4. Materials and Methods

4.1. Ethical Considerations

The study was approved by the Bioethical Committee at Poznan University of Medical Sciences and it is in accordance with the 1964 Declaration of Helsinki and its later amendments. Informed consent was obtained from the parents or guardians prior to initiating the study.

4.2. Children's Characteristics

Our study included 24 children, aged 3–18 years, with nephrotic syndrome treated with MMF and steroids in the Department of Pediatric Nephrology and Hypertension, Poznan University of Medical Sciences, Poland. MMF was administered orally twice a day at the same dose. On the day of blood collection, 18 children were in remission whereas six children had trace proteinuria. MMF was given under fasting conditions, 30 min before breakfast. The exclusion criteria were cyclosporine (CsA) co-administration, MMF dosing at unequal morning and evening doses, administration of MMF shorter than 1 month and too low number of blood samples. Blood samples were collected into EDTA tubes before MMF administration (C_0) and subsequently 1 h (C_1), 2 h (C_2), 3 h (C_3), 4 h (C_4), 6 h (C_6), 9 h (C_9), and 12 h (C_{12}) after its administration. The samples were centrifuged to obtain plasma, then immediately frozen and kept at $-20\ °C$ until analysis. The demographic and biochemical characteristics of the children are presented in Table 5.

4.3. Analytical Methods

MPA and fMPA concentrations were determined in the Department of Physical Pharmacy and Pharmacokinetics at Poznan University of Medical Sciences, Poland.

MPA plasma concentrations were determined using the HPLC method with ultraviolet detection. The analytical method for MPA determination was described elsewhere [11,57]. The calibration curve was linear, and within the range 0.25–40.0 µg/mL. The mean between-day coefficient of variation and average accuracy were 2.7% (range 0.5–6.1%) and 98.8% (range 93.8–103.0%), respectively [11].

Free MPA (fMPA) was determined using the HPLC method with fluorescence detection described previously [5,11]. The calibration curve was linear, and within the range of 0.0025–1.0 µg/mL. The mean between-day coefficient of variation and average accuracy were 6.5% (range 1.4–12.7%) and 99.9% (range 94.3–107.6%), respectively [11].

Table 5. Demographic and biochemical characteristics of the study group.

Parameter	Mean ± SD	Range
24 children	10 boys/14 girls	
age	11 ± 4	3–18
body weight	36.9 ± 14.7	17.7–66.5
body surface (m^2)	1.20 ± 0.32	0.70–1.85
MMF daily dose (mg)	Number of children	
500/600/700/800/1000/1200/1500/2000	2/1/1/1/10/1/7/1	
MMF daily dose (mg/m^2)	933 ± 218	505–1250
duration of MMF treatment (months)	12 ± 7	2–29
Protein concentration (g/dL)	6.60 ± 0.53	5.52–7.54
Glomerular filtration rate (mL/min/1.73 m^2)	133 ± 23	101–183
Creatinine concentration (mg/dL)	0.45 ± 0.13	0.25–0.72
Leukocytes count (10^9/L)	6.75 ± 2.34	3.46–13.88
Erythrocytes count (10^{12}/L)	4.65 ± 0.31	4.07–5.54
Hemoglobin (g/dL)	13.0 ± 1.1	11.1–15.5
Hematocrit (%)	37.8 ± 2.8	33.6–44.3
Alanine aminotransferase (U/L)	13 ± 4	5–25
Aspartate aminotransferase (U/L)	26 ± 6	17–45

MMF, mycophenolate mofetil; SD, standard deviation.

4.4. The Literature Data Search

We comprehensively searched the literature in December 2020 using the PUBMED database using the combination of 'mycophenolic acid' or 'mycophenolate mofetil' and the terms: 'limited sampling strategy', 'limited sampling strategies', 'limited sampling', 'optimal sampling', 'sparse sampling', and 'minimal sampling'. We included English written studies determining LSS based on MLR calculations for adult and pediatric patients receiving MMF after solid organ transplantation or with autoimmune diseases, and identified those LSSs which covered the same blood sampling times as in our study. We included LSSs which were established based on HPLC and EMIT MPA determinations. We excluded articles describing LSS for EC-MPS as there is an evident difference in MPA pharmacokinetics for the two formulations MMF and EC-MPS (unpredictable absorption profile after EC-MPS administration) [58]. We also excluded studies using previously established LSSs, those with Bayesian estimators and with different than twice daily MMF dosing schedules.

4.5. Pharmacokinetic Calculations and Statistical Analyses

For children with nephrotic syndrome, firstly, we calculated the MPA AUC_{total} using the linear trapezoidal rule. Secondly, based on the results of the literature data search, we calculated the AUC_{pred} for these children using the MLR formulae found in the literature. We applied LSSs established using MPA concentrations determined with HPLC, EMIT, and PETINIA to evaluate LSS usefulness. Due to the 15–20% higher MPA concentrations established with EMIT [16] and the similar magnitude of the MPA overestimation found for PETINIA when compared with EMIT [53], we multiplied the MPA concentration determined in the children included in this study with the HPLC method by 1.175, and applied the re-calculated AUC_{total} to the evaluation of the LSSs based on EMIT or PETINIA MPA determination. The multiplier of 1.175 was achieved by assuming that MPA concentrations established with EMIT are on average 17.5% higher than those determined with HPLC.

To assess the predictive performance of LSSs available in the literature, we calculated r^2 as well as the bias and precision for AUC_{pred} as the mean relative prediction error (%MPE) and the percentage of the mean absolute relative prediction error (%MAE), respectively, both with 95% confidence intervals. According to the literature, precision and bias ±15% were considered acceptable [22,59,60], although some authors defined the clinical acceptance as ±20% [18] or even as ±33% [61]. Although it does not translate into clinical practice, lower percentages of precision and bias result in more accurate calculations. We also calculated the percentage of the AUC_{pred} within ±15% of the AUC_{total} for each equation to analyze the agreement between the AUC_{pred} and the AUC_{total}. The equations used in the analysis were as follows [51,62]:

$$\%MPE = \frac{1}{N}\Sigma \frac{(AUC_{pred} - AUC_{total})}{AUC_{total}} \times 100 \qquad (1)$$

$$MAE = \frac{1}{N}\Sigma \frac{|AUC_{pred} - AUC_{total}|}{AUC_{total}} \times 100 \qquad (2)$$

Statistical analyses were performed using STATISTICA 13.0 software (StatSoft, Inc., Tulsa, OK, USA). For the best matched MLR LSSs, the Bland–Altman method was used to assess the agreement between the AUC_{pred} and the AUC_{total}. To compare the HPLC and EMIT/PETINIA predictive performance results, the Mann–Whitney test was applied.

5. Conclusions

We concluded that the optimal MPA LSS for children with nephrotic syndrome should include C_1, C_2, and C_6, as these time points coincide with MPA C_{max} and C_{max2}. MPA LSSs established using MPA concentrations determined with EMIT or PETINIA may be used in LSSs based on HPLC-determined MPA concentrations after multiplying the latter by 1.175. The MLR LSS which predicted MPA AUC the best in children with nephrotic syndrome was developed for MMF-treated renal transplant recipients. MPA binding with plasma protein is high in children with nephrotic syndrome, which suggests there are different fMPA pharmacokinetics in this group of patients than in renal, liver, and hematopoietic stem cell recipients treated with MMF. MPA LSSs may facilitate TDM in the case of MMF, however, more studies of fMPA LSS are required for children with nephrotic syndrome.

Author Contributions: Conceptualization, J.S.; methodology, J.S., M.R. and M.C.; validation, J.S. and M.R.; formal analysis, J.S.; investigation, J.S. and M.R.; resources, J.S., J.Z. and D.O.-N.; data curation, J.S.; writing—original draft preparation, J.S.; writing—review and editing, M.C., J.Z. and D.O.-N.; visualization, J.S.; supervision, M.C., J.Z. and D.O.-N.; project administration, J.S.; funding acquisition, J.S. All authors have read and agreed to the published version of the manuscript.

Funding: This research was funded by Poznan University of Medical Sciences, grant number 502-14-03306413-10156. The APC was funded by Poznan University of Medical Sciences.

Institutional Review Board Statement: The study was conducted according to the guidelines of the Declaration of Helsinki, and approved by the Institutional Review Board (or Ethics Committee) of Poznan University of Medical Sciences (protocol code 542/16 and date of approval 5 May 2016).

Informed Consent Statement: Informed consent was obtained from all subjects involved in the study.

Data Availability Statement: The data presented in this study are available on request from the corresponding author.

Acknowledgments: The authors would like to acknowledge the nursing staff that participated in the samples collection.

Conflicts of Interest: The authors declare no conflict of interest.

Sample Availability: Not applicable.

References

1. Staatz, C.E.; Tett, S.E. Pharmacology and Toxicology of Mycophenolate in Organ Transplant Recipients: An Update. *Arch. Toxicol.* **2014**, *88*, 1351–1389. [CrossRef]
2. Ostalska-Nowicka, D.; Malinska, A.; Silska, M.; Perek, B.; Zachwieja, J.; Nowicki, M. Mycophenolate Mofetil (MMF) Treatment Efficacy in Children with Primary and Secondary Glomerulonephritis. *Arch. Med. Sci.* **2011**, *7*, 1042–1048. [CrossRef]
3. Kiang, T.K.; Ensom, M.H. Population Pharmacokinetics of Mycophenolic Acid: An Update. *Clin. Pharm.* **2018**, *57*, 547–558. [CrossRef]
4. Dias-Polak, D.; Bergman, R.; Avitan-Hersh, E. Mycophenolate Mofetil Therapy in Adult Patients with Recalcitrant Atopic Dermatitis. *J. Dermatol. Treat.* **2019**, *30*, 49–51. [CrossRef] [PubMed]

5. Chen, B.; Gu, Z.; Chen, H.; Zhang, W.; Fen, X.; Cai, W.; Fan, Q. Establishment of High-Performance Liquid Chromatography and Enzyme Multiplied Immunoassay Technology Methods for Determination of Free Mycophenolic Acid and Its Application in Chinese Liver Transplant Recipients. *Ther. Drug Monit.* **2010**, *32*, 653–660. [CrossRef] [PubMed]
6. Filler, G.; Alvarez-Elías, A.C.; McIntyre, C.; Medeiros, M. The Compelling Case for Therapeutic Drug Monitoring of Mycophenolate Mofetil Therapy. *Pediatr. Nephrol.* **2017**, *32*, 21–29. [CrossRef] [PubMed]
7. Bruchet, N.K.; Ensom, M.H. Limited Sampling Strategies for Mycophenolic Acid in Solid Organ Transplantation: A Systematic Review. *Expert Opin. Drug Metab. Toxicol.* **2009**, *5*, 1079–1097. [CrossRef]
8. Filler, G. Abbreviated Mycophenolic Acid AUC from C0, C1, C2, and C4 Is Preferable in Children after Renal Transplantation on Mycophenolate Mofetil and Tacrolimus Therapy. *Transplant. Int.* **2004**, *17*, 120–125. [CrossRef]
9. David-Neto, E.; Araujo, L.M.; Sumita, N.M.; Mendes, M.E.; Castro, M.C.; Alves, C.F.; Kakehashi, E.; Romano, P.; Yagyu, E.M.; Queiroga, M.; et al. Mycophenolic Acid Pharmacokinetics in Stable Pediatric Renal Transplantation. *Pediatr. Nephrol.* **2003**, *18*, 266–272. [CrossRef] [PubMed]
10. Staatz, C.E.; Tett, S.E. Clinical Pharmacokinetics and Pharmacodynamics of Mycophenolate in Solid Organ Transplant Recipients. *Clin. Pharm.* **2007**, *46*, 13–58. [CrossRef]
11. Sobiak, J.; Resztak, M.; Ostalska-Nowicka, D.; Zachwieja, J.; Gąsiorowska, K.; Piechanowska, W.; Chrzanowska, M. Monitoring of Mycophenolate Mofetil Metabolites in Children with Nephrotic Syndrome and the Proposed Novel Target Values of Pharmacokinetic Parameters. *Eur. J. Pharm. Sci.* **2015**, *77*, 189–196. [CrossRef]
12. Hackl, Á.; Cseprekál, O.; Gessner, M.; Liebau, M.C.; Habbig, S.; Ehren, R.; Müller, C.; Taylan, C.; Dötsch, J.; Weber, L.T. Mycophenolate Mofetil Therapy in Children with Idiopathic Nephrotic Syndrome: Does Therapeutic Drug Monitoring Make a Difference? *Ther. Drug Monit.* **2016**, *38*, 274–279. [CrossRef]
13. Tellier, S.; Dallocchio, A.; Guigonis, V.; Saint-Marcoux, F.; Llanas, B.; Ichay, L.; Bandin, F.; Godron, A.; Morin, D.; Brochard, K.; et al. Mycophenolic Acid Pharmacokinetics and Relapse in Children with Steroid–Dependent Idiopathic Nephrotic Syndrome. *CJASN* **2016**, *11*, 1777–1782. [CrossRef] [PubMed]
14. Hibino, S.; Nagai, T.; Yamakawa, S.; Ito, H.; Tanaka, K.; Uemura, O. Pharmacokinetics of Mycophenolic Acid in Children with Clinically Stable Idiopathic Nephrotic Syndrome Receiving Cyclosporine. *Clin. Exp. Nephrol.* **2017**, *21*, 152–158. [CrossRef]
15. Gellermann, J.; Weber, L.; Pape, L.; Tönshoff, B.; Hoyer, P.; Querfeld, U. Mycophenolate Mofetil versus Cyclosporin A in Children with Frequently Relapsing Nephrotic Syndrome. *JASN* **2013**, *24*, 1689–1697. [CrossRef] [PubMed]
16. Abd Rahman, A.N.; Tett, S.E.; Staatz, C.E. How Accurate and Precise Are Limited Sampling Strategies in Estimating Exposure to Mycophenolic Acid in People with Autoimmune Disease? *Clin. Pharm.* **2014**, *53*, 227–245. [CrossRef]
17. Katsuno, T.; Ozaki, T.; Ozeki, T.; Hachiya, A.; Kim, H.; Kato, N.; Ishimoto, T.; Kato, S.; Kosugi, T.; Tsuboi, N.; et al. Investigation on the Benefits of Mycophenolate Mofetil and Therapeutic Drug Monitoring in the Treatment of Japanese Patients with Lupus Nephritis. *Clin. Exp. Nephrol.* **2018**, *22*, 1341–1350. [CrossRef]
18. Zicheng, Y.; Xianghui, W.; Peijun, Z.; Da, X.; Weixia, Z.; Hongzhuan, C. Evaluation of the Practicability of Limited Sampling Strategies for the Estimation of Mycophenolic Acid Exposure in Chinese Adult Renal Recipients. *Ther. Drug Monit.* **2007**, *29*, 600–606. [CrossRef] [PubMed]
19. Saint-Marcoux, F.; Guigonis, V.; Decramer, S.; Gandia, P.; Ranchin, B.; Parant, F.; Bessenay, L.; Libert, F.; Harambat, J.; Bouchet, S.; et al. Development of a Bayesian Estimator for the Therapeutic Drug Monitoring of Mycophenolate Mofetil in Children with Idiopathic Nephrotic Syndrome. *Pharmacol. Res.* **2011**, *63*, 423–431. [CrossRef]
20. Ting, L.S.; Partovi, N.; Levy, R.D.; Ignaszewski, A.P.; Ensom, M.H. Performance of Limited Sampling Strategies for Predicting Mycophenolic Acid Area under the Curve in Thoracic Transplant Recipients. *J. Heart Lung Transplant.* **2008**, *27*, 325–328. [CrossRef]
21. Prabha, R.; Mathew, B.; Jeyaseelan, V.; Kumar, T.; Agarwal, I.; Fleming, D. Development and Validation of Limited Sampling Strategy Equation for Mycophenolate Mofetil in Children with Systemic Lupus Erythematosus. *Indian J. Nephrol.* **2016**, *26*, 408–412. [CrossRef] [PubMed]
22. Al-Khatib, M.; Shapiro, R.J.; Partovi, N.; Ting, L.S.; Levine, M.; Ensom, M.H. Limited Sampling Strategies for Predicting Area under the Concentration-Time Curve of Mycophenolic Acid in Islet Transplant Recipients. *Ann. Pharm.* **2010**, *44*, 19–27. [CrossRef] [PubMed]
23. Tong, K.; Mao, J.; Fu, H.; Shen, H.; Liu, A.; Shu, Q.; Du, L. The Value of Monitoring the Serum Concentration of Mycophenolate Mofetil in Children with Steroid-Dependent/Frequent Relapsing Nephrotic Syndrome. *Nephron* **2016**, *132*, 327–334. [CrossRef]
24. Neuberger, M.; Sommerer, C.; Böhnisch, S.; Metzendorf, N.; Mehrabi, A.; Stremmel, W.; Gotthardt, D.; Zeier, M.; Weiss, K.H.; Rupp, C. Effect of Mycophenolic Acid on Inosine Monophosphate Dehydrogenase (IMPDH) Activity in Liver Transplant Patients. *Clin. Res. Hepatol. Gastroenterol.* **2020**, *44*, 543–550. [CrossRef]
25. Benz, M.; Ehren, R.; Kleinert, D.; Müller, C.; Gellermann, J.; Fehrenbach, H.; Schmidt, H.; Weber, L. Generation and Validation of a Limited Sampling Strategy to Monitor Mycophenolic Acid Exposure in Children with Nephrotic Syndrome. *Ther. Drug Monit.* **2019**, *41*, 696–702. [CrossRef]
26. Cai, W.; Cai, Q.; Xiong, N.; Qin, Y.; Lai, L.; Sun, X.; Hu, Y. Limited Sampling Strategy for Estimating Mycophenolic Acid Exposure on Day 7 Post-Transplant for Two Mycophenolate Mofetil Formulations Derived from 20 Chinese Renal Transplant Recipients. *Transplant. Proc.* **2018**, *50*, 1298–1304. [CrossRef]

27. Chaabane, A.; Aouam, K.; Fredj, N.B.; Hammouda, M.; Chadly, Z.; May, M.E.; Boughattas, N.; Skhiri, H. Limited Sampling Strategy of Mycophenolic Acid in Adult Kidney Transplant Recipients: Influence of the Post-Transplant Period and the Pharmacokinetic Profile. *J. Clin. Pharmacol.* **2013**, *53*, 925–933. [CrossRef]
28. Chen, H.; Gu, Z.; Chen, B.; Mao, H.; Zhang, W.; Fan, Q. Models for the Prediction of Mycophenolic Acid Area under the Curve Using a Limited-Sampling Strategy and an Enzyme Multiplied Immunoassay Technique in Chinese Patients Undergoing Liver Transplantation. *Clin. Ther.* **2008**, *30*, 2387–2401. [CrossRef]
29. Chen, H.; Peng, C.; Yu, Z.; Shen, B.; Deng, X.; Qiu, W.; Fei, Y.; Shen, C.; Zhou, G.; Yang, W.; et al. Pharmacokinetics of Mycophenolic Acid and Determination of Area under the Curve by Abbreviated Sampling Strategy in Chinese Liver Transplant Recipients. *Clin. Pharmacokinet.* **2007**, *46*, 175–185. [CrossRef]
30. Enokiya, T.; Nishikawa, K.; Muraki, Y.; Iwamoto, T.; Kanda, H.; Sugimura, Y.; Okuda, M. Usefulness of Limited Sampling Strategy for Mycophenolic Acid Area under the Curve Considering Postoperative Days in Living-Donor Renal Transplant Recipients with Concomitant Prolonged-Release Tacrolimus. *J. Pharm. Health Care Sci.* **2017**, *3*, 17. [CrossRef]
31. Fatela-Cantillo, D.; Hinojosa-Pérez, R.; Peralvo-Rodríguez, M.I.; Serrano-Díaz Canedo, J.; Gómez-Bravo, M.A. Pharmacokinetic Evaluation of Mycophenolic Acid Profiles during the Period Immediately Following an Orthotopic Liver Transplant. *Transplant. Proc.* **2006**, *38*, 2482–2485. [CrossRef] [PubMed]
32. Filler, G.; Feber, J.; Lepage, N.; Weiler, G.; Mai, I. Universal Approach to Pharmacokinetic Monitoring of Immunosuppressive Agents in Children. *Pediatr. Transplant.* **2002**, *6*, 411–418. [CrossRef] [PubMed]
33. Filler, G.; Mai, I. Limited Sampling Strategy for Mycophenolic Acid Area under the Curve. *Ther. Drug Monit.* **2000**, *22*, 169–173. [CrossRef] [PubMed]
34. Gota, V.; Purohit, V.; Gurjar, M.; Nayak, L.; Punatar, S.; Gokarn, A.; Bonda, A.; Bagal, B.; Vora, C.S.; Patil, A.; et al. A Limited Sampling Strategy for Therapeutic Drug Monitoring of Mycophenolate Mofetil for Prophylaxis of Acute Graft-Versus-Host Disease in Allogeneic Stem Cell Transplantation. *Cell Transplant.* **2020**, *29*, 0963689720912925. [CrossRef]
35. Gu, Z.; Chen, B.; Song, Y.; Shen, B.; Zhu, Z.; Zhang, W.; Xie, J.; Deng, X.; Peng, C.; Fan, Q.; et al. Pharmacokinetics of Free Mycophenolic Acid and Limited Sampling Strategy for the Estimation of Area under the Curve in Liver Transplant Patients. *Eur. J. Pharm. Sci.* **2012**, *47*, 636–641. [CrossRef] [PubMed]
36. Jiao, Z.; Zhong, J.; Zhang, M.; Shi, X.; Yu, Y.; Lu, W. Total and Free Mycophenolic Acid and Its 7-O-Glucuronide Metabolite in Chinese Adult Renal Transplant Patients: Pharmacokinetics and Application of Limited Sampling Strategies. *Eur. J. Clin. Pharm.* **2007**, *63*, 27–37. [CrossRef]
37. Johnson, A.G.; Rigby, R.J.; Taylor, P.J.; Jones, C.E.; Allen, J.; Franzen, K.; Falk, M.C.; Nicol, D. The Kinetics of Mycophenolic Acid and Its Glucuronide Metabolite in Adult Kidney Transplant Recipients. *Clin. Pharmacol. Ther.* **1999**, *66*, 492–500. [CrossRef]
38. Karimani, A.; Abedi, H.; Nazemian, F.; Poortaji, A.; Pour, A.H. Estimation of Abbreviated Mycophenolic Acid Area under the Concentration-Time Curve during Stable Post-Transplant Period by Limited Sampling Strategy. *Curr. Clin. Pharm.* **2020**. [CrossRef]
39. Kuriata-Kordek, M.; Boratynska, M.; Falkiewicz, K.; Porazko, T.; Urbaniak, J.; Wozniak, M.; Patrzalek, D.; Szyber, P.; Klinger, M. The Influence of Calcineurin Inhibitors on Mycophenolic Acid Pharmacokinetics. *Transplant. Proc.* **2003**, *35*, 2369–2371. [CrossRef]
40. Guellec, C.L.; Büchler, M.; Giraudeau, B.; Meur, Y.L.; Gakoué, J.E.; Lebranchu, Y.; Marquet, P.; Paintaud, G. Simultaneous Estimation of Cyclosporin and Mycophenolic Acid Areas under the Curve in Stable Renal Transplant Patients Using a Limited Sampling Strategy. *Eur. J. Clin. Pharm.* **2002**, *57*, 805–811. [CrossRef]
41. Miura, M.; Satoh, S.; Niioka, T.; Kagaya, H.; Saito, M.; Hayakari, M.; Habuchi, T.; Suzuki, T. Limited Sampling Strategy for Simultaneous Estimation of the Area under the Concentration-Time Curve of Tacrolimus and Mycophenolic Acid in Adult Renal Transplant Recipients. *Ther. Drug Monit.* **2008**, *30*, 52–59. [CrossRef]
42. Ng, J.; Rogosheske, J.; Barker, J.; Weisdorf, D.; Jacobson, P.A. A Limited Sampling Model for Estimation of Total and Unbound Mycophenolic Acid (MPA) Area under the Curve (AUC) in Hematopoietic Cell Transplantation (HCT). *Ther. Drug Monit.* **2006**, *28*, 394–401. [CrossRef] [PubMed]
43. Poulin, E.; Greanya, E.D.; Partovi, N.; Shapiro, R.J.; Al-Khatib, M.; Ensom, M.H. Development and Validation of Limited Sampling Strategies for Tacrolimus and Mycophenolate in Steroid-Free Renal Transplant Regimens. *Ther. Drug Monit.* **2011**, *33*, 50–55. [CrossRef] [PubMed]
44. Ting, L.S.; Partovi, N.; Levy, R.D.; Riggs, K.W.; Ensom, M.H. Limited Sampling Strategy for Predicting Area under the Concentration-Time Curve of Mycophenolic Acid in Adult Lung Transplant Recipients. *Pharmacother. J. Hum. Pharmacol. Drug Ther.* **2006**, *26*, 1232–1240. [CrossRef] [PubMed]
45. Wada, K.; Takada, M.; Kotake, T.; Ochi, H.; Morishita, H.; Komamura, K.; Oda, N.; Mano, A.; Hanatani, A.; Nakatani, T. Limited Sampling Strategy for Mycophenolic Acid in Japanese Heart Transplant Recipients Comparison of Cyclosporin and Tacrolimus Treatment. *Circ. J.* **2007**, *7*, 1022–1028. [CrossRef]
46. de Winter, B.C.; Neumann, I.; van Hest, R.M.; Gelder, T.; Mathot, R. Limited Sampling Strategies for Therapeutic Drug Monitoring of Mycophenolate Mofetil Therapy in Patients with Autoimmune Disease. *Ther. Drug Monit.* **2009**, *31*, 382–390. [CrossRef] [PubMed]
47. Yamaguchi, K.; Fukuoka, N.; Kimura, S.; Watanabe, M.; Tani, K.; Tanaka, H.; Sofue, T.; Kosaka, S.; Inui, M.; Kakehi, Y.; et al. Limited Sampling Strategy for the Estimation of Mycophenolic Acid Area under the Concentration–Time Curve Treated in

Japanese Living-Related Renal Transplant Recipients with Concomitant Extended-Release Tacrolimus. *Biol. Pharm. Bull.* **2013**, *36*, 1036–1039. [CrossRef]
48. Yeung, S.; Tong, K.L.; Tsang, W.K.; Tang, H.L.; Fung, K.S.; Chan, H.W.; Chan, A.Y.; Chan, L. Determination of Mycophenolate Area under the Curve by Limited Sampling Strategy. *Transplant. Proc.* **2001**, *33*, 1052–1053. [CrossRef]
49. Czyrski, A.; Kondys, K.; Szałek, E.; Karbownik, A.; Grześkowiak, E. The Pharmacokinetic Interaction between Levofloxacin and Sunitinib. *Pharm. Rep.* **2015**, *67*, 542–544. [CrossRef]
50. Danielak, D.; Sobiak, J.; Wachowiak, J.; Główka, F.; Chrzanowska, M. Development of a Limited Sampling Strategy for the Estimation of Exposure to High-Dose Etoposide after Intravenous Infusion in Pediatric Patients. *Ther. Drug Monit.* **2017**, *39*, 138–144. [CrossRef]
51. Pawinski, T.; Luszczynska, P.; Durlik, M.; Majchrzak, J.; Baczkowska, T.; Chrzanowska, M.; Sobiak, J.; Glyda, M.; Kuriata-kordek, M.; Kamińska, D.; et al. Development and Validation of Limited Sampling Strategies for the Estimation of Mycophenolic Acid Area under the Curve in Adult Kidney and Liver Transplant Recipients Receiving Concomitant Enteric-Coated Mycophenolate Sodium and Tacrolimus. *Ther. Drug Monit.* **2013**, *35*, 760–769. [CrossRef]
52. Sobiak, J.; Resztak, M.; Pawiński, T.; Żero, P.; Ostalska-Nowicka, D.; Zachwieja, J.; Chrzanowska, M. Limited Sampling Strategy to Predict Mycophenolic Acid Area under the Curve in Pediatric Patients with Nephrotic Syndrome: A Retrospective Cohort Study. *Eur. J. Clin. Pharm.* **2019**, *75*, 1249–1259. [CrossRef]
53. Kunicki, P.K.; Pawinski, T.; Boczek, A.; Was, J.; Bodnar-Broniarczyk, M. A Comparison of the Immunochemical Methods, PETINIA and EMIT, with That of HPLC-UV for the Routine Monitoring of Mycophenolic Acid in Heart Transplant Patients. *Ther. Drug Monit.* **2015**, *37*, 311–318. [CrossRef]
54. Abd Rahman, A.N.; Tett, S.E.; Staatz, C.E. Clinical Pharmacokinetics and Pharmacodynamics of Mycophenolate in Patients with Autoimmune Disease. *Clin. Pharm.* **2013**, *52*, 303–331. [CrossRef]
55. Zhou, P.J.; Xu, D.; Yu, Z.C.; Wang, X.H.; Shao, K.; Zhao, J.P. Pharmacokinetics of Mycophenolic Acid and Estimation of Exposure Using Multiple Linear Regression Equations in Chinese Renal Allograft Recipients. *Clin. Pharm.* **2007**, *46*, 389–401. [CrossRef] [PubMed]
56. Downing, H.J.; Pirmohamed, M.; Beresford, M.W.; Smyth, R.L. Paediatric Use of Mycophenolate Mofetil. *Br. J. Clin. Pharm.* **2013**, *75*, 45–59. [CrossRef] [PubMed]
57. Elbarbry, F.A.; Shoker, A.S. Liquid Chromatographic Determination of Mycophenolic Acid and Its Metabolites in Human Kidney Transplant Plasma: Pharmacokinetic Application. *J. Chromatogr. B* **2007**, *859*, 276–281. [CrossRef] [PubMed]
58. Jia, Y.; Peng, B.; Li, L.; Wang, J.; Wang, X.; Qi, G.; Rong, R.; Wang, L.; Qiu, J.; Xu, M.; et al. Estimation of Mycophenolic Acid Area under the Curve with Limited-Sampling Strategy in Chinese Renal Transplant Recipients Receiving Enteric-Coated Mycophenolate Sodium. *Ther. Drug Monit.* **2017**, *39*, 29–36. [CrossRef]
59. Gaies, E.; Ben Sassi, M.; El Jebari, H.; Jebabli, N.; Charfi, R.; Chokri, I.; Salouage, I.; Klouz, A.; Trabelsi, S. Limited Sampling Strategy for the Estimation of Mycophenolic Acid Area under the Curve in Tunisian Renal Transplant Patients. *Nephrol. Ther.* **2017**, *13*, 460–462. [CrossRef]
60. Barraclough, K.; Isbel, N.; Franklin, M.; Lee, K.; Taylor, P.; Campbell, S.; Petchey, W.; Staatz, C. Evaluation of Limited Sampling Strategies for Mycophenolic Acid after Mycophenolate Mofetil Intake in Adult Kidney Transplant Recipients. *Ther. Drug Monit.* **2010**, *32*, 723–733. [CrossRef]
61. Hao, C.; Erzheng, C.; Anwei, M.; Zhicheng, Y.; Baiyong, S.; Xiaxing, D.; Weixia, Z.; Chenghong, P.; Hongwei, L. Validation of Limited Sampling Strategy for the Estimation of Mycophenolic Acid Exposure in Chinese Adult Liver Transplant Recipients. *Liver Transplant.* **2007**, *13*, 1684–1693. [CrossRef] [PubMed]
62. Miura, M.; Satoh, S.; Niioka, T.; Kagaya, H.; Saito, M.; Hayakari, M.; Habuchi, T.; Suzuki, T. Early Phase Limited Sampling Strategy Characterizing Tacrolimus and Mycophenolic Acid Pharmacokinetics Adapted to the Maintenance Phase of Renal Transplant Patients. *Ther. Drug Monit.* **2009**, *31*, 467–474. [CrossRef] [PubMed]

Review

Metabolic Profiling and Quantitative Analysis of Cerebrospinal Fluid Using Gas Chromatography–Mass Spectrometry: Current Methods and Future Perspectives

Alisa Pautova [1,*], Natalia Burnakova [2] and Alexander Revelsky [2]

1. Federal Research and Clinical Center of Intensive Care Medicine and Rehabilitology, Laboratory of Human Metabolism in Critical States, Negovsky Research Institute of General Reanimatology, Petrovka str. 25-2, 107031 Moscow, Russia
2. Laboratory of Mass Spectrometry, Chemistry Department, Lomonosov Moscow State University, GSP-1, Leninskie Gory, 1-3, 119991 Moscow, Russia; nburnakova@me.com (N.B.); sorbent@yandex.ru (A.R.)
* Correspondence: alicepau@mail.ru; Tel.: +79-057-738-981

Abstract: Cerebrospinal fluid is a key biological fluid for the investigation of new potential biomarkers of central nervous system diseases. Gas chromatography coupled to mass-selective detectors can be used for this investigation at the stages of metabolic profiling and method development. Different sample preparation conditions, including extraction and derivatization, can be applied for the analysis of the most of low-molecular-weight compounds of the cerebrospinal fluid, including metabolites of tryptophan, arachidonic acid, glucose; amino, polyunsaturated fatty and other organic acids; neuroactive steroids; drugs; and toxic metabolites. The literature data analysis revealed the absence of fully validated methods for cerebrospinal fluid analysis, and it presents opportunities for scientists to develop and validate analytical protocols using modern sample preparation techniques, such as microextraction by packed sorbent, dispersive liquid–liquid microextraction, and other potentially applicable techniques.

Keywords: metabolomics; targeted analysis; nontargeted analysis; sample preparation; derivatization; validation; biomarkers

Citation: Pautova, A.; Burnakova, N.; Revelsky, A. Metabolic Profiling and Quantitative Analysis of Cerebrospinal Fluid Using Gas Chromatography–Mass Spectrometry: Current Methods and Future Perspectives. *Molecules* 2021, 26, 3597. https://doi.org/10.3390/molecules26123597

Academic Editors: Victoria Samanidou and Natasa Kalogiouri

Received: 17 May 2021
Accepted: 9 June 2021
Published: 11 June 2021

Publisher's Note: MDPI stays neutral with regard to jurisdictional claims in published maps and institutional affiliations.

Copyright: © 2021 by the authors. Licensee MDPI, Basel, Switzerland. This article is an open access article distributed under the terms and conditions of the Creative Commons Attribution (CC BY) license (https://creativecommons.org/licenses/by/4.0/).

1. Introduction

Modern differential diagnostics of a wide variety of diseases and pathologies is not complete without analyzing the composition of biological fluids of the body. The most available and studied fluids are blood and urine. Less studied, as well as less available, is cerebrospinal fluid (CSF). CSF performs a number of important physiological functions in the brain and spinal cord, providing metabolic processes between the blood and the brain. Chemical compounds of different structures can penetrate from the blood through the blood–brain barrier (BBB) into the CSF, and then into the brain cells, subsequently affecting the functioning of the central nervous system (CNS) [1].

The routine laboratory study of the CSF composition is usually aimed at the diagnosis of acute infectious diseases of the CNS such as meningitis and encephalitis. The current interest in the study of the CSF composition is due to the rapidly increasing number of neurodegenerative, mental, and other slowly progressive—and in most cases, incurable—diseases. Despite the advances in science and medicine, treatment of these diseases is directed at reducing the symptoms, but not at eliminating the cause of the disease, since the etiologies of most of the common diseases, such as multiple sclerosis [2–7], Parkinson's disease [8], Alzheimer's disease [9–14], and others, remain not fully understood.

One of the modern concepts in the medical community is the concept of "microbiota–gut–brain" connection. Previously, it was thought that the brain affects the functioning of the body "unidirectionally", but recent research studies clearly indicate that the composition and function of the gut microbiota have an equal impact on the brain and the CNS.

Numerous studies have shown that the composition of the microbiota in healthy volunteers differs from that in patients with various mental, neurodegenerative, and chronic diseases [15,16]. The gut microbiota synthesize and utilize a large number of biologically active substances, including fatty acids, amino acids, neurotransmitters, and others that can penetrate the BBB; thus, the microbiota are becoming a new potential target for the monitoring and treatment of various CNS diseases [17–19].

The main fact that limits the active study of the composition of the CSF is its inaccessibility. A lumbar puncture is required to obtain the CSF that carries the risk of side complications [20]. The most common one is a postdural puncture syndrome, which is accompanied by headaches and symptoms of meningism, as well as less common complications (spinal or epidural bleeding, adhesive arachnoiditis, and trauma to the spinal cord) [21]. The technique of lumbar puncture requires professionalism and experience from the specialist who performs it; this determines its nonprevalence along with the above-mentioned side effects. Lumbar puncture for diagnosis of subarachnoid hemorrhage, hydrocephalus, or infectious diseases of the CNS is frequently carried out in cases in which computed tomography or magnetic resonance imaging is not possible (with the exception of meningitis) [20].

During the diagnostic lumbar puncture, the volume of the sampled CSF is 2.0–5.0 mL (protocol approved by the regional guidelines approved by the Ministry of Health of the Russian Federation, 4 March 2004, #4.2.1887-04). Subsequent routine analysis includes the determination of cytosis, microbiological studies, and biochemical analysis (determination of the level of glucose, chlorides, etc.). Since the total volume of the CSF in an adult is 125–150 mL, any actions aimed at the additional CSF collection or procedure must be justified and approved by the local ethics committee of the institution. For scientific purposes, carrying the CSF analysis out in the residual volume of the sample after clinical and laboratory studies, which is usually about 0.5–1 mL, is highly desirable. Despite all mentioned limitations, the investigation of the CSF composition—in particular, of that of the healthy donors—is an important and crucial issue for the discovering of the new biomarkers.

The limited available sample volume requires the use of modern efficient sample preparation techniques and sensitive analytical methods for the determination of metabolites, most often at the trace level. Such methods are chromatographic ones, namely gas (GC) and high-performance liquid chromatography (HPLC) in combination with different mass-selective (MS) detectors (single quadrupole (Q), time-of-flight (TOF), including high resolution equipment and tandem MS/MS or Q-TOF) [22].

Nontargeted metabolic profiling analysis is the initial stage in the search for metabolites that distinguish patients with pathology from patients of the control group/healthy volunteers. The most common task is to identify as many chromatographic peaks in the sample as possible using the total ion current mode [8,10,23,24]. The subsequent steps should be directed at clearly establishing the formulas of the most promising markers, and statistically processing the quantitative data, which will answer the question of whether the found metabolite or several metabolites are promising biomarkers [25–27]. An important stage is the validation of the method for determining the target components, which currently needs to be carried out in accordance with the Food and Drug Administration (FDA) or the European Medicines Agency (EMA) guidelines [28]. Since the volume of biological samples for full validation is about 40–50 mL (for plasma samples), CSF validation requirements can be reduced because of its specificity. However, validation for the CSF metabolites is justified for several compounds simultaneously.

Based on the information provided, the following requirements for the CSF analysis method can be formulated:

- the use of the minimum (less than 0.5 mL) volume of the CSF for one analysis;
- the quantitative analysis of several target components at the trace level in one analysis.

Gas chromatography is the most common method for the analysis of volatile compounds in biological samples. Correctly selected sample preparation conditions lead to the quantitative determination of different classes of chemical compounds in one chromatographic analysis, which meets the above requirements. When comparing the capabilities of GC and HPLC, the lower cost of the GC equipment and the simpler selection of the parameters of the analysis are worth noting. Despite the wider possibilities of the HPLC in the analysis of nonvolatile high-molecular-weight compounds, gas chromatography is the "gold standard" for the tasks of metabolomics—in particular, the determination of low-molecular-weight compounds [29]. Thus, the aim of this review was to analyze the original articles published since 2000 on the study of the CSF composition using GC–MS and to describe the most promising modern methods of sample preparation, which are potentially suitable for studying CSF composition. The PubMed, Science Direct, and Google Scholar database platforms were used for the search. The keywords "cerebrospinal fluid" and "gas chromatography–mass spectrometry" or "GC–MS" were used in combination in the search list.

2. The Human Cerebrospinal Fluid Metabolome

A large-scale study of the CSF composition (Table 1) was carried out by a group of scientists who are the creators of the Human Metabolome Database. The main publication describing the result of their research provided a link to the created resource www.csfmetabolome.ca (accessed on 11 June 2021). From a total of 308 metabolites detected, 53 were identified using nuclear magnetic resonance (NMR) spectroscopy, 41 using GC–MS, and 17 using LC–MS [22]. Later work by the same authors described 476 metabolites [30]. At the time of preparing the present review, there were 445 metabolites in the metabolite catalog, 443 of which were listed as qualitatively and quantitatively measured. The main classes of low molecular weight metabolites found in CSF are the compounds that can be determined by GC–MS, namely amino acids, fatty acids, including short-chain ones, steroids and their derivatives, hydroxy acids, dicarboxylic acids, and nucleosides. Moreover, most of the identified compounds are neurotransmitters or their metabolites [22]. As noted by the authors, despite the greater number of compounds determined by the NMR, the potential remains with GC–MS when using selective methods of sample preparation and derivatization, as well as when using TOF mass spectrometry.

Table 1. The CSF metabolic profiling using GC-MS methods (CSF volume, samples preparation, and type of capillary column).

Aim	GC-MS Method, Capillary Column	CSF Sampling	Compounds	Sample Volume	Sample Preparation	Reference
Presenting a catalog of detectable metabolites (including their concentrations and disease associations) that can be found in human CSF. This catalog was assembled using a combination of both experimental (NMR, GC-MS, LC-FTMS) and literature-based research.	GC-MS: DB-5 column	Patients screened for meningitis (n = 50).	41 metabolites: amino acids, fatty acids, steroids, carbohydrates, et al.	200 μL	CSF + 800 μL 8:1 HPLC-grade MeOH-deionized water + vortexing + centrifugation + 200 μL of the supernatant was evaporated to dryness + 40 μL methoxyamine hydrochloride + incubation 90 min at 30 °C + 40 μL MSTFA + 20 μL proline IS + incubation at 30 °C for 45 min.	[22]
Update of the CSF metabolome database. Determination of metabolites using different methods, including NMR, GC-MS, LC-FTMS, direct flow injection-MS/MS and ICP-MS.	GC-MS: DB-5 column	Patients screened for meningitis (n = 7).	The same as in [22]	200 μL	The same as in [22].	[30]
Analysis of protein and metabolite abundances in CSF by multiple analytical platforms. Integration of metabolomics and proteomics to present biological variations in metabolite and protein abundances and compare these with technical variations with the currently used analytical methods.	GC-MS: 30 m × 0.25 mm × 0.25 μm, HP5-MS	Subjects (n = 9), the validation sample set (n = 28), and the experimental sample sets (n = 36 for proteomics and n = 42 for metabolomics).	93 metabolites: amino acids, organic acids, nucleosides, fatty acids, mono- and disaccharides, et al.	60–100 μL	60 μL CSF + 250 μL MeOH + centrifugation. 100 μL CSF samples from the validation sample set + 400 μL MeOH + drying under N₂ + derivatization with MSTFA in pyridine. The final volume was 45 μL for the original sample set and 135 μL for the validation sample set.	[31]
To conduct a global metabolomics analysis to provide an overview of the postprandial alterations in CSF and plasma metabolites and to facilitate the application of CSF for biomarker screening (using metabolomics).	GC-MS/MS: 30 m × 0.25 mm × 0.25 μm, DB-5MS-DG.	Healthy subjects (n = 9). CSF collected both preprandial and postprandial CSF. The postprandial time was set at 1.5 h (n = 3), 3 h (n = 3), and 6 h (n = 3).	150 metabolites: amino acids, fatty acids, indoles, carbohydrates, et al.	500 μL	CSF + 9 volumes MeOH + centrifugation + IS + drying under N₂ + oximation and trimethylsilylation.	[32]
Study on the effects of preanalytical factors on the porcine CSF proteome and metabolome using a variety of techniques comprising LC-MS, GC-MS, and MALDI-FT-ICP-MS.	GC-MS: 30 mm × 0.25 mm × 0.25 μm, HP5-MS.	Conventional pigs (n = 5).	49 metabolites: amino acids, sugars, hydroxy acids, et al.	100 μL	Lyophilization + derivatization with ethoxyamine hydrochloride in pyridine + derivatization with MSTFA.	[33]
Characterization of the metabolites present in CSF and comparison of the metabolite levels in patient-matched setting to those found in serum.	GC × GC-TOFMS: 10 m × 0.18 mm × 0.18 μm, Rxi-5ms; 1.5 m × 0.1 mm × 0.1 μm, BPX-50.	Healthy subjects (n = 53).	1280 metabolites, quantitatively determined 21 compounds	25 μL	25 μL CSF + 10 μL IS + 400 μL MeOH + vortexing + centrifugation + 30 min at −20 °C + drying under N₂ + derivatization 25 μL O-methylhydroxyamine hydrochloride + incubation 60 min, 45 °C + 25 μL MSTFA + incubation 60 min at 45 °C + hexane.	[34]
A GC-MS-based metabolomic analysis of CSF samples from glioma patients.	GC-MS: 30 m × 0.25 mm × 1.0 μm, DB-5, 30 m × 0.25 mm × 0.25 μm, CP-SIL 8 CB low bleed/MS.	Patients with intracranial glial tumors (n = 32).	45 metabolites, quantitatively determined 16 compounds.	50 μL	50 μL CSF + 250 μL MeOH-water-chloroform (2.5:1:1) + IS + vortexing + centrifugation + 250 μL of supernatant + 200 μL distilled water + vortexing + centrifugation + 250 μL of supernatant was lyophilized + 40 μL 20 mg/mL methoxyamine hydrochloride in pyridine + 20 μL MSTFA + centrifugation.	[23]
Exploring potential biomarkers and improving understanding of biochemical features of CSF-mediated autoimmune inflammatory diseases of the CNS.	GC-TOFMS: RTX-5Sil MS.	Patients suspected to have inflammatory demyelinating diseases (n = 145), control subjects without medical or neurological illness (n = 12)	962 metabolic signatures, quantitatively determined 85: sugars and sugar alcohols (24%), amino acids (28%), fatty acids (15%), organic acids (15%), amines (2%) et al.	100 μL	100 μL CSF on ice at 4 °C + 650 μL MeOH-isopropanol-water, 3:3:2, v/v/v + centrifugation + 700 μL of supernatant + drying + storage at −80 °C until derivatization + 5 μL 40 mg/mL methoxyamine hydrochloride in pyridine + incubation 90 min at 200 rpm, 30 °C + 2 μL IS + 45 μL MSTFA+1% TMCS + incubation 1 h at 200 rpm, 37 °C.	[26]

Table 1. Cont.

Aim	GC–MS Method, Capillary Column	CSF Sampling	Compounds	Sample Volume	Sample Preparation	Reference
CSF metabolome study in a group of patients with different clinical and genetic subtypes of amyotrophic lateral sclerosis using GC–TOFMS.	GC–TOFMS: 10 m × 0.18 mm × 0.18 μm DB 5-MS.	amyotrophic lateral sclerosis patients (n = 78), healthy subjects (for control)	120 peaks, 40 identified.	100 μL	100 μL CSF + 900 μL IS and MeOH–water, 1:9 + 11 IS + beadmill for 1 min (90 Hz), 2 h on ice + centrifugation + 200 μL evaporated + storage at −80 °C + 30 μL (15 μg/μL) methylhydroxylamine hydrochloride 98% in pyridine + vortexing + heating 70 °C, 1 h + 16 h at room temperature + 30 μL MSTFA+1% TMCS, room temperature, 1 h + 30 μL heptane.	[25]
A detailed analytical evaluation of GC–APCI–TOFMS. In addition to the detailed examination of the analytical performance (repeatability, reproducibility, linearity, and detection limits), the applicability of this technique for metabolic profiling of CSF was demonstrated.	GC–APCI–TOFMS 30 m × 0.25 mm × 0.25 μm HP-5-MS.	Human CSF	300 compounds, 21 identified	250 μL	250 μL CSF + 600 μL MeOH + centrifugation + evaporation + 100 μL methoxyamine:pyridine mixture, 40 °C, 60 min +100 μL BSTFA or MSTFA containing 1% TMCS, 40 °C, 30 min + 2 h equilibration.	[35]
GC–MS/MS-based metabolome analysis of the CSF in pediatric patients with and without epilepsy.	GC–MS/MS	Patients with epilepsy (n = 34), patients without epilepsy (n = 30)	180 metabolites	50 μL (from reference)	(From reference): 50 μL serum + 250 μL MeOH–water–chloroform (2.5:1:1) + shaking, 30 min at 37 °C + centrifugation + 225 μL supernatant + 200 μL distilled water + centrifugation + 250 μL + 40 μL 20 mg/mL methoxyamine hydrochloride in pyridine + shaking + 20 μL MSTFA + incubation 30 min, 1200 rpm at 37 °C + centrifugation.	[36]
Testing the hypothesis that fatty acid metabolism in Alzheimer's disease or mild cognitive impairment is altered compared to cognitively healthy study participants, and that details of the changes could be revealed by study of the brain-derived nanoparticles and supernatant fluid fractions of CSF.	GC–MS: 30 m × 0.25 mm × 0.50 μm, Phenomenex Zebron ZB-1MS.	Total participants (n = 139): cognitively healthy (n = 70), mild cognitive impairment (n = 40), Alzheimer's disease (n = 29).	20 fatty acids (6 saturated, 6 monounsaturated, 8 polyunsaturated fatty acids)	1 mL	1 mL CSF + 100 ng IS + formic acid (0.9%, 3 drops) + lipid extraction + 0.5 mL chloroform:MeOH solution (1:1, v/v), 0.5 mg/mL butylated hydroxytoluene + vortexing + storage −40 °C + PFBBr in MeCN solution (1:19 v/v, 50 μL) and diisopropyl ether in MeCN solution (1.9 v/v, 50 μL), 20 min at 45 °C + drying under N_2 + 1 mL hexane.	[10]
Investigation of CSF and plasma metabolomic profiles for prediction of Parkinson's disease progression.	GC–MS	Participants with relatively mild Parkinsonism. Donors (n = 49) were randomly selected placebo-treated participants.	383 biochemicals	No data	No exact information, except references to the earlier publications. Brief information describes extraction and derivatization with BSTFA.	[8]
Investigation of the metabolomics profile of patients affected by relapsing–remitting multiple sclerosis and primary progressive multiple sclerosis, in order to find potential biomarkers to distinguish between the two forms.	GC–MS: 30 m × 0.25 mm × 0.25 μm, TG-5MS.	Patients (relapsing–remitting multiple sclerosis n = 22, primary progressive multiple sclerosis n = 12)	Different classes of compounds (data most discussed in this article were obtained from LC-MS and flow injection-MS analysis).	200 μL	200 μL + lyophilization + drying + 50 μL methoxyamine in pyridine (10 mg/mL), 70 °C + 100 μL MSTFA, room temperature, 1 h + 100 μL hexane.	[24]

Table 1. Cont.

Aim	GC–MS Method, Capillary Column	CSF Sampling	Compounds	Sample Volume	Sample Preparation	Reference
Comprehensive analysis of the absorbed constituents in the plasma and CSF of rabbits after intranasal administration of Asari Radix et Rhizoma by headspace solid-phase microextraction–GC–MS and HPLC–atmospheric pressure chemical ionization–ion trap-time of flight-multistage mass spectrometry (HPLC–APCI–ion-trap-TOF-MSn).	GC–MS: 30 m × 0.25 mm × 0.25 μm, Rxi-5MS.	Rabbits (n = 15)	25 metabolites	500 μL	500 μL CSF in 10 mL headspace vial + 0.10 g NaCl + polydimethylsiloxane/divinylbenzene fiber was exposed to the headspace at 70 °C, 40 min + fiber was withdrawn into the needle + desorption at 250 °C for 3 min into the GC injection port.	[37]
Investigation of CSF metabolomics in an acute experimental autoimmune encephalomyelitis rat model using targeted LC–MS and GC–MS.	GC–MS: 30 m × 0.25 mm × 0.25 μm, HP5-MS.	Rats (n = 84)	14 amino acids and related compounds	30 μL	30 μL CSF + 250 μL MeOH + centrifugation + drying under N_2 + derivatization with MSTFA in pyridine. The end volume was 45 μL.	[2]
A nontargeted metabolomic analysis using GC–MS was conducted to identify differentially expressed metabolites between naturally occurring depressive and control macaques.	GC–MS: 30 m × 0.25 mm × 0.25 μm, HP-5MS.	Naturally occurring depressive female macaques (n = 10) and age- and gender-matched healthy controls (n = 12)	663 variables, 37 metabolites	15 μL	~15 μL CSF + 10 μL IS + vortexing + 90 μL MeOH + centrifugation + 95 μL of supernatant + drying under N_2 + 30 μL methoxyamine hydrochloride (20 mg/mL pyridine) + incubation 37 °C, 90 min + 30 μL of BSTFA + 1% TMCS 70 °C, 60 min + cooling to room temperature.	[38]
Report on an analytical method that can be used for metabolomics studies when only a limited amount of sample volume is available.	GC–MS: 30 m × 0.25 mm × 0.25 μm, HP-5MS.	Rats (n = 60, total number of CSF samples n = 90)	93 metabolites, 73 identified: fatty acids, amino acids, tricarboxylic cycle acids, carbohydrates, polyols, purine/pyrimidine bases, et al.	10 μL	10 μL CSF + 40 μL MeOH + centrifugation 10 min, 11,800 rpm + drying under N_2 + 10 μL ethoxyamine·HCl (c = 56 mg/mL (0.58M)) in pyridine), 90 min, 40 °C + 20 μL MSTFA, 50 min, 40 °C. The final volume was 50 μL.	[39]
Presentation of an analytical method using in-liner silylation coupled to GC–MS that is suitable for metabolic profiling in ultrasmall sample volumes of 2 μL down to 10 nL.	GC–TOFMS: 30 m × 0.25 mm × 0.25 μm HP5-MS.	Mouse and human CSF samples.	342 peaks, 52 identified in human CSF: amino acids, organic acids, fatty acids, sugars, et al.	2 μL	The microvials containing the dried sample were placed inside the PTV injection liner + 1 μL IS + 3 μL MSTFA.	[40]
Application of untargeted metabolomics using GC–TOFMS to the CSF of aneurysmal subarachnoid hemorrhage patients to determine global metabolic changes and metabolite predictors of long-term outcome that are independent of vasospasm status.	GC–MS: 30 m × 0.25 mm × 0.25 μm, Rtx-5Sil MS.	Patients with aneurysmal subarachnoid hemorrhage (n = 15).	97 metabolites	5 μL	5 μL of CSF + 1.0 mL MeCN, isopropanol, and water in proportion 3:3:2 + vortexing + centrifugation + drying + 450 μL degassed 50% MeCN + centrifugation + derivatization.	[41]

The authors also noted the presence of biological variability in the concentration of metabolites between individuals in average ±50%, and in some cases ±100%. This phenomenon should be considered by researchers who study potential biomarkers in the CSF [22]. A large-scale metabolomic–proteomic investigation was directed to the study of the biological variability of the concentrations of the CSF metabolites using normal human CSF from patients undergoing routine, non-neurological surgical procedures. As a result of the metabolomic study, which included nontargeted GC–MS analysis, 93 of 108 detected metabolites were identified, including amino acids, organic and fatty acids, nucleosides, mono- and disaccharides, the biological variability of which was 15–85% (analytical variability less than 20%) [31]. Another study was devoted to the characterization of postprandial effects on the CSF metabolic profile of healthy volunteers ($n = 9$), which was performed using GC–MS/MS. Individual plots of postprandial samples of 150 CSF hydrophilic metabolites were positioned similar to the corresponding plots of preprandial samples. The postprandial effects had a far lower impact compared with interindividual variations [32]. Thus, biological interindividual variations appear to have more significant impact on the CSF metabolite profile than food intake.

Important results were obtained describing different effects of preanalytical factors on stability of the proteomic and metabolomics profiles of the CSF. These factors were a 30/120 min delayed storage after the CSF collection at room temperature as the potential delays in the clinic, storage at 4 °C as the time that samples remain in the cooled autosampler, and repeated freeze–thaw cycles. The delayed storage factor led to the increased levels of 49 metabolites, which were analyzed using nontargeted GC–MS, and explained by metabolic processes that occurred because of the remaining white blood cells. The author's recommendations are to remove white blood cells by the CSF centrifugation immediately after collection, use liquid nitrogen for the snap-freeze of the supernatant for storage at −80°C, and avoid freeze/thaw cycles. Samples should not be left in the autosampler for more than 24 h [33].

The chemical composition of low-molecular-weight CSF metabolites justifies its study using GC–MS in case of compliance with the necessary requirements for sampling, storage, and selection of appropriate and sensitive methods of sample preparation, which will provide quantitative determination of the target components at the required level, taking their potential biological variability among individuals into account.

3. Metabolic Profiling of the Cerebrospinal Fluid Using Advanced GC–MS Technologies

Different groups of authors have made attempts for the metabolic profiling of the CSF using more sensitive types of MS detectors than those described in Section 2 (Table 1). The most abundant types of mass analyzers are scanning single and triple-quadrupoles (MS/MS) and time-of-flight (TOF). TOF analyzers are more suitable for metabolic profiling because of their high speed, resolving power, sensitivity, and high quality of identification achieved by retention time, combination of accurate mass, and isotopic distribution. At the same time, the sample preparation approach for the metabolic profiling remains nonselective and includes liquid–liquid extraction (most often using methanol for protein precipitation and as extraction solvent) and widespread two-step derivatization using oxymation (the first step) and silylation (the second step) with different types of reagents to form the volatile derivatives. The most abundant type of the capillary column for the GC is a 30 m × 0.32 mm × 0.25 μm column with a phase of 5% phenyl/95% dimethyl polysiloxane crosslinked polymer, which is characterized by low bleed of the stationary phase, resistance to active compounds, and high temperature stability. This column is suitable for the determination of a wide range of compounds and produced as HP-5MS, TR-5MS, and under other trade names. However, more specific types of columns are also applied.

Application of two-dimensional GC–TOFMS led to the identification of 91 metabolites out of over 1200 detected. Sensitivity was achieved using cryogenic modulation, which concentrated analyte fractions transferred from the first (a 10 m × 0.18 mm I.D. Rxi-5ms column with film thickness of 0.18 μm) to the second column (a 1.5 m × 0.1 mm i.d. BPX-50 column with film thickness of 0.1 μm) [34]. The method was based on two types of capillary columns and two types of mass-selective detectors (DB-5 column (30 mm × 0.25 mm i.d., film thickness 1.0 μm) and GCMS-QP2010 Plus; a fused silica capillary column CP-SIL 8 CB low bleed/MS (30 m × 0.25 mm i.d., film thickness 0.25 μm) and GCMS-QP 2010 Ultra) for targeted and nontargeted analysis led to the detection of 61 metabolites, where 45 metabolites were identified with a nontargeted semiquantitative analysis. Sixteen metabolites involved in the tricarboxylic acid cycle, glycolysis, and amino acids were identified quantitatively: succinic acid, fumaric acid, malic acid, aconitic acid, isocitric acid, citric acid, alanine, valine, leucine, isoleucine, proline, serine, threonine, methionine, phenylalanine, and tyrosine [23].

Two groups of authors explored a similar multicomponent biomarker approach [25,26] using classical two-step derivatization and GC–TOFMS metabolite profiling coupled to a multiplex bioinformatics approach. The first research resulted in 40 identified compounds of 120 peaks; the second resulted in 85 structurally identified and quantified compounds of 962 metabolic signatures. The identified metabolites in the second publication were classified as sugars and sugar alcohols (24%), amino acids (28%), fatty acids (15%), organic acids (15%), and amines (2%) [26].

Careful attention should be paid to the results of the study, in which metabolic profiling was performed using not common configuration of GC with an orthogonal-accelerated TOFMS with atmospheric pressure chemical ionization interface. A distinctive feature of the chemical ionization (CI) is a softer ionization at the energy not exceeding 5 eV, compared to the electron ionization at 70 eV commonly used in GC–MS systems, which provides less fragmentation of the precursor ion. Evaluation of the analytical parameters (repeatability, reproducibility, linearity, and detection limits) using model solutions led to the successful determination of the 25 different compounds (valine, alanine, sarcosine, leucine, proline, isoleucine, benzoic acid, glycine, serine, threonine, methionine, aspartic acid, glutamic acid, phenylalanine, Phenyl-Gly, hippuric acid, caffeine, theophylline, lysine, tyrosine, 4-methyldopamine, dopamine, uric acid, 5-hydroxyindole-3-acetic, and nortriptyline). The applicability of this technique for the CSF metabolic profiling was demonstrated and resulted in 21 identified compounds from more than 300 detected [35].

Tandem GC–MS/MS was applied for the metabolic analysis in pediatric patients and revealed 180 metabolite derivatives in the CSF samples. The main metabolites were 2-ketoglutaric acid, pyridoxamine, tyrosine, 2-propyl-5-hydroxypentanoic acid, 1,5-anhydroglucitol, 2-aminobutiric acid, 2-ketoisocaproic acid, 4-hydroxyproline, acetyl-glycine, methionine, N-acetylserine, and serine [36]. Fatty acid analysis was performed using GC–MS/MS and resulted in identification of 20 compounds (6 saturated, 6 monoun-saturated, and 8 polyunsaturated fatty acids) [10].

Metabolomic research studies are also in demand—specifically, studies in animals. A highly sensitive headspace solid-phase microextraction–GC–MS technique was successfully put into practice for the detection of 25 volatile constituents in rabbit CSF after intranasal administration of Asari Radix et Rhizoma frequently used in traditional Chinese medicine [37]. Numerous amino acids and related compounds (glycine, L-alanine, L-asparagine, L-glutamic acid, L-glutamine, L-isoleucine, L-leucine, L-lysine, L-methionine, L-phenylalanine, L-proline, L-serine, L-threonine, and O-phosphoethanolamine) were detected using GC–MS in the rat CSF [2]. GC–MS metabolomics profiling for the macaque CSF samples produced 663 variables across the two different groups of animals with depression, which were used in the subsequent multivariate analysis. In total, 37 metabolites responsible for discriminating these two groups were identified (propanoic acid, acetic acid, hydroxylamine, propanedioic acid, butanoic acid, proline, methanamine, glycine, isothiourea, nonanoic acid, carbamic acid, threonine, β-alanine, threitol, erythronic acid, L-aspartic acid, xylitol, ribitol, 2-keto-D-gluconic acid, 1,4-butanediamine, D-fructose, my-oinositol, glucaric acid, hexadecanoic acid, scyllitol, gulose, heptadecanoic acid, linolelaidic acid, trans-9-octadecenoic acid, oleic acid, octadecanoic acid, N-acetyl-D-glucosamine,

D-glycero-D-galactoheptitol, galactitol, 5-phenylvaleric acid, benzeneacetic acid, and 1H-indole-2-carboxylic acid) [38].

Most of the described investigations required from 25–50 µL [2,23,34,36] to 100–500 µL [25,26,37] of the CSF (even 1 mL [10]). After different steps of sample preparation, especially during derivatization and subsequent dilution with organic solvent, the final volume of the mixture is usually 100–200 µL. As the volume of CSF is usually limited, especially in cases of experimental rat or mouse models of diseases, approaches requiring extremely small amounts of the CSF and sample for the GC–MS analysis are of great interest.

Three investigations appeared to be promising in solving these issues. The analysis with modified vial design and sample workup procedure became applicable to small volume of the CSF (10 µL), and 50 µL of the final mixture was used for GC–MS analysis. The modified vial design reduced the required volume of the insert from 500 to 200 µL, and the smaller amount of derivatizing agent resulted in a reduction to 50 µL of the total volume of the mixture for the analysis. This approach had similar number of metabolites as in the analysis of >100 µL of the CSF, i.e., 73 identified compounds from 93 detected peaks, and was successfully applied for the metabolic profiling of the rat CSF [39]. Application of non-targeted metabolomics using 5 µL of the CSF for GC–TOFMS seemed to be interesting for investigation, as it resulted in 97 metabolites (including phenylalanine, leucine, threonine, valine, tryptophan, serine, glycerol, 1,5-anhydroglucitol, methionine, β-mannosylglycerate, asparagine, tyrosine, lysine, glutamine, isoleucine, proline, 2-hydroxyglutarate, tryptophan, glycine, proline, isoleucine, and alanine) being identified. Unfortunately, there was no complete information about sample preparation, particularly on derivatization and total sample volume for the analysis [41]. An analytical method based on in-liner silylation in the programmed temperature vaporizer (PTV) injector at 70 °C coupled to GC–TOFMS used only 0.01–2 µL of the CSF and was subsequently applied for metabolic profiling of the human and mouse CSF. A total of 342 peaks were found in both human and mouse profiles and 52 metabolites were identified in the human CSF (amino acids, organic acids, fatty acids, sugars, and others) [40]. The described methods for 0.01–2 and 10 µL of CSF demonstrated promising results and could be recommended for the metabolomics studies, although the method for 0.01–2 µL of CSF requires more expensive equipment (PTV injector and TOFMS) compared to those for 10 µL (GC–MS).

Metabolic profiling resulted in most cases in a number of compounds or groups of compounds that successfully distinguished the compared groups of patients or patients and healthy donors. These compounds are candidate biomarkers for a various type of diseases, including inflammatory demyelinating, neurodegenerative, oncological, infectious, mental, genetic, vascular, and epilepsies (Table 2).

Table 2. Candidate low-molecular-weight biomarkers, discovered using GC–MS methods, for different types of diseases.

Disease Classification	Diagnosis	Candidate Biomarkers	Reference
Inflammatory demyelinating	Multiple sclerosis	5 amino acids, O-phosphoethanolamin	[2]
		sorbitol, fructose	[3]
		homocysteine	[4]
		N-acetylaspartic acid	[5]
		2 metabolites of arachidonic acid	[6]
		quinolinic acid, picolinic acid	[7]
		3 neuroactive steroids	[42]
		4 endocannabinoids	[43]
	Neuromyelitis optica spectrum disorder and idiopathic transverse myelitis data	2 monoglycerides, salicylaldehyde, 4 organic acids, inosine, threose, butane-2,3-diol, hypoxanthine, glutamine	[26]

Table 2. Cont.

Disease Classification	Diagnosis	Candidate Biomarkers	Reference
Neurodegenerative	Amyotrophic lateral sclerosis	amino acids, organic acids	[25]
	Alzheimer's disease	2 steroids	[9]
		fatty acids	[10]
		8,12-iso-iPF2α-VI	[11]
		total isoprostane iPF2α-VI	[12]
		polyunsaturated fatty acids	[13]
		F2-IsoPs	[14]
	Multiple system atrophy	7 polyamines	[44]
		eicosapentaenoic acid	[45]
Oncological	Glioma	citric and iso-citric acid	[23]
	Leukemia	5-hydroxytryptamine, 5-hydroxyindole acetic acid	[46]
Infectious	Meningitis	5 amino acids	[47]
		prostaglandins, thromboxane B_2	[48]
		muramic acid	[49]
	Malaria	quinolinic acid, picolinic acid	[50]
			[51]
	HIV-associated impaired prospective memory	quinolinic acid	[52]
	Subacute sclerosing panencephalitis	quinolinic acid	[53]
Mental	Mood disorders	sorbitol	[54]
		fatty acids, amino acids	[38]
	Major depressive disorder	nervonic acid	[55]
	Post-traumatic stress disorder	allopregnanolone, pregnanolone	[56]
		allopregnanolone, pregnanolone	[57]
	Diagnosis of suicidal behavior	5-hydroxyindolacetic acid	[58]
		homovanillic acid	[59]
		5-hydroxyindolacetic acid, homovanillic acid	[60]
Genetic	Pyruvate carboxylase deficiency	free-gamma-aminobutyric acid, glutamine, C5 ketone bodies	[61]
	Combined sepiapterin reductase and methylmalonyl-CoA epimerase deficiency	2 polyunsaturated fatty acids	[62]
	Guanidinoacetate methyltransferase (GAMT) and creatine transporter deficiency	guanidinoacetate	[63]
Vascular	Aneurysmal subarachnoid hemorrhage	free amino acids	[41]
Epilepsies	Epilepsy	2-ketoglutaric acid, pyridoxamine, tyrosine, 1,5-anhydro-glucitol	[36]

4. Quantitative Analysis of Different Groups of the Cerebrospinal Fluid Metabolites Using GC–MS

As mentioned in the Introduction section, after metabolic profiling and various statistical and bioinformatics approaches, it is necessary to develop a selective method for the detection of a specific compound or group of related compounds. At this stage, modifications of sample preparation techniques are required to reduce the matrix effect and coextraction of nontargeted compounds.

Validation of the developed analytical methods for the quantitative evaluation of the potential biomarkers in a biological matrix is critical for the successful conduct of nonclinical and clinical studies, and it ensures that the obtained data are reliable. Validation includes the evaluation of a number of parameters, such as linearity, lower limit of detection and quantitation (LLOD and LLOQ), selectivity and specificity, sensitivity, accuracy, precision, recovery, and stability of the analyte in the matrix. However, full validation requires a large amount of the biological matrix. In the case of rare matrices, the validation can be performed on a pooled sample from several persons or a model solution that has a chemical composition similar to that of the matrix. We did not manage to find fully validated methods for the determination of the analytes in the CSF, which were performed according to the FDA or EMA guidelines. Several methods, such as recovery, LOQ and LOD values, and linearity will be discussed in this section (Table 3). However, most of the published methods do not provide enough validation data, and they will be mentioned in passing.

Table 3. The analytical methods for the determination of the low-molecular-weight compounds in the CSF using GC–MS methods.

Compounds	GC–MS Method, Capillary Column	CSF Sampling	Sample Volume	Sample Preparation	Method Validation	Concentration	Reference
Glycine, sarcosine, L-forms: alanine, valine, leucine, isoleucine, serine, threonine, methionine, aspartic acid, proline, cysteine, glutamic acid, phenylalanine, asparagine, lysine.	GC–MS: 30 m × 0.25 mm × 0.25 μm Rtx-5MS.	Artificial CSF, patients (n = 16).	200 μL	200 μL CSF + 800 μL MeOH at −10 °C + vortexing + centrifugation + 200 μL supernatant + evaporation to dryness at room temperature under N$_2$ + 15 μL methoxyamine in pyridine (20 mg/mL) + 35 μL BSTFA+TMCS (99:1 v/v) + vortexing + derivatization under microwave irradiation, 210 W, 3 min.	Recovery: 88–129%. LOD: 0.01–4.24 μM. LOQ: 0.02–7.07 μM. Intraday (RSD): 4.1–15.6%. Interday (RSD): 6.4–18.7%. Lin. 0.1–133.0 μM (R^2 = 0.99 for amino acids except cysteine).	Median (n = 16), μM: 6.9/4.9/19.9/9.3/6.7/ 4.2/7.9/7.6/10.4/6.2/ 375.0/1046.7/13.0/4.8/ 1.1/10.4.	[64]
5-hydroxyindole ethanol (5-HTOL), 5-hydroxyindole acetic acid (5-HIAA), 5-hydroxytryptophan (5-HTP), 5-hydroxytryptamine (5-HT).	GC–MS: 30 m × 0.25 mm × 0.25 μm, DB-5.	Children with acute lymphoblastic leukemia without chemotherapy (n = 36), control group (n = 24)	3 mL	3 mL CSF + SPE + washing + elution 1 mL MeOH + 0.5% formic acid + drying under N$_2$ + 70 μL BSTFA +1% TMCS + 30 μL pyridine + 2.5 μL MeOH + incubating for 1 h at 95 °C.	Matrix effect: 92.3–106.2% (no significant ME). Lin. 0.5–200.0 μg/L (5-HTOL, 5-HIAA, $R^2 \geq$ 0.9924) and 2.0–800.0 μg/L (5-HTP, 5-HT, $R^2 \geq$ 0.9918). LOD: 0.1–0.4 μg/L. LOQ: 0.5 (5-HTOL, 5-HIAA) and 2.0 (5-HTP, 5-HT) μg/L. Intraday recovery: 94.6–105.6% (CV 1.4–4.5%). Interday recovery: 93.0–106.9% (CV 1.8–4.5%).	Children with acute lymphoblastic leukemia without chemotherapy 4.3/61.0/5.3/3. Control group: 4.5/88.9/5.8/6.5	[46]
Indole-3-carboxylic (3ICA), indole-3-acetic (3IAA), indole-3-propionic (3IPA), indole-3-lactic (3ILA), 5-hydroxyindole-3-acetic (5-HIAA) acids.	GC–MS: 30 m × 0.25 mm × 0.25 μm, TR-5ms.	CSF (n = 3) samples of different patients with CNS diseases.	40 μL	40 μL CSF + 40 μL distilled water + MEPS + elution with diethyl ether + drying + 40 μL BSTFA/MTBSTFA + incubation 30 min at 90 °C + cooling 30 min at 4 °C + 350 μL of hexane.	Recovery: 40–80% (for pooled CSF). LOD: 0.2–0.4 μM. LOQ: 0.4–0.5 μM. Precision (RSD): <20%. Accuracy (the relative error, RE): <±20% (at the LOQ concentrations). Lin: 0.4–7 μM ($R^2 \geq$ 0.9949).	3IAA, μM: 0.42 ± 0.08; 0.6 ± 0.1; 0.43 ± 0.03	[65]
Quinolinic, picolinic, nicotinic acids.	GC–ECNI-MS: 30 m × 0.25 mm, HP-5MS (i) 0.25 μm or (ii) 1.0 μm stationary-phase film thickness.	Human CSF samples, artificial CSF	20–50 μL	CSF + evaporation to dryness + 100 μL trifluoroacetic anhydride + 100 μL hexafluoroisopropanol + heating at 60 °C for 30 min + dissolving in 1 mL toluene + washing with 1 mL 5% NaHCO$_3$ + 1 mL water + ~500 mg anhydrous Na$_2$SO$_4$.	On-column LOQ < 1 fmol (S/N 10:1). Lin.: 0–5 pmol on column. Slope: for nicotinic acid 5.8; for picolinic acid 25.8 ($R^2 >$ 0.996). Precision (RSD): 0.5–4.3%. Accuracy: 94.0–105.5%. Interday precision: 1.0–8.9%. Interday accuracy: 96.7%–104.0%.	Nicotinic acid: 2.0 (prehydrolysis) and 56.2 (after hydrolysis) μM.	[66]

Table 3. Cont.

Compounds	GC–MS Method, Capillary Column	CSF Sampling	Sample Volume	Sample Preparation	Method Validation	Concentration	Reference
Benzoic (BA), phenylpropionic (PhPA), phenyllactic (PhLA), 4-hydroxybenzoic (p-HBA), 4-hydroxyphenylacetic (p-HPhAA), 4-hydroxyphenylpropionic (p-HPhPA), homovanillic (HVA), 4-hydroxyphenyllactic (p-HPhLA).	GC–MS: 30 m × 0.25 mm × 0.25 μm, TR-5ms.	CSF samples (n = 138) from neurosurgical patients (n = 84), pooled CSF for validation.	40 μL (MEPS) 200 μL (LLE).	MEPS: 40 μL CSF + 40 μL distilled water + MEPS + elution with diethyl ether + drying + 40 μL BSTFA, 30 min at 90 °C + cooling 30 min at 4 °C + 350 μL of hexane. LLE: 200 μL CSF + 800 μL distilled water + 0.3–0.5 g solid NaCl + 15 μL concentrated sulfuric acid + diethyl ether + extraction 2 × 1 mL + evaporation at 40 °C + derivatization as for MEPS.	Recovery: 40–90%. LOD: 0.1–0.3 μM. LOQ: 0.4–0.7 μM. Precision (the reproducibility, RSD): <20%. Accuracy (the relative error, RE): ±±20%. Lin.: over 0.4–10 μM ($R^2 \geq 0.99$).	Median (BA/PhPA/PhLA/p-HBA/p-HPhAA/HVA/p-HPhLA), μM: 0.7/<LOQ/0.1/nd/<LOQ/0.3/0.7/2.5.	[67]
Guanidinoacetate, creatine	Stable isotope dilution GC–MS: SGE BPX-70.	GAMT-deficient patients (n = 8) and SLC6A8-deficient patients (n = 8)	100 μL	100 μL CSF + 50 μL NaHCO$_3$ + 50 μL hexafluoroacetylacetone + 500 μL toluene + heating 2 h to 80 °C + 300 μL toluene phase + drying under N$_2$ + 10 μL triethylamine + 100 μL 7% PFBBr in MeCN (v/v), 15 min + 200 μL 0.5N HCl + 1 mL hexane + extraction.	Linearity: 0.5–10 nmol and 0.05–0.5 nmol LOD (S/N = 5): 0.01 and 0.0012 μM. LOQ (S/N = 10): 0.02 and 0.0024 μM. Intra-assay (n = 10): 0.25 ± 0.02 (CV 6.0%) and 57 ± 3 (CV 6.0%) μM. Interassay (n = 5): 0.25 ± 0.01 (CV 4.0%) and 62 ± 3.7 (CV 6.0%) μM	Control (n = 25): 0.036—0.22 μM and 24–66 μM GAMT deficient: 14–15 μM and not detected SLC6A8 deficient creatine levels: 56–62 μM.	[63]
Gamma-hydroxybutyric acid (GHB)	GC–MS: 30 m × 0.25 mm × 0.25 μm, VF-5 ms.	From autopsy cases (n = 21)	50 μL	50 μL CSF + IS + 200 μL 0.1M HCl + 1 mL ethyl acetate + centrifugation + evaporation to dryness at 20 °C, 2 mbar in a vacuum centrifuge + 50 μL MeCN + 25 μL BSTFA+1% TMCS + mixing.	Inter- and intraday accuracy: >91%. Imprecision: <9%. LOD: 0.5 mg/L. LOQ: 0.6 mg/L. Cal. curve: 1.0 mg/L, 10 mg/L, 40 mg/L, 80 mg/L and 100 mg/L.	Range concentrations after immediate analysis/after storage for 14 days at 4 °C/20 °C, mg/L: 1.1–10.4/0.6–13.2/<0.5–21.6.	[68]
Gamma-aminobutyric acid (GABA)	Isotope-dilution GC–ECNI-MS: 25 m × 0.32 mm, CPSil 88.	CSF samples of a patient before and during Vigabatrin treatment, control samples.	50, 500 μL	Free GABA: 500 μL CSF + 800 μL 1M phosphate buffer, pH 11.5 + 50 μL methylchloroformate + 150 μL 6 M HCl + 4 mL ethyl acetate + drying under N$_2$ at 40 °C + 100 μL 7% PFBBr in MeCN + 10 μL triethylamine + 150 μL 0.5 M HCl + 1 mL hexane + drying under N$_2$ at 40 °C + 50 μL hexane. Total GABA: 50 μL CSF + 450 μL water + 250 μL 20% sulphosalicylic acid + hydrolysis 24 h at 110 °C.	LOD: <0.005 μM. Free GABA. Intra-assay: 0.188 ± 0.004 μM (1.9% SD). Interassay: 0.177 ± 0.013 μM (7.3% SD). Total GABA. Intra-assay: 3.00 ± 0.05 μM (1.8% SD). Interassay: 3.57 ± 0.33 μM (9.2% SD).	Free/total GABA, μM. Control: 0.029–0.127/4.72–11.8. Before therapy: 0.153/13.2. During therapy: 0.274/24.1.	[69]

Table 3. Cont.

Compounds	GC–MS Method, Capillary Column	CSF Sampling	Sample Volume	Sample Preparation	Method Validation	Concentration	Reference
Pipecolic acid	Isotope-dilution GC–ECNI-MS: 30 m × 0.25 mm, DB-19.	Pediatric CSF samples	500 µL	CSF/aqueous standard solution + 1.0 mL 1 M potassium phosphate–sodium carbonate buffer, pH 11.5 + 50 µL methyl chloroformate + 0.15 mL 6 M HCl, pH 2 + 4 mL ethyl acetate + drying under N_2 + 5 µL PFBBr + 10 µL triethylamine in 50 µL MeCN + 1 mL hexane + washing with 0.5 mL 100 mM HCl + 50 µL hexane.	Lin. 0.05–5 nmol ($R > 0.999$). LOD: 1.6F-6 nmol of PA (~0.5 nM). Recovery (CV): 97.3%–101.2% (4.0%–7.2%)	Patients: 0.93–4.53 µM. Control: 0.010–0.120 µM.	[70]
Androsterone, dihydrotestosterone, testosterone, allopregnanolone, isopregnanolone, pregnenolone.	GC-ECNI-MS: 15 m × 0.25 mm × 0.05 µm, HP 5890.	Normal volunteers, cisternal monkey.	1–2 mL	1–2 mL CSF + 100 mg C18 SPE + 50 µL 0.2% carboxymethoxylamine hemihydrochloride in pyridine + incubation 45 min at 60 °C + drying in N_2+ 100 µL 1.25% pentafluorobenzyl bromide + 2.5% di-isopropylethylamine in MeCN + drying in N_2 + 100 µL 50% BSTFA in MeCN + drying + 5 µL hexane.	Recovery: 78.2–99.5%. Reproducibility (RSD): 4.6–35.0%. Lin.: 10–1000 pg/ml ($R^2 > 0.996$). Two-month variation <10%.	Human/monkey CSF, pg/mL: androsterone—52.8/24.7; testosterone—158.3/73.7; allopregnanolone—44.1/6.3; pregnenolone—52.8/16.7.	[71]
Pregnenolone, dehydroepiandrosterone, progesterone, androstenedione, testosterone, allopregnanolone, isopregnanolone, androsterone, epiandrosterone, 7α-hydroxy-dehydroepiandrosterone, 7β-Hydroxy-dehydroepiandrosterone, 5-androstene-3β,7α,17β-triol, 5-androstene-3β,7β,17β-triol, 16α-hydroxy-pregnenolone, 16α-hydroxy-dehydroepiandrosterone, 16α-hydroxy-progesterone.	GC–MS: 15 m × 0.25 mm × 0.1 µm, RESTEK Rxi.	Patients that underwent an endoscopic third ventriculostomy because of obstructive hydrocephalus ($n = 15$).	1 mL	1 mL CSF + 3 mL of diethyl ether + drying at 37 °C + 1 mL MeOH–water (4:1) + 1 mL pentane + drying of the polar phase 2 h in the vacuum centrifuge at 60 °C + 50 µL methoxylamine–hydrochloride solution in pyridine (2%) + incubation 1 h at 60 °C + drying in the N_2 + 50 µL Sylon B + incubating 1 h at 90 °C + drying in the N_2 + 20 µL isooctane.	Lin.: 10–1000 pg Slope: 0.96–1.33. R: >0.995. CV: 1.0–5.1%. LOD: 0.04–11.3 pM. Recovery: 75–104%.	Median, nM: 0.060/0.078/0.235/0.208/ 0.231/0.008/0.040/0.005/ 0.004/0.300/0.037/0.007/ 0.012/0.001/0.006/0.072.	[72]
Indomethacin	GC-NiCI-MS: 30 m × 0.25 mm × 0.25 µm, HP-5MS.	Children ($n = 31$).	250 µL	250 µL CSF + acidification + C18 SPE + evaporation + 200 µL PFBBr (3.5% v/v, in MeCN) + 50 µL di-isopropylethylamine + extraction (water and toluene).	LOQ: 0.1 ng/sample. Accuracy: 98–122%. Recovery: 85–87%. Intraday (RSD, $n = 3$): 3–34%. Lin.: 0.1–5 ng/sample.	0.2–5.0 ng/mL (median, 1.4 ng/mL)	[73]

Table 3. Cont.

Compounds	GC–MS Method, Capillary Column	CSF Sampling	Sample Volume	Sample Preparation	Method Validation	Concentration	Reference
Scyllo-Inositol (Elnd005)	GC–MS	Healthy adults (n = 8).	No data	Extraction from human CSF diluted 1:1 with blank human plasma by protein precipitation/derivatization/LLE.	LLOQ: 0.4 µg/mL. Linearity 0.4–80 µg/mL. Precision for QC samples: 1.7–2.3%. Accuracy at all concentrations: 0.5% to +3.0%.	Prior to administration of ELND005 1.4–1.5 µg/mL.	[74]
6-monoacetyl morphine, morphine, codeine	GC–MS: 30 m × 0.25 mm × 0.25 µm, HP-5.	Deceased individuals (n = 25).	3 mL	3 mL CSF + 18 mL 0.1M phosphate buffer, pH 6.0 + SPE + 100 µL toluene + analysis of codeine + toluene evaporation + 50 µL BSTFA + 1% TMCS 10 min at 70 °C.	Linearity: to 1.7 mg/L. LOQ: 0.002 mg/L for morphine and 0.001 mg/L for codeine and 6-monoacetyl morphine. Precision: 0.115–0.121 mg/L (CV 4.4–7.2%).	0.001–0.406/0.01–0.38/<0.01–0.04 mg/L	[75]
Diethylene glycol, ethylene glycol, glycolic, oxalic, diglycolic, hydroxyethoxy acetic acids.	GC–NICI-MS: 30 m × 0.25 mm ID × 0.50 µm, ZB-5 ms.	Control CSF	250 µL	250 µL CSF + 1.0 mL water + 100 µL 5 N NaOH + 500 µL toluene + 50 µL PFBCl	LOQ: 0.05–1.0 µg/mL. Lin.: for diethylene glycol and ethylene glycol 0.02–2 (R^2 0.9984–0.9989) µg/mL; for other 0.05–5 (0.9990), 0.5–25 (0.9907), 0.5–100 (0.9985), 1–100 (0.9922) µg/mL. Accuracy: ≤15%.	No data	[76]

4.1. Amino Acids

The presence and levels of free amino acids in the CSF can be indicators of neurological diseases [4,47,64,77]. Silylation is commonly used for the amino acids and an alternative derivatization using a microwave-assisted derivatization was described [64]. A 200 µL aliquot of the artificial CSF (contains 127 µM NaCl, 2 µM KCl, 1.2 µM KH_2PO_4, 26 µM $NaHCO_3$, 2 µM $MgSO_4$, 2 µM $CaCl_2$, 10 µM 4-(2-hydroxyethyl)-1-piperazineethanesulfonic acid (HEPES) and 10 mM glucose and bubbled with a carbogenic mixture (95% v/v O_2 and 5% v/v CO_2)) was used for the validation of this method for 16 amino acids (glycine, sarcosine, L-forms: alanine, valine, leucine, isoleucine, serine, threonine, methionine, aspartic acid, proline, cysteine, glutamic acid, phenylalanine, asparagine, and lysine) according to the Eurachem guidelines. Arginine and histidine were not analyzed because of the thermal instability of their derivatives. The evaluated analytical parameters, such as LOD 0.01–4.24 µM, LOQ 0.02–7.07 µM, intraday and interday precision values, recoveries, and linearity allowed the authors to determine all 16 amino acids in the human CSF samples (n = 16) at the level higher than LOD values, which indicated that the developed analytical method is applicable to solving the task of the quantitative determination of free amino acids in the CSF.

4.2. Tryptophan Metabolites

Tryptophan is one of the most important amino acid for CNS function. Its metabolism occurs in two main ways: the indole and kynurenine pathways. The indole metabolism is divided into the serotonin (5-hydroxytryptamine, 5-HT) via 5-hydroxytryptophan (5-HTP) and the microbial pathways. 5-Hydroxyindole-3-acetic acid (5-HIAA) or 5-hydroxyindole-3-ethanol (5-HTOL) appears as a result of the serotonin metabolism. The microbial pathway leads to the formation of metabolites containing an indole ring, for example, tryptamine and several indole-containing acids (indole-3-acetic (3IAA), indole-3-propanoic (3IPA), indole-3-carboxylic (3ICA), and indole-3-lactic (3ILA) acids).

5-HT is a neurotransmitter, and its related indole derivatives from serotonin pathway are involved in physiological and pathological responses, which are associated with many neurological diseases [46,58–60]. A method for detecting 5-HTOL, 5-HIAA, 5-HTP and 5-HT using solid-phase extraction (SPE) with Cleanert PEP-2 column was developed. This method required 3 mL of CSF, which seemed to be too much in the case of children, who were the participants of the study. However, many important analytical parameters (without reference to the FDA or EMA guidelines) were evaluated (matrix effect, linearity, LOD 0.1–0.4 µg/L and LOQ 0.5–2.0 µg/L, intraday and interday precision values, recovery, and coefficient of variation); these allowed the authors to obtain the statistically significant data about the changes in the concentration of the target compounds between children with acute lymphoblastic leukemia and the control group [46].

In contrast to the described method with classical SPE, a modern microextraction by packed sorbent (MEPS) with C18 was applied for the determination of the indole-containing acids (5-HIAA, 3IAA, 3IPA, 3ICA, and 3ILA) using only 40 µL of the CSF. The pooled CSF samples were used for the validation and the following parameters were evaluated (according to the FDA guidelines): linearity, recovery, LOD 0.2–0.4 µM and LOQ 0.4–0.5 µM, accuracy, precision, selectivity, and carryover effects. Despite the satisfactory results, only 3 IAA was detected in CSF samples of the patients with the CNS diseases [65].

Several studies describe an importance of the changes in the concentration of the kynurenine pathway metabolites, particularly pyridine-containing quinolinic, picolinic, and nicotinic acids, which are involved in the inflammatory and apoptotic processes associated with the CNS neuronal cell damage and death [50–53,66]. These metabolites were detected in the CSF after derivatization with trifluoroacetic anhydride and hexafluoroisopropanol, and electron-capture negative-ion chemical ionization (ECNI) GC–MS. One of these studies describes an almost fully validated method (LOQ less than 1 fmol for each of the analytes, linearity, precision, and accuracy) using for the concurrent quantification of quinolinic, picolinic, and nicotinic acids in 20–50 µL of model solutions and an artificial CSF [66].

4.3. Organic Acids

Phenyl-containing acids (benzoic, 3-phenylpropionic, 3-phenyllactic, 4-hydroxybenzoic, 2-(4-hydroxyphenyl)acetic, homovanillic, and 3-(4-hydroxyphenyl)lactic acids), which are mostly microbial metabolites of the tyrosine and phenylalanine, were detected in the CSF samples ($n = 138$) from neurosurgical patients ($n = 84$) with different CNS pathology using MEPS and traditional liquid–liquid extraction (LLE). The validation (linearity, recovery, LOD 0.1–0.3 µM and LOQ 0.4–0.7 µM, accuracy, precision, selectivity, and carryover effects) according to FDA guidelines was performed for both MEPS and LLE, demonstrating the equal possibilities of these sample preparation techniques. Similar results were achieved using 40 µL of the CSF sample for MEPS instead of 200 µL for LLE [67].

The detection of creatine, an N-containing acid, and its precursor guanidinoacetate is crucial in cases of creatine deficiency syndromes, a group of inherited metabolic disorders that are caused by abnormalities in creatine biosynthesis and/or transport [78]. A sample preparation technique included LLE from 100 µL of the CSF and derivatization with subsequent stable isotope dilution (SID) GC–MS. This method provides LOD and LOQ 0.0012 and 0.0024 µM for creatine, and 0.01 and 0.02 µM for guanidinoacetate, respectively. Linearity, interassay, and intra-assay variability were also evaluated. The reference values for creatine and guanidinoacetate were revealed and ranged from 17 to 78 µM and 0.02 to 0.56 for µM, respectively [63].

Gamma-hydroxybutyric acid (GHB) is a naturally occurring neurotransmitter and a precursor to gamma-aminobutyric acid (GABA), glutamate, and glycine in certain brain areas. The postmortem examination of the influence of temperature and time storage to in vivo production of GHB was evaluated using traditional LLE and silylation. The validation was performed using 50 µL of the CSF and resulted in LOD 0.5 mg/L, LOQ 0.6 mg/L, and interday and intraday accuracy $\geq 91\%$. GHB concentration changes were affected both during postmortem interval in the dead body and during in vitro storage [68].

GABA is the chief inhibitory neurotransmitter and plays an important role in various neurological and mental disorders, in which both elevated and decreased concentrations in CSF may occur. A sensitive, selective, and accurate SID GC-ECNI-MS method for the determination of free and total GABA was developed using 500 µL of the CSF, derivatization in aqueous solution with methylchloroformate, extraction with ethyl acetate, and derivatization of the dried residue with pentafluorobenzylbromide in acetonitrile and triethylamine (for free GABA). Total GABA determination included hydrolysis with sulphosalicylic acid during 24 h. The following analytical parameters were evaluated: LOD <0.005 µM and interassay and intra-assay variability for both free and total GABA. The applicability of the method was successfully demonstrated for the determination of free and total GABA in patients suffering from succinic semialdehyde dehydrogenase deficiency before and during specific treatment [69]. This method was used for the evaluation of free GABA in the CSF samples in a patient with pyruvate carboxylase deficiency [61] and for the determination of the pipecolic acid [70], a carboxylic acid of piperidine and one of the biomarkers of the pyridoxine dependent epilepsy [79].

Different polyunsaturated fatty acids are the components of neuronal and glial membrane phospholipids and participate in the development of Parkinson's [45] and Alzheimer's diseases [13]. Although the determination of these compounds includes traditional LLE and silylation, there are no validated methods for the CSF [13,45,62]. Furthermore, there are no validated methods for the determination of the nervonic acid, a candidate biomarker for depressive and manic symptoms [55], and N-acetylaspartic acid, a neuron-specific marker that is identified in multiple sclerosis [5].

4.4. Neuroactive Steroids

Neuroactive steroids are steroids synthesized de novo in the CNS and play a central role in neuronal processes [80]. Allopregnanolone and related neurosteroids (androsterone, dihydrotestosterone, testosterone, isopregnanolone, and pregnenolone) were detected as carboxymethoxime, pentafluorobenzyl, and trimethylsilyl derivatives using GC–ECNI-MS. The sample preparation included SPE with C18 sorbent from 1–2 mL of the CSF. Linearity, LOD 0.2–1.2 µg/L, recovery, and reproducibility were evaluated. This method was successfully applied for the analysis of the human and monkey CSF [56]. Another study was devoted to the evaluation of the correlations between peripheral and CSF steroids using a wide spectrum of bioactive steroids, their precursors and metabolites. Unconjugated steroids (pregnenolone, dehydroepiandrosterone, progesterone, androstenedione, testosterone, allopregnanolone, isopregnanolone, androsterone, epiandrosterone, 7α-hydroxy-dehydroepiandrosterone, 7β-Hydroxy-dehydroepiandrosterone, 5-androstene-3β, 7α, 17β-triol, 5-androstene-3β, 7β, 17β-triol, 16α-hydroxy-pregnenolone, 16α-hydroxy-dehydroepiandrosterone, and 16α-hydroxy-progesterone) were extracted from 1 mL of the CSF and derivatized in a common two-step procedure. LOD from 0.04 (for 5-androstene-3β, 7β, 17β-triol) to 11.3 pM (for androstenedione) were measured together with other analytical parameters. Significant correlations between some steroids in serum and CSF were revealed, particularly between the 7α/β-hydroxy-metabolites of dehydroepiandrosterone and androstenediol [72]. Another study applied similar sample preparation for the detection of free dehydroepiandrosterone and its 7-hydroxylated derivatives: 7α-hydroxy-dehydroepiandrosterone, 7β-hydroxy-dehydroepiandrosterone, 5-androstene-3β, 7α, 17β-triol, and 5-androstene-3β, 7β, 17β-triol [81]. Different neuroactive steroids were evaluated using GC–MS preceded by HPLC purification in Alzheimer's disease [9], relapsing–remitting multiple sclerosis [42], and post-traumatic stress disorder [56].

4.5. Arachidonic Acid Metabolites

F_2-isoprostanes (F_2-IsoPs) and F4-neuroprotanes (F_4-NPs) are compounds formed in vivo from the nonenzymatic free-radical-catalyzed peroxidation of essential fatty acids, primarily arachidonic and docosahexaenoic acids, respectively. Since CNS is characterized by a high level of polyunsaturated fatty acids and significant oxygen demand, considering its weak antioxidant defenses, it is also rather liable to oxidative damage caused by reactive oxygen or nitrogen species. Imbalance between free radicals and antioxidants, so-called "oxidative stress", plays a crucial role in neurodegenerative disorders. F_2-IsoPs and F_4-NPs, being products of lipid peroxidation, can be biomarkers of oxidative stress and neurodegenerative diseases. The literature indicates that levels of F_2-IsoPs and F_4-NPs in CSF and brain tissue are elevated in case of such disorders as Alzheimer's disease [11,14,82–84] and equine neuroaxonal dystrophy [85]. For sensitive quantification of these compounds, which are present in the CSF samples in low concentrations, GC–MS methods are applied [86], and their validation is required not only for urine and serum [87], but for CSF as well.

Prostanoids, which include prostaglandins and thromboxanes, are the metabolites of the enzymatic pathways of arachidonic acid. These compounds have similar chemical structures but different biological and therapeutic effects. Simultaneous assay of these compounds in the CSF was developed, which included extraction with octadecyl silica gel and two-step purification with silicic acid gel chromatography [48]. Some representatives of this class of compounds were elevated in patients with multiple sclerosis [6].

The validated analytical procedures for different metabolites of arachidonic acid are required because of the high demand in their evaluation in different neurodegenerative disorders.

4.6. Glucose Metabolites

Abnormalities in carbohydrate metabolism are of interest in mood disorders studies, as a possible relationship between diabetes and major depression has been shown. Glucose, which serves as an energy source for cells, is conversed to fructose via the polyol pathway with sorbitol being an intermediate compound formed during this two-step process. The CSF sorbitol levels were investigated in patients with bipolar and unipolar mood disorder, and sorbitol concentrations were higher in the CSF of depressed subjects compared to normal controls [54]. Sorbitol levels along with fructose levels, both being glucose metabolites, were found to be elevated in the CSF of multiple sclerosis patients as well, while concentrations of myoinositol that is not produced via the polyol pathway did not differ significantly from its concentrations in the CSF of control subjects [3]. No validation data were demonstrated; thus, the development of the validated analytical method for the determination of the glucose metabolites is required.

4.7. Drugs and Toxic Metabolites

Indomethacin is a nonsteroidal anti-inflammatory drug, mainly known for its ability to inhibit cyclooxygenase, which is responsible for the prostaglandins production catalysis. Indomethacin is often prescribed to treat inflammation and pain caused by rheumatic and orthopedic diseases or surgery. To determine its concentrations in 250 µL of CSF, a SPE sample preparation technique was applied followed by derivatization with pentafluorobenzylbromide and GC–NICI–MS. The method provides moderate analytical characteristics (recoveries, accuracy, intraday precision) with LOQ 0.1 ng/sample. The CSF indomethacin levels in healthy children were found to be 0.2 and 5.0 ng/mL after administering it intravenously [73].

GC–MS was also used for determination of another drug: ELND005 (scyllo-inositol), an endogenous inositol stereoisomer. This drug could be used for Alzheimer' disease treatment, and its pharmacokinetic behavior in the CSF after oral administration was of interest. A traditional combination of protein precipitation, LLE, and derivatization was used for sample treatment, and validation resulted in LLOQ 0.4 µg/mL (linearity, precision, and accuracy were also evaluated) [74].

Analysis of the CSF could also be beneficial for toxicology studies. For instance, morphine can be found in human biological samples and tissues due to the ingestion of heroin or codeine, since these compounds are both metabolized to morphine, or because of exposure to morphine itself. To distinguish whether it was heroine or morphine administering, determination of 6-monoacetylmorphine is often used. However, it converses to morphine rather rapidly, and its concentrations in blood may be lower than the LOD of the method used for 6-monoacetylmorphine determination. Several studies suggested that 6-monoacetylmorphine persists in the CSF and some other human biological samples when compared to blood. 6-Monoacetylmorphine, free morphine, and free codeine levels were investigated in the CSF samples in 25 heroin deaths. The sample pretreatment procedure included such steps as SPE and derivatization and was combined with GC–MS analysis. The method's LOQ for 6-monoacetylmorphine was 0.001 mg/L; linearity and precision were evaluated. 6-Monoacetylmorphine levels were 6.6 times higher on average in the CSF samples than in blood [75].

Analysis of the diethylene glycol and its potential metabolites (ethylene glycol, glycolic acid, oxalic acid, diglycolic acid, and hydroxyethoxy acetic acid) is required because of the human poisoning during misformulation into pharmaceutical products. Sample preparation for acid metabolites from 100 µL of the CSF included traditional LLE and silylation. Sample preparation for diethylene and ethylene glycol from 250 µL of the CSF included extraction and derivatization with pentafluorobenzoyl chloride with the following analysis by GC–NICI–MS. The LOQ values were 0.05–1.0 µg/mL; accuracy and linearity were evaluated [76].

Different low-molecular-weight compounds are required to be detected in the CSF for the diagnosis of the CNS diseases. The literature data analysis revealed the absence of fully validated methods, and it presents opportunities for scientists to develop and validate analytical protocols using modern sample preparation techniques.

5. Miniaturization in Sample Preparation Techniques for the GC–MS Analysis

A small sample volume is one of the main criteria for the analysis of CSF. There are interesting approaches to sample preparation of biological fluids for GC–MS analysis, which use the principle of miniaturization.

The method of homogenous liquid–liquid microextraction (HLLME), compared to the classical LLE using significant volumes of organic solvents, is based on the extraction of polar organic compounds from aqueous matrices, including biofluids, with small volumes (microliters) of water-miscible organic solvents. For the phase separation, the salting-out effect is often used, followed by centrifugation. The volumes of biological fluids are usually up to hundreds of microliters and the volumes of polar solvents are often several times less; the weighed portion of the salting-out agent is tenths of a gram. The achievable LOD of analytes are nanograms and tenths of nanograms per milliliter [88]. The method could be combined with HPLC–MS [89], but a combination with GC–MS is also possible. To determine volatile analytes without derivatization, it is necessary to dry the extract by adding anhydrous sodium sulfate. Analytes can be derivatized both directly in the extract (addition of chloroformates) and in the dried extract (silylation in acetonitrile).

Dispersive liquid–liquid microextraction (DLLME) is based on the extraction of analytes with a microemulsion followed by the phase separation. Usually, the volume of a mixture of an extracting solvent and a dispersing solvent is hundreds of times less than the volume of the analyzed solution. Analytes are extracted quantitatively at high preconcentration factors. Extraction equilibrium is established in minutes [90]. Various methods of dispersion have been proposed and various extracting solvents have been studied for a large number of different matrices, including biological ones [91,92]. The DLLME method is applicable for the analysis of small volumes of analyzed solutions and could be combined with derivatization procedure for polar and/or nonvolatile analytes with small volumes of reagents. Different derivatizing agents are used, e.g., for analysis of human urine, ethyl chloroformate in pyridine was utilized to convert 20 amino acids into their volatile carbamate esters, which were further analyzed using GC–MS. The derivatization process was carried out simultaneously with DLLME using trichloroethylene and acetonitrile as extracting and dispersing solvent, respectively. The range of LOD was 0.4–3.7 µg/L [93].

An interesting solution is to combine the capabilities of DLLME and injector port silylation technique for the determination of polar analytes in biological matrices. Aliquots of the extract and derivatizing reagent are injected simultaneously or sequentially into a heated GC–MS injector. A gas phase reaction occurs between the silylating reagent and polar analytes at the injector temperature. This approach reduces the derivatization time (less than a minute), the possibility of derivatives decomposition, and the amount of toxic reagent and solvents used for the process is smaller [94].

QuEChERS (the name is formed from "quick, easy, cheap, effective, rugged, and safe") is a two-step process involving liquid extraction (usually using acetonitrile) and dispersive SPE (dSPE) using (more commonly) primary secondary amines, C_{18}, and/or graphitized carbon black sorbents to eliminate significantly interfering matrix components (for example, humic acids, lipids, etc.). QuEChERS, originally developed for the extraction of acidic and basic pesticides from food [95], is also used for analysis of blood plasma samples. A method for the simultaneous extraction of acidic, basic, neutral, and amphiphilic analytes from blood plasma using a micro version of QuEChERS (micro-QuEChERS) is proposed. It reduces the volumes of samples (200 µL of plasma compared to 1.5 mL required in the nonmicro version) and reagents by 8 times. The method allowed for the extraction of analytes from blood plasma with high (65 to 80%) recoveries and low matrix effect. The developed approach is considered to be a fast and clean alternative to "dilute and shoot" approaches or the protein precipitation procedure, which are used for high-throughput clinical diagnostics (including analysis of the CSF samples), coupled with HPLC–MS. However, it is of interest to study a possible combination of micro-QuEChERS with GC–MS [96].

Micro-QuEChERS was used to analyze 148 avian blood samples collected in an environmental field study of the impact of rodenticides (applied for the treatment of common vole plague) on the wildlife. The volume of each sample was 250 µL. In combination with GC–MS/MS, this method detected the desired analytes at the level of 1.5 ng/mL [97]. Although micro-QuEChERS is not yet widely used for sample preparation of biological matrices [95–98], the ability to vary components for the extraction and extracts purification procedures applying dSPE, reduce the cost and time of analysis, and analyze relatively small sample volumes (even smaller than those described in the publications above) presents broad prospects for the analysis of biological fluids, including CSF [96–98].

Solid-phase microextraction (SPME) is a solvent-free sample preparation technique [99]. This method is widely used for analyzing the vapor phase of various biological fluids in order to determine volatile organic compounds and to extract analytes directly from the liquid phase [100–102]. Substances are absorbed by a polymer film or solid sorbent covering a piece of fiber (a piece of fused silica capillary). The capillary is placed inside a needle connected to a syringe-like device. During sorption and desorption, the capillary moves out of the needle. The metabolomic composition of the circulating blood of laboratory mice was investigated [100], and for selective recovery of hydrophilic and hydrophobic analytes with respect to high molecular weight matrix components, an SPME fiber coated with mixed-mode polymers (phenylboronic acid and polystyrenedivinylbenzene) was used. The fiber was placed in an injection needle, and it absorbed the metabolites directly from the bloodstream. The vapor phase over saliva samples (sample volume 500 µL) was investigated. The LOD for the 20 detected volatile metabolites ranged from 0.008 to 1 µM [101]. The vapor phase over urine samples was studied in order to identify biomarkers of cancer; 82 metabolites were found, and the sample volume was 4 mL [102]. The SPME capabilities allow the researchers to vary the volumes of the studied samples, depending on the aims of the research.

Stir bar sorptive extraction was also developed as another solvent-free sample pretreatment technique, and it is actively used for isolation of low-molecular-weight components of different polarity from biological fluids [103–105]. The device used for stir bar sorptive extraction is a glass tube with a metal rod inside (magnetic stirrer), often coated with polydimethylsiloxane. The main difference between stir bar sorptive extraction and SPME is the larger amount of stationary phase covering the surface of the mixer (up to 25 µL) compared to the capillary cut (0.5 µL), which increases the extraction efficiency. Desorption of analytes is carried out either by solvent re-extraction or thermal desorption. For polar components in biological matrices, it is possible to combine simultaneous deconjugation and extraction of analytes (in situ deconjugation) or derivatization and extraction (in situ derivatization) followed by thermal desorption into the GC injector. Derivatization could also be carried out after extraction (postextraction derivatization), both during thermal desorption of analytes and after their re-extraction. Sample volumes are typically on the order of 1 mL, with LOD being attainable from ng/mL to pg/mL.

A method of amino acids microextraction from biological fluids (including CSF), based on a combination of hollow fiber SPME and extraction with stir bar sorptive extraction—hollow fiber–stir bar sorptive extraction—was proposed [106]. Hollow polymer fibers are obtained using the coaxial electrospinning technology when an electrostatic field acts on an electrically charged jet of a polymer solution or its melt. It is also possible to obtain a hollow fiber membrane with specified properties (average pore diameter, membrane thickness), providing a semipermeable barrier (analytes pass through the pores of the membrane, the matrix components remain in solution). For the first time, the hollow fiber membrane was used as a SPME fiber coating for the extraction of BTEX (benzene, toluene, ethylbenzene, and xylenes) from aqueous matrices, and was based on a polypropylene coated copper wire [107].

To extract amino acids, hollow polyvinylidene fluoride fiber was used. A piece of polyvinylidene fluoride hollow fiber was sealed on the one side, and a steel rod was placed inside the fiber. A dispersion system of 0.1 g of silica microspheres in ethanol was introduced into the fiber. After removal (including evaporation) of ethanol, the fiber was sealed on the other side. Before use, the resulting hollow fiber–stir bar sorptive extraction device was washed with acetone and dried. An aliquot of the biological fluid was mixed with ethanol in a 3:1 ratio in order to reduce the surface tension of the sample to facilitate the penetration of analytes into the membrane micropores. A hollow fiber–stir bar was placed in a vial with the sample and extraction was performed while stirring under the influence of a magnetic field for a chosen time. During this time, amino acids selectively with respect to biological macromolecules penetrated from the solution of biological fluid through the pores of the membrane and were absorbed on the surface of silica microspheres due to the formation of hydrogen bonds. Next, the hollow fiber–stir bar was removed from the sample vial, dried until moisture was completely evaporated, placed in another vial, and 0.1 mL of BSTFA and 0.9 µL of a nonpolar solvent were added to extract amino acid derivatives, which should have been formed as a result of silylation. BSTFA molecules also penetrated the pores of the membrane and interacted with amino acid molecules. Derivatization was carried out in a microwave field. The resulting derivatives were extracted with a nonpolar solvent and analyzed using GC–MS. Before the next extraction, the hollow fiber–stir bar was conditioned in distilled water and acetone. The LOD of the studied amino acids ranged from 3×10^{-4} to 6×10^{-3} µg/mL. Recoveries from the rat CSF samples ranged from 71.8 to 101.2%. The resulting extraction device could be used 30 times without loss of analyte sensitivity [106]. Due to the miniaturization of the device, it can be used for small volumes of biological fluids. One of the advantages of this method is that there is no need in sample cleanup.

One of the perspective sample preparation techniques is a MEPS [108] that is also aimed at miniaturizing SPE. The method is based on the use of a small amount of sorbent (1–4 mg) and multiple passage of the test sample through this sorbent layer located in the extension of the syringe needle. The method requires small amounts of sample (tens, hundreds of microliters), and solvent volumes (up to tens of microliters) which eluate analytes. This technique minimizes the dead volume as well. The syringe is placed in an automatic dispenser, and the speed and number of cycles of passing the sample through the sorbent could be programmed. It is also possible to connect the dispenser to an HPLC or GC system. A large number of needles with sorbents, which are common for classical SPE (C_{18}, C_8, silica gel), have been developed and are commercially available. A method for the determination of low-molecular-weight metabolites using a hand-operated device with hypercrosslinked polystyrene was described [109]. MEPS has found wide application in analytical chemistry, including the field of biological fluids analysis [110–113]—in particular, for the determination of low-molecular-weight microbial metabolites in the CSF (see Section 4) [65,67].

The CSF samples are characterized not only by low levels of the analytes of interest, but also by small sample volumes available for analysis. Miniaturization of classic extraction methods, such as LLE and SPE which are often used for analysis of biological fluids, coupled with GC–MS presents new perspectives in the metabolic analysis of the CSF.

6. Conclusions

GC–MS plays an important role in the development of the metabolic analysis of biological fluid samples. High separation efficiency and detection sensitivity, stable retention times, and reproducible mass spectra of analytes make it possible to analyze multicomponent mixtures of low-molecular-weight organic compounds of complex composition and to perform nontargeted and targeted analysis. Sample preparation based on the selective extraction of analytes with respect to interfering matrix components is often required to detect a wide range of components by this method. The complexity of the CSF analysis is caused not only by the low content of the target analytes, but also by the small volume of samples available for analysis. The miniaturization of the traditional LLE and SPE methods in combination with various options for the derivatization of polar analytes presents new possibilities in the metabolic analysis of the CSF using GC–MS.

Author Contributions: Conceptualization, A.P. and A.R.; writing—original draft preparation, A.P., N.B. and A.R.; writing—review and editing, A.P., N.B. and A.R.; visualization, A.P. and N.B.; supervision, A.P.; funding acquisition, A.P. All authors have read and agreed to the published version of the manuscript.

Funding: This research was funded by Ministry of Science and Higher Education (Russian Federation) on the state assignment number 0563-2019-0020, and by The Council on Grants of The President of The Russian Federation, grant number MK-627.2020.7.

Acknowledgments: The authors are grateful to Maria Getsina, Anastasia Megley, and Maksim Golubev for the help in editing the manuscript.

Conflicts of Interest: The authors declare no conflict of interest.

Abbreviations

APCI	atmospheric pressure chemical ionization
BA	benzoic acid
BBB	blood-brain barrier
BSTFA	N, O-bis(trimethylsilyl)trifluoroacetamide
CI	chemical ionization
CNS	central nervous system
CSF	cerebrospinal fluid
CV	coefficient of variation
DLLME	dispersive liquid–liquid microextraction
dSPE	dispersive solid-phase extraction
ECNI	electron-capture negative-ion chemical ionization
EMA	European Medicines Agency
FDA	Food and Drug Administration
FTMS	Fourier transform mass spectrometry
F_2-IsoPs	F_2-isoprostanes
F_4-NPs	F_4-neuroprotanes
GABA	gamma-aminobutyric acid
GAMT	guanidinoacetate methyltransferase
GC	gas chromatography
GHB	gamma-hydroxybutyric acid
HEPES	4-(2-hydroxyethyl)-1-piperazineethanesulfonic acid
HIV	human immunodeficiency virus
HLLME	homogenous liquid–liquid microextraction
HPLC	high-performance liquid chromatography
5-HT	5-hydroxytryptamine
5-HTP	5-hydroxytryptophan
HVA	homovanillic acid
5-HIAA	5-hydroxyindole-3-acetic acid
5-HTOL	5-hydroxyindole-3-ethanol
ICP	inductively coupled plasma
IS	internal standard
3IAA	indole-3-acetic acid
3ICA	indole-3-carboxylic acid
3ILA	indole-3-lactic acid
3IPA	indole-3-propionic acid
LC	liquid chromatography
LLE	Liquid–liquid extraction
LLOD	lower limit of detection
LLOQ	lower limit of quantitation

LOD	limit of detection
LOQ	limit of quantitation
MALDI	matrix-assisted laser desorption/ionization
ME	matrix effect
MEPS	microextraction by packed sorbent
MS	mass spectrometry
MS/MS	tandem mass spectrometry
MSn	multistage mass spectrometry
MSTFA	N-methyl-N-(trimethylsilyl)trifluoroacetamide
MTBSTFA	N-(tert-butyldimethylsilyl)-N-methyltrifluoroacetamide
NMR	nuclear magnetic resonance
PFBBr	2,3,4,5,6-pentafluorobenzyl bromide
PFBCl	2,3,4,5,6-pentafluorobenzoyl chloride
p-HBA	4-hydroxybenzoic acid
PhLA	phenyllactic acid
PhPA	phenylpropionic acid
p-HPhAA	4-hydroxyphenylacetic acid
p-HphLA	4-hydroxyphenyllactic acid
p-HphPA	4-hydroxyphenylpropionic acid
PTV	programmed temperature vaporizer
Q	single quardupole
QuEChERS	"quick, easy, cheap, effective, rugged, and safe"
RE	relative error
RSD	relative standard deviation
SID	stable isotope dilution
SPE	solid-phase extraction
SPME	solid-phase microextraction
TBDMS	tert-butyldimethylsilyl
TMCS	trimethylchlorosilane
TOF	time-of-flight

References

1. Segal, M.B. Extracellular and cerebrospinal fluids. *J. Inherit. Metab. Dis.* **1993**, *16*, 617–638. [CrossRef]
2. Noga, M.J.; Dane, A.; Shi, S.; Attali, A.; van Aken, H.; Suidgeest, E.; Tuinstra, T.; Muilwijk, B.; Coulier, L.; Luider, T.; et al. Metabolomics of cerebrospinal fluid reveals changes in the central nervous system metabolism in a rat model of multiple sclerosis. *Metabolomics* **2012**, *8*, 253–263. [CrossRef]
3. Regenold, W.T.; Phatak, P.; Makley, M.J.; Stone, R.D.; Kling, M.A. Cerebrospinal fluid evidence of increased extra-mitochondrial glucose metabolism implicates mitochondrial dysfunction in multiple sclerosis disease progression. *J. Neurol. Sci.* **2008**, *275*, 106–112. [CrossRef]
4. Obeid, R.; Kasoha, M.; Knapp, J.P.; Kostopoulos, P.; Becker, G.; Fassbender, K.; Herrmann, W. Folate and methylation status in relation to phosphorylated tau protein(181P) and β-amyloid(1-42) in cerebrospinal fluid. *Clin. Chem.* **2007**, *53*, 1129–1136. [CrossRef]
5. Jasperse, B.; Jakobs, C.; Eikelenboom, M.J.; Dijkstra, C.D.; Uitdehaag, B.M.J.; Barkhof, F.; Polman, C.H.; Teunissen, C.E. N-acetylaspartic acid in cerebrospinal fluid of multiple sclerosis patients determined by gas-chromatography-mass spectrometry. *J. Neurol.* **2007**, *254*, 631–637. [CrossRef]
6. Mattsson, N.; Yaong, M.; Rosengren, L.; Blennow, K.; Månsson, J.-E.; Andersen, O.; Zetterberg, H.; Haghighi, S.; Zho, I.; Pratico, D. Elevated cerebrospinal fluid levels of prostaglandin E2 and 15-(S)-hydroxyeicosatetraenoic acid in multiple sclerosis. *J. Intern. Med.* **2009**, *265*, 459–464. [CrossRef]
7. Lim, C.K.; Bilgin, A.; Lovejoy, D.B.; Tan, V.; Bustamante, S.; Taylor, B.V.; Bessede, A.; Brew, B.J.; Guillemin, G.J. Kynurenine pathway metabolomics predicts and provides mechanistic insight into multiple sclerosis progression. *Sci. Rep.* **2017**, *7*, 1–9. [CrossRef]
8. Lewitt, P.A.; Li, J.; Lu, M.; Guo, L.; Auinger, P. Metabolomic biomarkers as strong correlates of Parkinson disease progression. *Neurology* **2017**, *88*, 862–869. [CrossRef]
9. Naylor, J.C.; Hulette, C.M.; Steffens, D.C.; Shampine, L.J.; Ervin, J.F.; Payne, V.M.; Massing, M.W.; Kilts, J.D.; Strauss, J.L.; Calhoun, P.S.; et al. Cerebrospinal fluid dehydroepiandrosterone levels are correlated with brain dehydroepiandrosterone levels, elevated in Alzheimer's disease, and related to neuropathological disease stage. *J. Clin. Endocrinol. Metab.* **2008**, *93*, 3173–3178. [CrossRef]
10. Fonteh, A.N.; Cipolla, M.; Chiang, J.; Arakaki, X.; Harrington, M.G. Human cerebrospinal fluid fatty acid levels differ between supernatant fluid and brain-derived nanoparticle fractions, and are altered in Alzheimer's disease. *PLoS ONE* **2014**, *9*, e100519. [CrossRef]

11. Praticò, D.; Clark, C.M.; Liun, F.; Lee, V.Y.M.; Trojanowski, J.Q. Increase of Brain Oxidative Stress in Mild Cognitive Impairment. *Arch. Neurol.* **2002**, *59*, 972–976. [CrossRef]
12. Yao, Y.; Clark, C.M.; Trojanowski, J.Q.; Lee, V.M.Y.; Praticò, D. Elevation of 12/15 lipoxygenase products in AD and mild cognitive impairment. *Ann. Neurol.* **2005**, *58*, 623–626. [CrossRef]
13. Fonteh, A.N.; Cipolla, M.; Chiang, A.J.; Edminster, S.P.; Arakaki, X.; Harrington, M.G. Polyunsaturated Fatty Acid Composition of Cerebrospinal Fluid Fractions Shows Their Contribution to Cognitive Resilience of a Pre-symptomatic Alzheimer's Disease Cohort. *Front. Physiol.* **2020**, *11*. [CrossRef]
14. Montine, T.J.; Kaye, J.A.; Montine, K.S.; McFarland, L.; Morrow, J.D.; Quinn, J.F. Cerebrospinal fluid Aβ42, tau, and F2-isoprostane concentrations in patients with Alzheimer disease, other dementias, and in age-matched controls. *Arch. Pathol. Lab. Med.* **2001**, *125*. [CrossRef]
15. Chernevskaya, E.; Beloborodova, N.; Klimenko, N.; Pautova, A.; Shilkin, D.; Gusarov, V.; Tyakht, A.; Tyakht, A. Serum and fecal profiles of aromatic microbial metabolites reflect gut microbiota disruption in critically ill patients: A prospective observational pilot study. *Crit. Care* **2020**, *24*, 1–13. [CrossRef]
16. Chernevskaya, E.; Klimenko, N.; Pautova, A.; Buyakova, I.; Tyakht, A.; Beloborodova, N. Host-microbiome interactions mediated by phenolic metabolites in chronically critically ill patients. *Metabolites* **2021**, *11*, 122. [CrossRef]
17. Carabotti, M.; Scirocco, A.; Maselli, M.A.; Severi, C. The gut-brain axis: Interactions between enteric microbiota, central and enteric nervous systems. *Ann. Gastroenterol.* **2015**, *28*, 203.
18. Kuwahara, A.; Matsuda, K.; Kuwahara, Y.; Asano, S.; Inui, T.; Marunaka, Y. Microbiota-gut-brain axis: Enteroendocrine cells and the enteric nervous system form an interface between the microbiota and the central nervous system. *Biomed. Res.* **2020**, *41*, 199–216. [CrossRef]
19. Chen, M.X.; Wang, S.Y.; Kuo, C.H.; Tsai, I.L. Metabolome analysis for investigating host-gut microbiota interactions. *J. Formos. Med. Assoc.* **2019**, *118*, S10–S22. [CrossRef]
20. Doherty, C.M.; Forbes, R.B. Diagnostic lumbar puncture. *Ulster Med. J.* **2014**, *83*, 93–102.
21. Wang, L.P.; Schmidt, J.F. Central nervous side effects after lumbar puncture. A review of the possible pathogenesis of the syndrome of postdural puncture headache and associated symptoms. *Dan. Med. Bull.* **1997**, *44*, 79–81.
22. Wishart, D.S.; Lewis, M.J.; Morrissey, J.A.; Flegel, M.D.; Jeroncic, K.; Xiong, Y.; Cheng, D.; Eisner, R.; Gautam, B.; Tzur, D.; et al. The human cerebrospinal fluid metabolome. *J. Chromatogr. B Anal. Technol. Biomed. Life Sci.* **2008**, *871*, 164–173. [CrossRef]
23. Nakamizo, S.; Sasayama, T.; Shinohara, M.; Irino, Y.; Nishiumi, S.; Nishihara, M.; Tanaka, H.; Tanaka, K.; Mizukawa, K.; Itoh, T.; et al. GC/MS-based metabolomic analysis of cerebrospinal fluid (CSF) from glioma patients. *J. Neurooncol.* **2013**, *113*, 65–74. [CrossRef]
24. Murgia, F.; Lorefice, L.; Poddighe, S.; Fenu, G.; Secci, M.A.; Marrosu, M.G.; Cocco, E.; Atzori, L. Multi-Platform Characterization of Cerebrospinal Fluid and Serum Metabolome of Patients Affected by Relapsing–Remitting and Primary Progressive Multiple Sclerosis. *J. Clin. Med.* **2020**, *9*, 863. [CrossRef]
25. Wuolikainen, A.; Moritz, T.; Marklund, S.L.; Antti, H.; Andersen, P.M. Disease-Related Changes in the Cerebrospinal Fluid Metabolome in Amyotrophic Lateral Sclerosis Detected by GC/TOFMS. *PLoS ONE* **2011**, *6*, e17947. [CrossRef]
26. Park, S.J.; Jeong, I.H.; Kong, B.S.; Lee, J.E.; Kim, K.H.; Lee, D.Y.; Kim, H.J. Disease type- and status-specific alteration of CSF metabolome coordinated with clinical parameters in inflammatory demyelinating diseases of CNS. *PLoS ONE* **2016**, *11*. [CrossRef]
27. Gordon, S.M.; Srinivasan, L.; Taylor, D.M.; Master, S.R.; Tremoglie, M.A.; Hankeova, A.; Flannery, D.D.; Abbasi, S.; Fitzgerald, J.C.; Harris, M.C. Derivation of a metabolic signature associated with bacterial meningitis in infants. *Pediatr. Res.* **2020**, *88*, 184–191. [CrossRef]
28. Kraus, V.B. Biomarkers as drug development tools: Discovery, validation, qualification and use. *Nat. Rev. Rheumatol.* **2018**, *14*, 354–362. [CrossRef]
29. Koek, M.M.; Jellema, R.H.; van der Greef, J.; Tas, A.C.; Hankemeier, T. Quantitative metabolomics based on gas chromatography mass spectrometry: Status and perspectives. *Metabolomics* **2011**, *7*, 307–328. [CrossRef]
30. Mandal, R.; Guo, A.C.; Chaudhary, K.K.; Liu, P.; Yallou, F.S.; Dong, E.; Aziat, F.; Wishart, D.S. Multi-platform characterization of the human cerebrospinal fluid metabolome: A comprehensive and quantitative update. *Genome Med.* **2012**, *4*, 1–11. [CrossRef]
31. Stoop, M.P.; Coulier, L.; Rosenling, T.; Shi, S.; Smolinska, A.M.; Buydens, L.; Ampt, K.; Stingl, C.; Dane, A.; Muilwijk, B.; et al. Quantitative proteomics and metabolomics analysis of normal human cerebrospinal fluid samples. *Mol. Cell. Proteomics* **2010**, *9*, 2063–2075. [CrossRef]
32. Saito, K.; Hattori, K.; Andou, T.; Satomi, Y.; Gotou, M.; Kobayashi, H.; Hidese, S.; Kunugi, H. Characterization of postprandial effects on csf metabolomics: A pilot study with parallel comparison to plasma. *Metabolites* **2020**, *10*, 185. [CrossRef]
33. Rosenling, T.; Slim, C.L.; Christin, C.; Coulier, L.; Shi, S.; Stoop, M.P.; Bosman, J.; Suits, F.; Horvatovich, P.L.; Stockhofe-Zurwieden, N.; et al. The Effect of Preanalytical Factors on Stability of the Proteome and Selected Metabolites in Cerebrospinal Fluid (CSF). *J. Proteome Res.* **2009**, *8*, 5511–5522. [CrossRef]
34. Hartonen, M.; Mattila, I.; Ruskeepää, A.L.; Orešič, M.; Hyötyläinen, T. Characterization of cerebrospinal fluid by comprehensive two-dimensional gas chromatography coupled to time-of-flight mass spectrometry. *J. Chromatogr. A* **2013**, *1293*, 142–149. [CrossRef]

35. Carrasco-Pancorbo, A.; Nevedomskaya, E.; Arthen-Engeland, T.; Zey, T.; Zurek, G.; Baessmann, C.; Deelder, A.M.; Mayboroda, O.A. Gas chromatography/atmospheric pressure chemical ionization-time of flight mass spectrometry: Analytical validation and applicability to metabolic profiling. *Anal. Chem.* **2009**, *81*, 10071–10079. [CrossRef]
36. Akiyama, T.; Saigusa, D.; Hyodo, Y.; Umeda, K.; Saijo, R.; Koshiba, S.; Kobayashi, K. Metabolic Profiling of the Cerebrospinal Fluid in Pediatric Epilepsy. *Acta Med. Okayama* **2020**, *74*, 65–72. [CrossRef]
37. Li, C.; Xu, F.; Xie, D.M.; Jing, Y.; Shang, M.Y.; Liu, G.X.; Wang, X.; Cai, S.Q. Identification of absorbed constituents in the rabbit plasma and cerebrospinal fluid after intranasal administration of asari radix et rhizoma by HS-SPME-GC-MS and HPLC-APCI-IT-TOF-MSN. *Molecules* **2014**, *19*, 4857–4879. [CrossRef]
38. Deng, F.L.; Pan, J.X.; Zheng, P.; Xia, J.J.; Yin, B.M.; Liang, W.W.; Li, Y.F.; Wu, J.; Xu, F.; Wu, Q.Y.; et al. Metabonomics reveals peripheral and central shortchain fatty acid and amino acid dysfunction in a naturally occurring depressive model of macaques. *Neuropsychiatr. Dis. Treat.* **2019**, *15*, 1077–1088. [CrossRef]
39. Coulier, L.; Muilwijk, B.; Bijlsma, S.; Noga, M.; Tienstra, M.; Attali, A.; van Aken, H.; Suidgeest, E.; Tuinstra, T.; Luider, T.M.; et al. Metabolite profiling of small cerebrospinal fluid sample volumes with gas chromatography–mass spectrometry: Application to a rat model of multiple sclerosis. *Metabolomics* **2013**, *9*, 78–87. [CrossRef]
40. Koek, M.M.; Bakels, F.; Engel, W.; Van Den Maagdenberg, A.; Ferrari, M.D.; Coulier, L.; Hankemeier, T. Metabolic profiling of ultrasmall sample volumes with GC/MS: From microliter to nanoliter samples. *Anal. Chem.* **2010**, *82*, 156–162. [CrossRef]
41. Lu, A.Y.; Damisah, E.C.; Winkler, E.A.; Grant, R.A.; Eid, T.; Bulsara, K.R. Cerebrospinal fluid untargeted metabolomic profiling of aneurysmal subarachnoid hemorrhage: An exploratory study. *Br. J. Neurosurg.* **2018**, *32*, 637–641. [CrossRef]
42. Orefice, N.S.; Carotenuto, A.; Mangone, G.; Bues, B.; Rehm, R.; Cerillo, I.; Saccà, F.; Calignano, A.; Orefice, G. Assessment of neuroactive steroids in cerebrospinal fluid comparing acute relapse and stable disease in relapsing-remitting multiple sclerosis. *J. Steroid Biochem. Mol. Biol.* **2016**, *159*, 1–7. [CrossRef]
43. Di Filippo, M.; Pini, L.A.; Pelliccioli, G.P.; Calabresi, P.; Sarchielli, P. Abnormalities in the cerebrospinal fluid levels of endo-cannabinoids in multiple sclerosis. *J. Neurol. Neurosurg. Psychiatry* **2008**, *79*, 1224–1229. [CrossRef]
44. Paik, M.J.; Ahn, Y.H.; Lee, P.H.; Kang, H.; Park, C.B.; Choi, S.; Lee, G. Polyamine patterns in the cerebrospinal fluid of patients with Parkinson's disease and multiple system atrophy. *Clin. Chim. Acta* **2010**, *411*, 1532–1535. [CrossRef]
45. Lee, P.H.; Lee, G.; Paik, M.J. Polyunsaturated fatty acid levels in the cerebrospinal fluid of patients with Parkinson's disease and multiple system atrophy. *Mov. Disord.* **2008**, *23*, 309–310. [CrossRef]
46. Zhang, P.; Wang, B.; Sun, Y.; Gao, J.; Lian, K. Analysis of 5-hydroxytryptamine and its related indoles in cerebrospinal fluid of leukemic children by gas chromatography-mass spectrometry. *J. Lab. Med.* **2020**, *44*, 41–45. [CrossRef]
47. Mason, S.; Reinecke, C.J.; Solomons, R. Cerebrospinal Fluid Amino Acid Profiling of Pediatric Cases with Tuberculous Meningitis. *Front. Neurosci.* **2017**, *11*. [CrossRef]
48. Obata, T.; Nagakura, T.; Maeda, H.; Yamashita, K.; Maekawa, K. Simultaneous assay of prostaglandins and thromboxane in the cerebrospinal fluid by gas chromatography-mass spectrometry-selected ion monitoring. *J. Chromatogr. B Biomed. Sci. Appl.* **1999**, *731*, 73–81. [CrossRef]
49. Kozar, M.P.; Krahmer, M.T.; Fox, A.; Gray, B.M. Failure to detect muramic acid in normal rat tissues but detection in cerebrospinal fluids from patients with pneumococcal meningitis. *Infect. Immun.* **2000**, *68*, 4688–4698. [CrossRef]
50. Medana, I.M.; Hien, T.T.; Day, N.P.; Phu, N.H.; Mai, N.T.H.; Van Chu'ong, L.; Chau, T.T.H.; Taylor, A.; Salahifar, H.; Stocker, R.; et al. The Clinical Significance of Cerebrospinal Fluid Levels of Kynurenine Pathway Metabolites and Lactate in Severe Malaria. *J. Infect. Dis.* **2002**, *185*, 650–656. [CrossRef]
51. Medana, I.M.; Day, N.P.J.; Salahifar-Sabet, H.; Stocker, R.; Smythe, G.; Bwanaisa, L.; Njobvu, A.; Kayira, K.; Turner, G.D.H.; Taylor, T.E.; et al. Metabolites of the kynurenine pathway of tryptophan metabolism in the cerebrospinal fluid of Malawian children with malaria. *J. Infect. Dis.* **2003**, *188*, 844–849. [CrossRef]
52. Anderson, A.M.; Croteau, D.; Ellis, R.J.; Rosario, D.; Potter, M.; Guillemin, G.J.; Brew, B.J.; Woods, S.P.; Letendre, S.L. HIV, prospective memory, and cerebrospinal fluid concentrations of quinolinic acid and phosphorylated Tau. *J. Neuroimmunol.* **2018**, *319*, 13–18. [CrossRef]
53. Inoue, H.; Matsushige, T.; Ichiyama, T.; Okuno, A.; Takikawa, O.; Tomonaga, S.; Anlar, B.; Yüksel, D.; Otsuka, Y.; Kohno, F.; et al. Elevated quinolinic acid levels in cerebrospinal fluid in subacute sclerosing panencephalitis. *J. Neuroimmunol.* **2020**, *339*, 577088. [CrossRef]
54. Regenold, W.T.; Kling, M.A.; Hauser, P. Elevated sorbitol concentration in the cerebrospinal fluid of patients with mood disorders. *Psychoneuroendocrinology* **2000**, *25*, 593–606. [CrossRef]
55. Kageyama, Y.; Deguchi, Y.; Hattori, K.; Yoshida, S.; Goto, Y.; Inoue, K.; Kato, T. Nervonic acid level in cerebrospinal fluid is a candidate biomarker for depressive and manic symptoms: A pilot study. *Brain Behav.* **2021**, *11*. [CrossRef]
56. Kim, B.K.; Fonda, J.R.; Hauger, R.L.; Pinna, G.; Anderson, G.M.; Valovski, I.T.; Rasmusson, A.M. Composite contributions of cerebrospinal fluid GABAergic neurosteroids, neuropeptide Y and interleukin-6 to PTSD symptom severity in men with PTSD. *Neurobiol. Stress* **2020**, *12*, 100220. [CrossRef]
57. Rasmusson, A.M.; King, M.W.; Valovski, I.; Gregor, K.; Scioli-Salter, E.; Pineles, S.L.; Hamouda, M.; Nillni, Y.I.; Anderson, G.M.; Pinna, G. Relationships between cerebrospinal fluid GABAergic neurosteroid levels and symptom severity in men with PTSD. *Psychoneuroendocrinology* **2019**, *102*, 95–104. [CrossRef]

58. Jokinen, J.; Nordström, A.L.; Nordström, P. Cerebrospinal fluid monoamine metabolites and suicide. *Nord. J. Psychiatry* **2009**, *63*, 276–279. [CrossRef]
59. Engström, G.; Alling, C.; Blennow, K.; Regnéll, G.; Träskman-Bendz, L. Reduced cerebrospinal HVA concentrations and HVA/5-HIAA ratios in suicide attempters: Monoamine metabolites in 120 suicide attempters and 47 controls. *Eur. Neuropsychopharmacol.* **1999**, *9*, 399–405. [CrossRef]
60. Lindqvist, D.; Janelidze, S.; Hagell, P.; Erhardt, S.; Samuelsson, M.; Minthon, L.; Hansson, O.; Björkqvist, M.; Träskman-Bendz, L.; Brundin, L. Interleukin-6 Is Elevated in the Cerebrospinal Fluid of Suicide Attempters and Related to Symptom Severity. *Biol. Psychiatry* **2009**, *66*, 287–292. [CrossRef]
61. Mochel, F.; DeLonlay, P.; Touati, G.; Brunengraber, H.; Kinman, R.P.; Rabier, D.; Roe, C.R.; Saudubray, J.M. Pyruvate carboxylase deficiency: Clinical and biochemical response to anaplerotic diet therapy. *Mol. Genet. Metab.* **2005**, *84*, 305–312. [CrossRef]
62. Mazzuca, M.; Maubert, M.A.; Damaj, L.; Clot, F.; Cadoudal, M.; Dubourg, C.; Odent, S.; Benoit, J.F.; Bahi-Buisson, N.; Christa, L.; et al. Combined Sepiapterin Reductase and Methylmalonyl-CoA Epimerase Deficiency in a Second Patient: Cerebrospinal Fluid Polyunsaturated Fatty Acid Level and Follow-Up Under L-DOPA, 5-HTP and BH4 Trials. *JIMD Rep.* **2015**, *22*, 47–55. [CrossRef]
63. Almeida, L.S.; Verhoeven, N.M.; Roos, B.; Valongo, C.; Cardoso, M.L.; Vilarinho, L.; Salomons, G.S.; Jakobs, C. Creatine and guanidinoacetate: Diagnostic markers for inborn errors in creatine biosynthesis and transport. *Mol. Genet. Metab.* **2004**, *82*, 214–219. [CrossRef]
64. de Paiva, M.J.N.; Menezes, H.C.; Christo, P.P.; Resende, R.R.; Cardeal, Z.d.L. An alternative derivatization method for the analysis of amino acids in cerebrospinal fluid by gas chromatography-mass spectrometry. *J. Chromatogr. B Anal. Technol. Biomed. Life Sci.* **2013**, *931*, 97–102. [CrossRef]
65. Pautova, A.; Khesina, Z.; Getsina, M.; Sobolev, P.; Revelsky, A.; Beloborodova, N. Determination of Tryptophan Metabolites in Serum and Cerebrospinal Fluid Samples Using Microextraction by Packed Sorbent, Silylation and GC-MS Detection. *Molecules* **2020**, *85*, 3258. [CrossRef]
66. Smythe, G.A.; Braga, O.; Brew, B.J.; Grant, R.S.; Guillemin, G.J.; Kerr, S.J.; Walker, D.W. Concurrent Quantification of Quinolinic, Picolinic, and Nicotinic Acids Using Electron-Capture Negative-Ion Gas Chromatography–Mass Spectrometry. *Anal. Biochem.* **2002**, *301*, 21–26. [CrossRef]
67. Pautova, A.K.; Khesina, Z.B.; Litvinova, T.N.; Revelsky, A.I.; Beloborodova, N.V. Metabolic profiling of aromatic compounds in cerebrospinal fluid of neurosurgical patients using microextraction by packed sorbent and liquid–liquid extraction with gas chromatography–mass spectrometry analysis. *Biomed. Chromatogr.* **2021**, *35*, 1–11. [CrossRef]
68. Kietzerow, J.; Otto, B.; Wilke, N.; Rohde, H.; Iwersen-Bergmann, S.; Andresen-Streichert, H. The challenge of post-mortem GHB analysis: Storage conditions and specimen types are both important. *Int. J. Legal Med.* **2020**, *134*, 205–215. [CrossRef]
69. Kok, R.M.; Howells, D.W.; Van Den Heuvel, C.C.M.; Guérand, W.S.; Thompson, G.N.; Jakobs, C. Stable Isotope Dilution Analysis of GABA in CSF Using Simple Solvent Extraction and Electron-Capture Negative-Ion Mass Fragmentography. *J. Inherit. Metab. Dis.* **1993**, *16*, 508–512. [CrossRef]
70. Kok, R.M.; Kaster, L.; De Jong, A.P.; Poll-Thé, B.; Saudubray, J.-M.; Jakobs, C. Stable isotope dilution analysis of pipecolic acid in cerebrospinal fluid, plasma, urine and amniotic fluid using electron capture negative ion mass fragmentography. *Clin. Chim. Acta* **1987**, *168*, 143–152. [CrossRef]
71. Kim, Y.S.; Zhang, H.; Kim, H.Y. Profiling neurosteroids in cerebrospinal fluids and plasma by gas chromatography/electron capture negative chemical ionization mass spectrometry. *Anal. Biochem.* **2000**, *277*, 187–195. [CrossRef]
72. Kancheva, R.; Hill, M.; Novák, Z.; Chrastina, J.; Velíková, M.; Kancheva, L.; Říha, I.; Stárka, L. Peripheral neuroactive steroids may be as good as the steroids in the cerebrospinal fluid for the diagnostics of CNS disturbances. *J. Steroid Biochem. Mol. Biol.* **2010**, *119*, 35–44. [CrossRef]
73. Mannila, A.; Kumpulainen, E.; Lehtonen, M.; Heikkinen, M.; Laisalmi, M.; Salo, T.; Rautio, J.; Savolainen, J.; Kokki, H. Plasma and cerebrospinal fluid concentrations of indomethacin in children after intravenous administration. *J. Clin. Pharmacol.* **2007**, *47*, 94–100. [CrossRef]
74. Liang, E.; Garzone, P.; Cedarbaum, J.M.; Koller, M.; Tran, T.; Xu, V.; Ross, B.; Jhee, S.S.; Ereshefsky, L.; Pastrak, A.; et al. Pharmacokinetic profile of orally administered scyllo-inositol (Elnd005) in plasma, cerebrospinal fluid and brain, and corresponding effect on amyloid-beta in healthy subjects. *Clin. Pharmacol. Drug Dev.* **2013**, *2*, 186–194. [CrossRef]
75. Wyman, J.; Bultman, S. Postmortem Distribution of Heroin Metabolites in Femoral Blood, Liver, Cerebrospinal Fluid, and Vitreous Humor. *J. Anal. Toxicol.* **2004**, *28*, 260–263. [CrossRef]
76. Perala, A.W.; Filary, M.J.; Bartels, M.J.; McMartin, K.E. Quantitation of Diethylene Glycol and Its Metabolites by Gas Chromatography Mass Spectrometry or Ion Chromatography Mass Spectrometry in Rat and Human Biological Samples. *J. Anal. Toxicol.* **2014**, *38*, 184–193. [CrossRef]
77. Wardlaw, S.L.; Burant, C.F.; Klein, S.; Meece, K.; White, A.; Kasten, T.; Lucey, B.P.; Bateman, R.J. Continuous 24-hour leptin, proopiomelanocortin, and amino acid measurements in human cerebrospinal fluid: Correlations with plasma leptin, soluble leptin receptor, and amino acid levels. *J. Clin. Endocrinol. Metab.* **2014**, *99*, 2540–2548. [CrossRef]
78. Young, S.; Struys, E.; Wood, T. Quantification of creatine and guanidinoacetate using GC-MS and LC-MS/MS for the detection of cerebral creatine deficiency syndromes. *Curr. Protoc. Hum. Genet.* **2007**, *Chapter 17*, 1–18. [CrossRef]

79. Plecko, B.; Paul, K.; Paschke, E.; Stoeckler-Ipsiroglu, S.; Struys, E.; Jakobs, C.; Hartmann, H.; Luecke, T.; di Capua, M.; Korenke, C.; et al. Biochemical and Molecular Characterizationof 18 Patients With Pyridoxine-Dependent Epilepsyand Mutations of the Antiquitin (ALDH7A1) Gene. *Hum. Mutat.* **2007**, *28*, 19–26. [CrossRef]
80. Wang, Y.; Karu, K.; Griffiths, W.J. Analysis of neurosterols and neurosteroids by mass spectrometry. *Biochimie* **2007**, *89*, 182–191. [CrossRef]
81. Starka, L.; Hill, M.; Kancheva, R.; Novak, Z.; Chrastina, J.; Pohanka, M.; Morfin, R. 7-Hydroxylated derivatives of dehydroepiandrosterone in the human ventricular cerebrospinal fluid. *Neuroendocrinol. Lett.* **2009**, *30*, 368–372.
82. Miller, E.; Morel, A.; Saso, L.; Saluk, J. Isoprostanes and neuroprostanes as biomarkers of oxidative stress in neurodegenerative diseases. *Oxid. Med. Cell. Longev.* **2014**. [CrossRef]
83. Praticò, D.; Clark, C.M.; Lee, V.M.-Y.; Trojanowski, J.Q.; Rokach, J.; FitzGerald, G.A. Increased 8,12-iso-iPF 2-VI in Alzheimer's Disease: Correlation of a Noninvasive Index of Lipid Peroxidation with Disease Severity. *Ann. Neurol.* **2000**, *48*, 809–812.
84. Li, G.; Millard, S.P.; Peskind, E.R.; Zhang, J.; Yu, C.E.; Leverenz, J.B.; Mayer, C.; Shofer, J.S.; Raskind, M.A.; Quinn, J.F.; et al. Cross-sectional and longitudinal relationships between cerebrospinal fluid biomarkers and cognitive function in people without cognitive impairment from across the adult life span. *JAMA Neurol.* **2014**, *71*, 742–751. [CrossRef]
85. Finno, C.J.; Estell, K.E.; Winfield, L.; Katzman, S.; Bordbari, M.H.; Burns, E.N.; Miller, A.D.; Puschner, B.; Tran, C.K.; Xu, L. Lipid peroxidation biomarkers for evaluating oxidative stress in equine neuroaxonal dystrophy. *J. Vet. Intern. Med.* **2018**, *32*, 1740–1747. [CrossRef]
86. Yen, H.C.; Wei, H.J.; Chen, T.W. Analytical variables affecting analysis of Fisoprostanes and Fneuroprostanes in human cerebrospinal fluid by gas chromatography/mass spectrometry. *Biomed. Res. Int.* **2013**, *2013*. [CrossRef]
87. Milne, G.L.; Sanchez, S.C.; Musiek, E.S.; Morrow, J.D. Quantification of F2-isoprostanes as a biomarker of oxidative stress. *Nat. Protoc.* **2007**, *2*, 221–226. [CrossRef]
88. Dmitrienko, S.G.; Apyari, V.V.; Gorbunova, M.V.; Tolmacheva, V.V.; Zolotov, Y.A. Homogeneous Liquid–Liquid Microextraction of Organic Compounds. *J. Anal. Chem.* **2020**, *75*, 1371–1383. [CrossRef]
89. Gupta, M.; Jain, A.; Verma, K.K. Salt-assisted liquid-liquid microextraction with water-miscible organic solvents for the determination of carbonyl compounds by high-performance liquid chromatography. *Talanta* **2009**, *80*, 526–531. [CrossRef]
90. Rezaee, M.; Assadi, Y.; Milani Hosseini, M.R.; Aghaee, E.; Ahmadi, F.; Berijani, S. Determination of organic compounds in water using dispersive liquid-liquid microextraction. *J. Chromatogr. A* **2006**, *1116*, 1–9. [CrossRef]
91. Dmitrienko, S.G.; Apyari, V.V.; Tolmacheva, V.V.; Gorbunova, M.V. Dispersive Liquid–Liquid Microextraction of Organic Compounds: An Overview of Reviews. *J. Anal. Chem.* **2020**, *75*, 1237–1251. [CrossRef]
92. Zuloaga, O.; Olivares, M.; Navarro, P.; Vallejo, A.; Prieto, A. Dispersive liquid-liquid microextraction: Trends in the analysis of biological samples. *Bioanalysis* **2015**, *7*, 2211–2225. [CrossRef]
93. Mudiam, M.K.R.; Ratnasekhar, C. Ultra sound assisted one step rapid derivatization and dispersive liquid-liquid microextraction followed by gas chromatography-mass spectrometric determination of amino acids in complex matrices. *J. Chromatogr. A* **2013**, *1291*, 10–18. [CrossRef]
94. Jain, R.; Kumar, A.; Shukla, Y. Dispersive Liquid-liquid Microextraction-injector Port Silylation: A Viable Option for the Analysis of Polar Analytes using Gas Chromatography-Mass Spectrometry. *Austin J. Anal. Pharm. Chem.* **2015**, *2*, 1042.
95. Anastassiades, M.; Lehotay, S.J.; Štajnbaher, D.; Schenck, F.J. Fast and easy multiresidue method employing acetonitrile extraction/partitioning and "dispersive solid-phase extraction" for the determination of pesticide residues in produce. *J. AOAC Int.* **2003**, *86*, 412–431. [CrossRef]
96. Townsend, R.; Keulen, G.; Desbrow, C.; Godfrey, A.R. An investigation of the utility of QuEChERS for extracting acid, base, neutral and amphiphilic species from example environmental and clinical matrices. *Anal. Sci. Adv.* **2020**, *1*, 152–160. [CrossRef]
97. Rial-Berriel, C.; Acosta-Dacal, A.; Zumbado, M.; Luzardo, O.P. Micro QuEChERS-based method for the simultaneous biomonitoring in whole blood of 360 toxicologically relevant pollutants for wildlife. *Sci. Total Environ.* **2020**, *736*. [CrossRef]
98. Santana-Mayor, Á.; Socas-Rodríguez, B.; Herrera-Herrera, A.V.; Rodríguez-Delgado, M.Á. Current trends in QuEChERS method. A versatile procedure for food, environmental and biological analysis. *TrAC Trends Anal. Chem.* **2019**, *116*, 214–235. [CrossRef]
99. Arthur, C.L.; Pawliszyn, J. Solid Phase Microextraction with Thermal Desorption Using Fused Silica Optical Fibers. *Anal. Chem.* **1990**, *62*, 2145–2148. [CrossRef]
100. Vuckovic, D.; De Lannoy, I.; Gien, B.; Shirey, R.E.; Sidisky, L.M.; Dutta, S.; Pawliszyn, J. In vivo solid-phase microextraction: Capturing the elusive portion of metabolome. *Angew. Chemie Int. Ed.* **2011**, *50*, 5344–5348. [CrossRef]
101. Campanella, B.; Onor, M.; Lomonaco, T.; Benedetti, E.; Bramanti, E. HS-SPME-GC-MS approach for the analysis of volatile salivary metabolites and application in a case study for the indirect assessment of gut microbiota. *Anal. Bioanal. Chem.* **2019**, *411*, 7551–7562. [CrossRef]
102. Silva, C.L.; Passos, M.; Cmara, J.S. Investigation of urinary volatile organic metabolites as potential cancer biomarkers by solid-phase microextraction in combination with gas chromatography-mass spectrometry. *Br. J. Cancer* **2011**, *105*, 1894–1904. [CrossRef]
103. Tienpont, B.; David, F.; Desmet, K.; Sandra, P. Stir bar sorptive extraction-thermal desorption-capillary GC-MS applied to biological fluids. *Anal. Bioanal. Chem.* **2002**, *373*, 46–55. [CrossRef]
104. Nazyropoulou, C.; Samanidou, V. Stir bar sorptive extraction applied to the analysis of biological fluids. *Bioanalysis* **2015**, *7*, 2241–2250. [CrossRef]

105. Hasan, C.K.; Ghiasvand, A.; Lewis, T.W.; Nesterenko, P.N.; Paull, B. Recent advances in stir-bar sorptive extraction: Coatings, technical improvements, and applications. *Anal. Chim. Acta* **2020**, *1139*, 222–240. [CrossRef]
106. Li, J.; Qi, H.Y.; Wang, Y.B.; Su, Q.; Wu, S.; Wu, L. Hollow fiber–stir bar sorptive extraction and microwave assisted derivatization of amino acids in biological matrices. *J. Chromatogr. A* **2016**, *1474*, 32–39. [CrossRef]
107. Farajzadeh, M.A.; Matin, A.A. Determination of BTEX in water samples with an SPME hollow fiber coated copper wire. *Chromatographia* **2008**, *68*, 443–446. [CrossRef]
108. Abdel-Rehim, M. New trend in sample preparation: On-line microextraction in packed syringe for liquid and gas chromatography applications. *J. Chromatogr. B* **2004**, *801*, 317–321. [CrossRef]
109. Pautova, A.K.; Sobolev, P.D.; Revelsky, A.I. Microextraction of aromatic microbial metabolites by packed hypercrosslinked polystyrene from blood serum. *J. Pharm. Biomed. Anal.* **2020**, *177*. [CrossRef]
110. Silva, C.; Cavaco, C.; Perestrelo, R.; Pereira, J.; Câmara, J.S. Microextraction by packed Sorbent (MEPS) and solid-phase microextraction (SPME) as sample preparation procedures for the metabolomic profiling of urine. *Metabolites* **2014**, *4*, 71–97. [CrossRef]
111. Peters, S.; Kaal, E.; Horsting, I.; Janssen, H.G. An automated method for the analysis of phenolic acids in plasma based on ion-pairing micro-extraction coupled on-line to gas chromatography/mass spectrometry with in-liner derivatisation. *J. Chromatogr. A* **2012**, *1226*, 71–76. [CrossRef]
112. Bustamante, L.; Cárdenas, D.; von Baer, D.; Pastene, E.; Duran-Sandoval, D.; Vergara, C.; Mardones, C. Evaluation of microextraction by packed sorbent, liquid–liquid microextraction and derivatization pretreatment of diet-derived phenolic acids in plasma by gas chromatography with triple quadrupole mass spectrometry. *J. Sep. Sci.* **2017**, *40*, 3487–3496. [CrossRef]
113. Pautova, A.K.; Sobolev, P.D.; Revelsky, A.I. Analysis of phenylcarboxylic acid-type microbial metabolites by microextraction by packed sorbent from blood serum followed by GC–MS detection. *Clin. Mass Spectrom.* **2019**, *14*, 46–53. [CrossRef]

Article

Counteracting the Ramifications of UVB Irradiation and Photoaging with *Swietenia macrophylla* King Seed

Camille Keisha Mahendra [1], Syafiq Asnawi Zainal Abidin [2], Thet Thet Htar [1], Lay-Hong Chuah [1], Shafi Ullah Khan [1,3], Long Chiau Ming [4], Siah Ying Tang [5,6,7], Priyia Pusparajah [8,*] and Bey Hing Goh [1,9,10,*]

1. Biofunctional Molecule Exploratory Research Group, School of Pharmacy, Monash University Malaysia, Bandar Sunway 47500, Malaysia; camille.mahendra@monash.edu (C.K.M.); thet.thet.htar@monash.edu (T.T.H.); alice.chuah@monash.edu (L.-H.C.); shafi.khan1@monash.edu (S.U.K.)
2. Liquid Chromatography Mass Spectrometry (LCMS) Platform, Jeffrey Cheah School of Medicine and Health Sciences, Monash University Malaysia, Jalan Lagoon Selatan, Bandar Sunway 47500, Malaysia; syafiq.asnawi@monash.edu
3. Department of Pharmacy, Abasyn University, Peshawar 25000, Pakistan
4. PAP Rashidah Sa'adatul Bolkiah Institute of Health Sciences, Universiti Brunei Darussalam, Gadong BE1410, Brunei; longchiauming@gmail.com
5. Chemical Engineering Discipline, School of Engineering, Monash University Malaysia, Bandar Sunway 47500, Malaysia; patrick.tang@monash.edu
6. Advanced Engineering Platform, School of Engineering, Monash University Malaysia, Bandar Sunway 47500, Malaysia
7. Tropical Medicine and Biology Platform, School of Science, Monash University Malaysia, Bandar Sunway 47500, Malaysia
8. Medical Health and Translational Research Group, Jeffrey Cheah School of Medicine and Health Sciences, Monash University Malaysia, Bandar Sunway 47500, Malaysia
9. College of Pharmaceutical Sciences, Zhejiang University, 866 Yuhangtang Road, Hangzhou 310058, China
10. Health and Well-Being Cluster, Global Asia in the 21st Century (GA21) Platform, Monash University Malaysia, Bandar Sunway 47500, Malaysia
* Correspondence: priyia.pusparajah@monash.edu (P.P.); goh.bey.hing@monash.edu (B.H.G.)

Abstract: In this day and age, the expectation of cosmetic products to effectively slow down skin photoaging is constantly increasing. However, the detrimental effects of UVB on the skin are not easy to tackle as UVB dysregulates a wide range of molecular changes on the cellular level. In our research, irradiated keratinocyte cells not only experienced a compromise in their redox system, but processes from RNA translation to protein synthesis and folding were also affected. Aside from this, proteins involved in various other processes like DNA repair and maintenance, glycolysis, cell growth, proliferation, and migration were affected while the cells approached imminent cell death. Additionally, the collagen degradation pathway was also activated by UVB irradiation through the upregulation of inflammatory and collagen degrading markers. Nevertheless, with the treatment of *Swietenia macrophylla* (*S. macrophylla*) seed extract and fractions, the dysregulation of many genes and proteins by UVB was reversed. The reversal effects were particularly promising with the *S. macrophylla* hexane fraction (SMHF) and *S. macrophylla* ethyl acetate fraction (SMEAF). SMHF was able to oppose the detrimental effects of UVB in several different processes such as the redox system, DNA repair and maintenance, RNA transcription to translation, protein maintenance and synthesis, cell growth, migration and proliferation, and cell glycolysis, while SMEAF successfully suppressed markers related to skin inflammation, collagen degradation, and cell apoptosis. Thus, in summary, our research not only provided a deeper insight into the molecular changes within irradiated keratinocytes, but also serves as a model platform for future cosmetic research to build upon. Subsequently, both SMHF and SMEAF also displayed potential photoprotective properties that warrant further fractionation and in vivo clinical trials to investigate and obtain potential novel bioactive compounds against photoaging.

Keywords: photoaging; proteomics; genomics; *Swietenia macrophylla*; UV irradiation; keratinocytes; epidermal layer; cosmetics; natural product; LC-MS/MS

1. Introduction

The existence of ultraviolet radiation (UVR) in our lives is very much like a double-edged sword. On one hand, we cannot live without it, however, excessive exposure could also lead to our demise. This is especially true for UVB, the reasons being that through it, our bodies produce the much-needed vitamin D, however, conversely, among the two types of UVR in our atmosphere, it is the one that causes the most harm to our skin, with photodamage ranging from sunburns to skin carcinogenesis [1]. Another benefit of UVB, or more precisely narrowband UVB with an emission peak of 311 nm, is its wide use in phototherapy treatments against skin diseases such as psoriasis, mycosis fungoides, vitiligo, etc. It has been witnessed that UVB phototherapy often improved skin conditions of diseased skin, giving temporary periods of respite to the patients [2–4]. Nevertheless, regardless of its benefits, overexposure to UVB is more often than not the case for the general population. This is often showcased in the appearance of irregular pigmentation, fine lines, wrinkles, poor texture, sagging skin, etc., on our skin after prolonged exposure, which are key signs of skin photoaging [5]. Although there are 'intrinsic' factors like the natural generation of reactive oxygen species (ROS) and reactive nitrogen species (RNS) in our skin and 'extrinsic' factors such as lifestyles changes and environmental pollution that causes skin aging, it is still undeniable that the exposure of UVB plays a part in skin aging [6–8].

There are three categories of UVR based on their wavelengths: UVA, UVB, and UVC. Among the three kinds of UVR, only UVA and UVB can penetrate through the ozone, with UVA (320–400 nm) having a 95% penetration level and UVB (290–320 nm) with a maximum penetration level of 5% through the ozone layer [9]. Despite the wavelength of UVB being much shorter than UVA and therefore, mostly absorbed through the epidermal layer, its detrimental effect is not limited to the epidermal layer [10]. Direct penetration of UVB not only form cyclobutene pyrimidine dimers (CPD) and pyrimidine-pyrimidone (6–4) photoproducts in the DNA, but also incites the production of ROS and RNS in the skin [11–13]. This then increases the oxidative stress levels in the skin, which quickly depletes the skin's antioxidant defense, and initiates a cascade of pro-inflammatory and other intracellular signals like matrix metalloproteases (MMP) and melanogenic cytokines by keratinocyte cells [14–16]. Ultimately, this leads to the formation of unwanted irregular pigmentation through the activation of the tyrosinase family, and the degradation of our skin's extracellular matrixes, reducing the elasticity of the skin and forming wrinkles [15–17]. Thus, in this study, the impact of UVB on keratinocytes was investigated. The purpose of this was to not only produce a wider view, and therefore better understanding of the molecular changes and pathway influenced by UVB, but also to aid in the advancement of cosmetic products that can better counteract the photodamaging effects of UVB. To achieve this, the Nanoflow-Ultra High-Performance Liquid Chromatography-Tandem Mass Spectrometry (LC-MS/MS) platform was utilized in the proteomics analysis of this study. The LC-MS/MS is a high throughput platform that is capable of accurately measuring fold changes and identifying a large range of proteins and peptides with the aid of bioinformatics in a relatively short time while remaining cost-effective. Another benefit of LC-MS/MS is its ability to separate and distinguish structurally or chemically similar peptides and proteins from each other [18].

Furthermore, this research also evaluated the capabilities of *S. macrophylla* seed extract as a photoprotective agent. *S. macrophylla* is a timber tree from the Meliaceae family that can be found in the tropics of Central America, Southeast Asia, and Mexico [19–21]. Besides being well prized for its mahogany wood, its seeds, containing flavonoids, alkaloids, and saponins, are often used in traditional medicine to treat sicknesses such as diabetes, hypertension, and even physical pain [22,23]. To prove its medicinal claim, many studies had been conducted, and through them, it has been reported that the seed possesses anti-cancer, neuroprotection, anti-hyperglycemic, anti-inflammation, antioxidant, and anti-viral properties [21,23–28]. Recently, it was discovered that one of the limonoid compounds, swietenine, isolated from the seed were responsible for the seed's antioxidant and anti-inflammatory activity on LPSEc stimulated RAW264.7 murine macrophage. Not only was the compound able to significantly inhibit the production of nitric oxide, but it also engaged the nuclear factor erythroid 2 (NRF2)/heme oxygenase-1 (HO-1) antioxidant pathway while downregulating the production of pro-inflammatory markers like interleukin (IL)-1β, tumor necrosis factor (TNF)-α, interferon gamma (IFN-γ), IL-6, cyclooxygenase (COX-2), and nuclear factor-κB (NF-κB) [28]. On the other hand, its wound healing ability has also been evaluated by Nilugal et al. [29]. In their study, the application of *S. macrophylla* ethanolic seed extract ointment was seen to significantly speed up the healing process of the excised wounds on the rats [29]. Thus, based on these claims, especially those regarding its antioxidant, wound healing, and anti-inflammatory properties, it would prove interesting to investigate if the seed extract and fractions can act as a photoprotective reagent against UVB and therefore be a potential active ingredient in the formulation of photoprotective cosmetics given the reasons that those aforementioned properties are inherently important in counteracting UVB-induced photodamage.

2. Results and Discussion

2.1. Cytotoxicity Assessment of S. macrophylla Extract and Fractions

HaCaT cells were treated with various concentrations (0–100 μg/mL) of the extract and fractions for 24 h. According to the data obtained, *S. macrophylla* crude extract (SMCE) begins to induce a dose-dependent decrease in cell viability starting from the concentration of 12.5 μg/mL with cell viability of 87.5 ± 3% ($p \leq 0.01$). The cell viability then continues to decrease to 74.83 ± 4.94% ($p \leq 0.001$), 51.77 ± 3.96% ($p \leq 0.001$), and 44.36 ± 3.36% ($p \leq 0.001$) when treated with 25, 50, and 100 μg/mL SMCE, respectively. On the other hand, after fractionation, SMHF did not induce any significant decrease in cell viability, even at concentrations as high as 100 μg/mL. As for SMEAF, cell viability was significantly decreased dose-dependently instead at concentrations of 25, 50, and 100 μg/mL to 82.04 ± 5.4% ($p \leq 0.001$), 49.93 ± 3.63% ($p \leq 0.001$), 35.25 ± 7.76% ($p \leq 0.001$), respectively. Finally, *S. macrophylla* water fraction (SMWF) became cytotoxic toward HaCaT at 100 μg/mL with cell viability of 80.7 ± 6.15% ($p \leq 0.001$) compared to the untreated control cells. Among the four samples, SMCE had the highest cytotoxicity against HaCaT, but the cytotoxicity levels were reduced after fractionation. This demonstrates that synergistic cytotoxic compounds were most likely separated during fractionation, and therefore, the fractions had reduced cytotoxic levels compared to the crude extract itself. Following the data obtained, the concentration 6.25 μg/mL for SMCE, 100 μg/mL for SMHF, 12.5 μg/mL for SMEAF, and 50 μg/mL for SMWF were chosen as the non-cytotoxic concentration to treat the cells in the upcoming experiments.

2.2. The Dynamic Proteomics and Genomic Dysregulation in Keratinocyte Cells Effectuated by UVB and Its Attenuation by S. macrophylla

2.2.1. Analysis of UVB-Induced Protein Modifications and the Reversal Effect of *S. macrophylla*

To ascertain the effect of UVB irradiation, with or without the treatment of *S. macrophylla*, on the protein expression changes in HaCaT cells, a high throughput proteomics analysis using LC-MS/MS was conducted. For the controls, untreated HaCaT cells were either unexposed or exposed with 50 mJ/cm^2 UVB to obtain the unexposed (non-UVB) and exposed (UVB) controls, respectively. Similarly, cells treated with *S. macrophylla* extract and fractions were also irradiated with 50 mJ/cm^2 UVB and their protein samples were individually collected at the 24 h time point. After processing and identifying individual proteins with the Uniprot database, a comparison of protein expression (in ratio) was obtained via PEAKS Q, based on the area under the curve. Five heatmaps (Figure 1A–E) depicting the comparison between the two untreated (non-UVB and UVB) controls and the difference between the UVB control and *S. macrophylla* treated cells were also obtained from PEAKS Q. Next, to truly compare amongst the data obtained, a compiled list of significant ($p \leq 0.05$) differentially expressed proteins, from the controls and *S. macrophylla* treated cells, were drawn up in Table 1. Additional detailed data of each protein can subsequently be found in Supplementary Materials Table S1. In Table 1, proteins obtained from both non-UVB and UVB controls were first compared to determine the protein expression changes induced by UVB. This is then followed by the changes seen in the treated cells compared to the UVB control. Proteins of similar Uniprot ID or name were directly compared across all groups for a clear comparison. When there are no significant changes seen in the protein ratio between the two comparing groups, the label N/S (not significant) will be assigned. In total, 151 proteins were identified to be differentially expressed among the controls and treated cells. From these differentially expressed proteins, it can be seen that UVB exposure has detrimental effects on a wide range of molecular functions such as DNA maintenance and repair, RNA synthesis, protein synthesis and processing (biogenesis, folding, stabilizing, proteostasis, etc.), cell growth, glycolysis process, etc., which ultimately determines the survival of the cells. When treated with SMCE, only four proteins within the HaCaT cells (60 S ribosomal protein L18, fumarate hydratase, annexin A3, and filamin B β) showed significant changes in expression in comparison to the UVB control. Other than fumarate hydratase, the three other proteins were instead significantly decreased in their expression levels. In contrast, SMHF showed the most attenuation in the differentially expressed proteins compared to the UVB control. Of all the proteins, only three proteins, 60 S ribosomal protein L18, neuroblast differentiation-associated protein AHNAK (AHNAK), and peroxiredoxin (PRDX)-3, were downregulated by SMHF. Aside from that, SMEAF induced changes in only 16 proteins in which seven of them were upregulated and nine were downregulated in comparison to the UVB control. The proteins that were upregulated were histone H2A type 1-A, histone H1.2, keratin type I cytoskeletal 14, exportin-2, nucleolar and coiled-body phosphoprotein 1 (NOLC1), and protein kinase C substrate 80 K–H isoform. Vice versa, PRDX-3, protein disulfide-isomerase (PDI) A3, annexin A3, polyubiquitin-C, HNRPCL1 protein, fascin, cathepsin D, prothymosin alpha, and GTP-binding nuclear protein Ran were downregulated by SMEAF. Finally, SMWF displayed significant downregulation of all proteins except the receptor of activated protein C kinase 1 (RACK-1) was upregulated. To elucidate the impact of *S. macrophylla* extract and fractions on UVB irradiated HaCaT cells, the functions of each protein were further studied and classified accordingly. An overview of the proteins affected by UVB and treatment is depicted in Figures 2 and 3 based on the data obtained. Figure 2 depicts the changes in protein expression occurring in various systems such as the redox system, DNA maintenance and repair, RNA transcription to protein processing, and the glycolysis process. On the other hand, Figure 3 covers proteins involved in cell growth, proliferation, and migration.

Table 1. A proteomic study of molecular changes within HaCaT cells after UVB irradiation and treatment with *S. macrophylla* extract and fractions.

No.	Name of Protein (Uniprot ID)	Comparison Ratio of Untreated Controls	Comparison Ratio of UVB Control vs. *S. macrophylla* Treatment + UVB Samples			
		Non-UVB: UVB	UVB: SMCE + UVB	UVB: SMHF + UVB	UVB: SMEAF + UVB	UVB: SMWF + UVB
1	Chaperonin containing TCP1 subunit 2 β isoform CRA_b (CCT-β) (V9HW96)	1.00:0.49 (↓)	N/S[1]	1.00:1.41 (↑)	N/S	N/S
2	T-complex protein 1 subunit delta (CCT-δ) (A8K3C3)	1.00:0.47 (↓)	N/S	1.00:1.71 (↑)	N/S	1.00:0.52 (↓)
3	T-complex protein 1 subunit gamma (CCT-γ) (B3KX11)	N/S	N/S	1.00:1.58 (↑)	N/S	N/S
4	T-complex protein 1 subunit eta (CCT-η) (Q99832)	N/S	N/S	1.00:1.92 (↑)	N/S	N/S
5	T-complex protein 1 subunit epsilon (CCT-ε) (P48643)	N/S	N/S	1.00:2.39 (↑)	N/S	N/S
6	Ribosomal protein L12 variant (Q59FI9)	1.00:0.34 (↓)	N/S	N/S	N/S	N/S
7	Ribosomal protein S8 (Q9BS10)	N/S	N/S	1.00:1.90 (↑)	N/S	N/S
8	40S ribosomal protein SA (C9J9K3)	N/S	N/S	1.00:1.87 (↑)	N/S	N/S
9	40S ribosomal protein S5 (M0QZN2)	1.00:0.32 (↓)	N/S	N/S	N/S	1.00:0.32 (↓)
10	40S ribosomal protein S7 (P62081)	1.00:0.45 (↓)	N/S	N/S	N/S	1.00:0.29 (↓)
11	40S ribosomal protein S10 (P46783)	1.00:0.33 (↓)	N/S	N/S	N/S	1.00:0.31 (↓)
12	40S ribosomal protein S16 (M0R210 or M0R3H0)	1.00:0.36 (↓)	N/S	N/S	N/S	1.00:0.51 (↓)
13	40S ribosomal protein S25 (P62851)	1.00:0.48 (↓)	N/S	N/S	N/S	1.00:0.50 (↓)
14	60S ribosomal protein L6 (Q8N5Z7)	1.00:0.31 (↓)	N/S	N/S	N/S	N/S
15	60S ribosomal protein L18 (F8VUA6)	1.00:0.51 (↓)	1.00:0.22 (↓)	1.00:0.68 (↓)	N/S	1.00:0.44 (↓)
16	60S ribosomal protein L22 (K7EJT5)	1.00:0.37 (↓)	N/S	1.00:1.71 (↑)	N/S	N/S

Table 1. Cont.

No.	Name of Protein (Uniprot ID)	Comparison Ratio of Untreated Controls	Comparison Ratio of UVB Control vs. *S. macrophylla* Treatment + UVB Samples			
		Non-UVB: UVB	UVB: SMCE + UVB	UVB: SMHF + UVB	UVB: SMEAF + UVB	UVB: SMWF + UVB
17	60S ribosomal protein L24 (C9JXB8)	1.00:0.48 (↓)	N/S	N/S	N/S	N/S
18	60S ribosomal protein L29 (A0A024R326)	1.00:0.50 (↓)	N/S	N/S	N/S	1.00:0.39 (↓)
19	60s acidic ribosomal protein P0 (Q53HW2)	N/S	N/S	N/S	N/S	1.00:0.40 (↓)
20	60s acidic ribosomal protein P2 (P05387)	N/S	N/S	1.00:1.39 (↑)	N/S	1.00:0.55 (↓)
21	Albumin (F6KPG5)	1.00:2.66 (↑)	N/S	N/S	N/S	1.00:0.41 (↓)
22	chloride intracellular channel (A0A1U9 × 8Y4)	N/S	N/S	N/S	N/S	1.00:0.63 (↓)
23	chloride intracellular channel 1 (CLIC1) (O00299)	1.00:0.56 (↓)	N/S	1.00:1.55 (↑)	N/S	N/S
24	Cofilin 1 (V9HWI5)	1.00:0.61 (↓)	N/S	1.00:1.38 (↑)	N/S	1.00:0.63 (↓)
25	Elongation factor 1-alpha (Q6IPN6 or Q53HM9)	1.00:0.36 (↓)	N/S	1.00:1.88 (↑)	N/S	N/S
26	Elongation factor 1-delta (A0A087X1X7 or E9PN91)	1.00:0.33 (↓)	N/S	1.00:1.51 (↑)	1.00:1.49 (↑)	1.00:0.66 (↓)
27	Eukaryotic translation elongation factor 2 (eEF2) (Epididymis secretory sperm binding protein) (A0A384N6H1)	1.00:0.55 (↓)	N/S	1.00:1.67 (↑)	N/S	1.00:0.41 (↓)
28	Glycine-tRNA ligase (P41250)	1.00:0.28 (↓)	N/S	1.00:7.22 (↑)	N/S	N/S
29	Heat shock protein (HSP)-10 kDa (Chaperonin 10) (Epididymis secretory sperm binding protein) (A0A384N6A4)	N/S	N/S	1.00:1.48 (↑)	N/S	1.00:0.53 (↓)
30	HSP-60 kDa mitochondrial (P10809)	N/S	N/S	N/S	N/S	1.00:0.52 (↓)

Table 1. *Cont.*

No.	Name of Protein (Uniprot ID)	Comparison Ratio of Untreated Controls	Comparison Ratio of UVB Control vs. *S. macrophylla* Treatment + UVB Samples				
		Non-UVB: UVB	UVB: SMCE + UVB	UVB: SMHF + UVB	UVB: SMEAF + UVB	UVB: SMWF + UVB
31	HSP-70 protein 1A variant (Q59EJ3)	N/S	N/S	N/S	N/S	1.00:0.54 (↓)
32	HSP-70 kDa protein 4 (Q59GF8)	N/S	N/S	1.00:1.36 (↑)	N/S	N/S
33	HSP-70 family protein 5 (Epididymis secretory sperm binding protein Li 89n) (78 kDa glucose-regulated protein) (V9HWB4)	N/S	N/S	N/S	N/S	1.00:0.54 (↓)
34	HSP-70 kDa protein 8, isoform CRA_a (Epididymis luminal protein 33) (V9HW22)	1.00:0.52 (↓)	N/S	1.00:1.50 (↑)	N/S	N/S
35	HSP-70 kDa protein 9 (75 kDa glucose-regulated protein) (B7Z4V2)	N/S	N/S	1.00:1.35 (↑)	N/S	N/S
36	HSP 90α (P07900)	1.00:0.60 (↓)	N/S	1.00:1.51 (↑)	N/S	1.00:0.51 (↓)
37	Gluthathione S-transferase (GST)-pi (Epididymis secretory protein Li 22) (V9HWE9)	1.00:0.63 (↓)	N/S	1.00:1.45 (↑)	N/S	1.00:0.58 (↓)
38	Poly (RC) binding protein 1 (Epididymis secretory protein Li 85) (Q53SS8)	1.00:0.32 (↓)	N/S	1.00:2.10 (↑)	N/S	1.00:0.56 (↓)
39	Histone H2A type 1-A (Q96QV6)	N/S	N/S	N/S	1.00:1.41 (↑)	N/S
40	Histone H2A type 2-B (Q8IUE6)	1.00:2.75 (↑)	N/S	1.00:1.66 (↑)	N/S	1.00:0.56 (↓)
41	Histone H2B (A8K9J7 or B4DR52)	1.00:2.86 (↑)	N/S	N/S	N/S	1.00:0.40 (↓)
42	Histone H1.2 (P16403)	N/S	N/S	1.00:1.69 (↑)	1.00:1.41 (↑)	1.00:0.62 (↓)
43	Histone H1.5 (P16401)	N/S	N/S	1.00:2.01 (↑)	N/S	N/S
44	Histone H4 (B2R4R0)	N/S	N/S	N/S	N/S	1.00:0.50 (↓)

Table 1. Cont.

No.	Name of Protein (Uniprot ID)	Comparison Ratio of Untreated Controls	Comparison Ratio of UVB Control vs. *S. macrophylla* Treatment + UVB Samples			
		Non-UVB: UVB	UVB: SMCE + UVB	UVB: SMHF + UVB	UVB: SMEAF + UVB	UVB: SMWF + UVB
45	Myosin light polypeptide 6 (F8W1R7)	1.00:0.43 (↓)	N/S	N/S	N/S	N/S
46	Myosin 9 (P35579)	1.00:0.48 (↓)	N/S	N/S	N/S	1.00:0.52 (↓)
47	Nucleophosmin (Nucleolar phosphoprotein B23 numatrin) isoform CRA_f (A0A0S2Z4G7)	1.00:0.71 (↓)	N/S	1.00:1.34 (↑)	N/S	1.00:0.43 (↓)
48	Peptidyl-prolyl cis-trans isomerase (V9HWF5)	1.00:0.56 (↓)	N/S	N/S	N/S	1.00:0.43 (↓)
49	PRDX-1 (A0A384NPQ2 or A0A0A0MSI0)	1.00:0.57 (↓)	N/S	1.00:1.67 (↑)	N/S	1.00:0.43 (↓)
50	PRDX-2 (B4DF70)	N/S	N/S	N/S	N/S	1.00:0.51 (↓)
51	PRDX-3 (Thioredoxin-dependent peroxide reductase mitochondrial) (P30048)	N/S	N/S	1.00:0.58 (↓)	1.00:0.73 (↓)	N/S
52	PRDX-6 (P30041)	1.00:0.58 (↓)	N/S	N/S	N/S	1.00:0.42 (↓)
53	Phosphoglycerate kinase 1 (P00558)	1.00:0.61 (↓)	N/S	1.00:1.58 (↑)	N/S	1.00:0.50 (↓)
54	Phosphoserine aminotransferase (Q9Y617)	1.00:0.56 (↓)	N/S	1.00:1.46 (↑)	N/S	1.00:0.39 (↓)
55	Profilin-1 (P07737)	1.00:0.63 (↓)	N/S	1.00:1.42 (↑)	N/S	1.00:0.56 (↓)
56	PDI (A0A024R8S5)	N/S	N/S	1.00:1.52 (↑)	N/S	1.00:0.63 (↓)
57	PDI-A3 (P30101)	N/S	N/S	1.00:1.66 (↑)	1.00:0.76 (↓)	N/S
58	PDI-A4 (P13667)	N/S	N/S	1.00:2.30 (↑)	N/S	1.00:0.60 (↓)
59	PDI-A6 (Endoplasmic reticulum protein 5) (Q15084)	1.00:1.74 (↑)	N/S	N/S	N/S	1.00:0.43 (↓)
60	Protein S100 (A0A590UJ49 or B2R5H0)	1.00:0.30 (↓)	N/S	1.00:1.47 (↑)	N/S	1.00:0.33 (↓)

Table 1. Cont.

No.	Name of Protein (Uniprot ID)	Comparison Ratio of Untreated Controls	Comparison Ratio of UVB Control vs. S. macrophylla Treatment + UVB Samples			
		Non-UVB: UVB	UVB: SMCE + UVB	UVB: SMHF + UVB	UVB: SMEAF + UVB	UVB: SMWF + UVB
61	Protein-arginine deiminase type-1 (Q9ULC6)	1.00:15.79 (↑)	N/S	N/S	N/S	1.00:0.21 (↓)
62	Pyruvate kinase PKM (P14618)	1.00:0.67 (↓)	N/S	1.00:1.33 (↑)	N/S	1.00:0.48 (↓)
63	RACK-1 (P63244 or D6RF23)	1.00:0.22 (↓)	N/S	N/S	N/S	1.00:9.32 (↑)
64	Signal recognition particle 14 kDa protein (SRP-14) (P37108)	1.00:0.33 (↓)	N/S	N/S	N/S	N/S
65	SYNCRIP protein (Q05CK9)	1.00:0.48 (↓)	N/S	N/S	N/S	N/S
66	Transketolase (B3KSI4)	1.00:0.52 (↓)	N/S	N/S	N/S	N/S
67	D-3-phosphoglycerate dehydrogenase (B3KSC3)	1.00:0.50 (↓)	N/S	N/S	N/S	N/S
68	Fumarate hydratase (Epididymis secretory sperm binding protein) (A0A0S2Z4C3)	N/S	1.00:2.02 (↑)	1.00:2.43 (↑)	N/S	N/S
69	Annexin A1 (Q5TZZ9)	N/S	N/S	1.00:1.46 (↑)	N/S	1.00:0.43 (↓)
70	Annexin A2 (A0A024R5Z7)	N/S	N/S	N/S	N/S	1.00:0.46 (↓)
71	Annexin A3 (P12429)	N/S	1.00:0.63 (↓)	N/S	1.00:0.52 (↓)	1.00:0.42 (↓)
72	Annexin A5 (P08758)	N/S	N/S	N/S	N/S	1.00:0.42 (↓)
73	Filamin A (Q60FE6)	N/S	N/S	1.00:1.54 (↑)	N/S	1.00:0.36 (↓)
74	Filamin B β (Actin binding protein 278) isoform CRA_a (A0A024R321)	N/S	1.00:0.44 (↓)	N/S	N/S	N/S
75	3-phosphoglycerate dehydrogenase (Q9UMY2 or Q9UMY3)	N/S	N/S	1.00:2.27 (↑)	N/S	1.00:0.26 (↓)

Table 1. Cont.

No.	Name of Protein (Uniprot ID)	Comparison Ratio of Untreated Controls	Comparison Ratio of UVB Control vs. S. macrophylla Treatment + UVB Samples				
		Non-UVB: UVB	UVB: SMCE + UVB	UVB: SMHF + UVB	UVB: SMEAF + UVB	UVB: SMWF + UVB
76	14-3-3 protein α/β (P31946)	N/S	N/S	N/S	N/S	1.00:0.60 (↓)
77	14-3-3 protein σ (P31947)	N/S	N/S	1.00:1.97 (↑)	N/S	1.00:0.47 (↓)
78	14-3-3 protein γ (P61981)	N/S	N/S	N/S	N/S	1.00:0.18 (↓)
79	14-3-3 protein ε (P62258)	N/S	N/S	N/S	N/S	1.00:0.38 (↓)
80	Ubiquitin-activating enzyme E1 (Testicular secretory protein Li 63 (A0A024R1A3)	N/S	N/S	1.00:1.84 (↑)	N/S	N/S
81	AHNAK (Desmoyokin) (Q09666)	N/S	N/S	1.00:0.55 (↓)	N/S	N/S
82	Keratin type I cytoskeletal 14 (P02533)	N/S	N/S	1.00:1.86 (↑)	1.00:1.88 (↑)	N/S
83	Keratin type II cytoskeletal 8 (P05787)	N/S	N/S	1.00:1.48 (↑)	N/S	1.00:0.52 (↓)
84	Heterogeneous nuclear ribonucleoprotein (hnRP) D0 (D6RF44) hnRP K	N/S	N/S	1.00:1.74 (↑)	N/S	N/S
85	Adenylyl cyclase-associated protein (CAP1) (B4DFF1 or B4DUQ1)	N/S	N/S	1.00:1.54 (↑)	N/S	1.00:0.51 (↓)
86	Reticulon-4 (B4DI38)	N/S	N/S	1.00:1.54 (↑)	N/S	1.00:0.38 (↓)
87	Malate dehydrogenase cytoplasmic (Q6IPN0 or Q9NQC3)	N/S	N/S	1.00:1.70 (↑)	N/S	1.00:0.64 (↓)
88	Transketolase (P40925)	N/S	N/S	1.00:1.70 (↑)	N/S	1.00:0.72 (↓)
89	Glucose-6-phosphate isomerase (A0A0B4J1R6)	N/S	N/S	1.00:1.62 (↑)	N/S	1.00:0.34 (↓)
90	(P06744)	N/S	N/S	1.00:1.61 (↑)	N/S	1.00:0.36 (↓)

Table 1. *Cont.*

No.	Name of Protein (Uniprot ID)	Comparison Ratio of Untreated Controls	Comparison Ratio of UVB Control vs. *S. macrophylla* Treatment + UVB Samples			
		Non-UVB: UVB	UVB: SMCE + UVB	UVB: SMHF + UVB	UVB: SMEAF + UVB	UVB: SMWF + UVB
91	Proteasome subunit alpha type (Q53GF5 or H0YLC2)	N/S	N/S	1.00:1.45 (↑)	N/S	1.00:0.32 (↓)
92	Proteasome activator complex subunit 2 (Q86SZ7)	N/S	N/S	1.00:1.63 (↑)	N/S	N/S
93	Proteasome subunit β type-2 (P49721)	N/S	N/S	N/S	N/S	1.00:0.42 (↓)
94	Proteasome subunit β type-3 (A0A087WXQ8)	N/S	N/S	1.00:1.59 (↑)	N/S	N/S
95	Glyceraldehyde-3-phosphate dehydrogenase (GAPDH) (P04406)	N/S	N/S	1.00:1.57 (↑)	N/S	1.00:0.47 (↓)
96	Proliferating cell nuclear antigen (PCNA) (P12004 or Q6FHF5)	N/S	N/S	1.00:1.55 (↑)	N/S	1.00:0.53 (↓)
97	Exportin-2 (P55060)	N/S	N/S	1.00:1.62 (↑)	1.00:1.33 (↑)	N/S
98	Serpin peptidase inhibitor clade B (Ovalbumin) member 5 isoform CRA_b (SERPINB5) (A0A024R2B6)	N/S	N/S	1.00:1.52 (↑)	N/S	1.00:0.51 (↓)
99	Collagen-binding protein (Serpin H1) (B4DN87)	N/S	N/S	1.00:1.80 (↑)	N/S	N/S
100	Ezrin (Q6NUR7)	N/S	N/S	1.00:1.51 (↑)	N/S	1.00:0.19 (↓)
101	Tubulin beta chain (A0A384NYT8)	N/S	N/S	1.00:1.47 (↑)	N/S	1.00:0.49 (↓)
102	Thioredoxin (P10599)	N/S	N/S	1.00:1.40 (↑)	N/S	1.00:0.54 (↓)
103	Thioredoxin domain-containing protein 17 (TXNDC17) (Testicular tissue protein Li 214) (A0A140VJY7)	N/S	N/S	1.00:1.47 (↑)	N/S	N/S
104	Alpha-enolase (P06733)	N/S	N/S	1.00:1.46 (↑)	N/S	1.00:0.46 (↓)
105	Triosephosphate isomerase (V9HWK1)	N/S	N/S	1.00:1.45 (↑)	N/S	1.00:0.44 (↓)

Table 1. Cont.

No.	Name of Protein (Uniprot ID)	Comparison Ratio of Untreated Controls	Comparison Ratio of UVB Control vs. S. macrophylla Treatment + UVB Samples				
		Non-UVB: UVB	UVB: SMCE + UVB	UVB: SMHF + UVB	UVB: SMEAF + UVB	UVB: SMWF + UVB
106	Alpha-actinin-1 (P12814)	N/S	N/S	1.00:1.44 (↑)	N/S	N/S
107	Calreticulin variant (Q53G71)	N/S	N/S	1.00:1.44 (↑)	N/S	1.00:0.42 (↓)
108	L-lactate dehydrogenase A (V9HWB9)	N/S	N/S	1.00:1.38 (↑)	N/S	1.00:0.35 (↓)
109	L-lactate dehydrogenase B (Q5U077)	N/S	N/S	1.00:1.39 (↑)	N/S	1.00:0.42 (↓)
110	ATP synthase subunit α, mitochondrial (V9HW26)	N/S	N/S	N/S	N/S	1.00:0.44 (↓)
111	ATP synthase subunit β, mitochondrial (P06576)	N/S	N/S	1.00:1.37 (↑)	N/S	1.00:0.59 (↓)
112	Endoplasmin (P14625)	N/S	N/S	1.00:1.36 (↑)	N/S	1.00:0.57 (↓)
113	Nucleosome assembly protein 1-like 1 (F8W020)	N/S	N/S	1.00:1.35 (↑)	N/S	N/S
114	Neutral amino acid transporter B(0) (Q15758)	N/S	N/S	1.00:1.38 (↑)	N/S	1.00:0.48 (↓)
115	LIM and SH3 domain protein 1 (LASP-1) (A8K1D2)	N/S	N/S	1.00:1.36 (↑)	N/S	N/S
116	Polyubiquitin-C (F5GYU3)	N/S	N/S	N/S	1.00:0.56 (↓)	1.00:0.33 (↓)
117	HNRPCL1 protein (Q6PKD2)	N/S	N/S	N/S	1.00:0.46 (↓)	N/S
118	Fascin (B3KTA3)	N/S	N/S	N/S	1.00:0.64 (↓)	1.00:0.36 (↓)
119	Cathepsin D (A0A1B0GW44)	N/S	N/S	N/S	1.00:0.70 (↓)	1.00:0.49 (↓)
120	Prothymosin alpha (Q15203)	N/S	N/S	N/S	1.00:0.20 (↓)	1.00:0.27 (↓)
121	GTP-binding nuclear protein Ran (P62826)	N/S	N/S	N/S	1.00:0.72 (↓)	1.00:0.41 (↓)
122	NOLC1 (B2RAU8)	N/S	N/S	N/S	1.00:1.39 (↑)	N/S

Table 1. Cont.

No.	Name of Protein (Uniprot ID)	Comparison Ratio of Untreated Controls	Comparison Ratio of UVB Control vs. S. macrophylla Treatment + UVB Samples				
		Non-UVB: UVB	UVB: SMCE + UVB	UVB: SMHF + UVB	UVB: SMEAF + UVB	UVB: SMWF + UVB
123	Protein kinase C substrate 80K-H isoform 1 (A0A0S2Z4D8)	N/S	N/S	N/S	1.00:1.35 (↑)	N/S
124	Actin cytoplasmic 1 (P60709)	N/S	N/S	N/S	N/S	1.00:0.25 (↓)
125	Actin cytoplasmic 2 (P63261)	N/S	N/S	N/S	N/S	1.00:0.47 (↓)
126	Rab GDP dissociation inhibitor β (P50395)	N/S	N/S	N/S	N/S	1.00:0.34 (↓)
127	Eukaryotic translation initiation factor 5A (I3L397)	N/S	N/S	N/S	N/S	1.00:0.39 (↓)
128	Phosphoglycerate mutase (Q6FHU2)	N/S	N/S	N/S	N/S	1.00:0.43 (↓)
129	Neutral alpha-glucosidase AB (Epididymis secretory sperm binding protein Li 164nA) (V9HWJ0)	N/S	N/S	N/S	N/S	1.00:0.33 (↓)
130	Adenosylhomocysteinase (P23526)	N/S	N/S	N/S	N/S	1.00:0.44 (↓)
131	4F2 cell-surface antigen heavy chain (F5GZS6)	N/S	N/S	N/S	N/S	1.00:0.44 (↓)
132	Protein SET (A0A0C4DFV9)	N/S	N/S	N/S	N/S	1.00:0.45 (↓)
133	Inorganic pyrophosphatase (Epididymis secretory sperm binding protein Li 66p) (V9HWB5)	N/S	N/S	N/S	N/S	1.00:0.42 (↓)
134	Aspartate aminotransferase (B3KUZ8)	N/S	N/S	N/S	N/S	1.00:0.47 (↓)
135	Plasminogen activator inhibitor 2 (P05120)	N/S	N/S	N/S	N/S	1.00:0.52 (↓)
136	Acidic leucine-rich nuclear phosphoprotein 32 family member A (ANP32B) (H0YN26)	N/S	N/S	N/S	N/S	1.00:0.52 (↓)
137	Complement component 1 Q subcomponent-binding protein mitochondrial (Q07021)	N/S	N/S	N/S	N/S	1.00:0.52 (↓)

Table 1. Cont.

No.	Name of Protein (Uniprot ID)	Comparison Ratio of Untreated Controls	Comparison Ratio of UVB Control vs. S. macrophylla Treatment + UVB Samples				
		Non-UVB: UVB	UVB: SMCE + UVB	UVB: SMHF + UVB	UVB: SMEAF + UVB	UVB: SMWF + UVB
138	Fructose-bisphosphate aldolase A (P04075)	N/S	N/S	N/S	N/S	1.00:0.53 (↓)
139	Transaldolase (A0A140VK56)	N/S	N/S	N/S	N/S	1.00:0.53 (↓)
140	Macrophage migration inhibitory factor (P14174)	N/S	N/S	N/S	N/S	1.00:0.55 (↓)
141	Endoplasmic reticulum resident protein 29 (P30040)	N/S	N/S	N/S	N/S	1.00:0.49 (↓)
142	Calmodulin-2 (P0DP24)	N/S	N/S	N/S	N/S	1.00:0.59 (↓)
143	RAN binding protein 1 isoform CRA_g (A0A140VK94)	N/S	N/S	N/S	N/S	1.00:0.59 (↓)
144	Vinculin isoform CRA_c (A0A024QZN4)	N/S	N/S	N/S	N/S	1.00:0.32 (↓)
145	Small ubiquitin-related modifier (A0A024R853)	N/S	N/S	N/S	N/S	1.00:0.37 (↓)
146	Calnexin (P27824)	N/S	N/S	N/S	N/S	1.00:0.61 (↓)
147	Nucleolin (P19338)	N/S	N/S	N/S	N/S	1.00:0.49 (↓)
148	RPLP1 protein (Q6ICQ4)	N/S	N/S	N/S	N/S	1.00:0.64 (↓)
149	FK506 binding protein 12 (Q1JUQ3)	N/S	N/S	N/S	N/S	1.00:0.12 (↓)
150	Lysosome-associated membrane glycoprotein 1 (B3KRY3)	N/S	N/S	N/S	N/S	1.00:0.26 (↓)
151	Transgelin-2 (P37802)	N/S	N/S	N/S	N/S	1.00:0.52 (↓)

[1] N/S represents data that was not significant, while the upregulation and downregulation of the protein are represented as (↑) and (↓), respectively, based on the ratio that was obtained. The data were compared individually with one another using PEAKS Q. N number is 3 for all controls and samples, whereas, the p-value was set at 0.05 significance level.

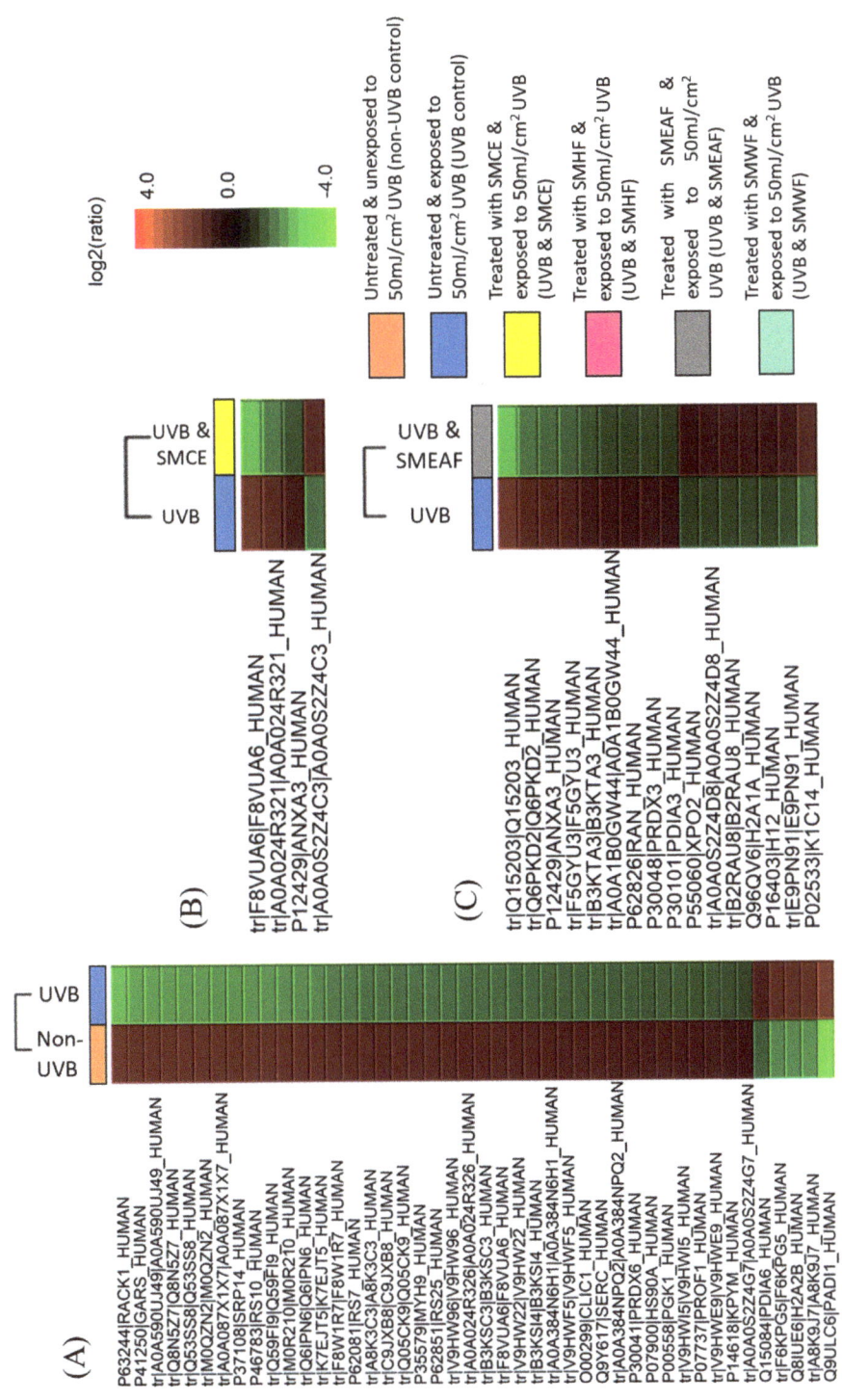

Figure 1. Cont.

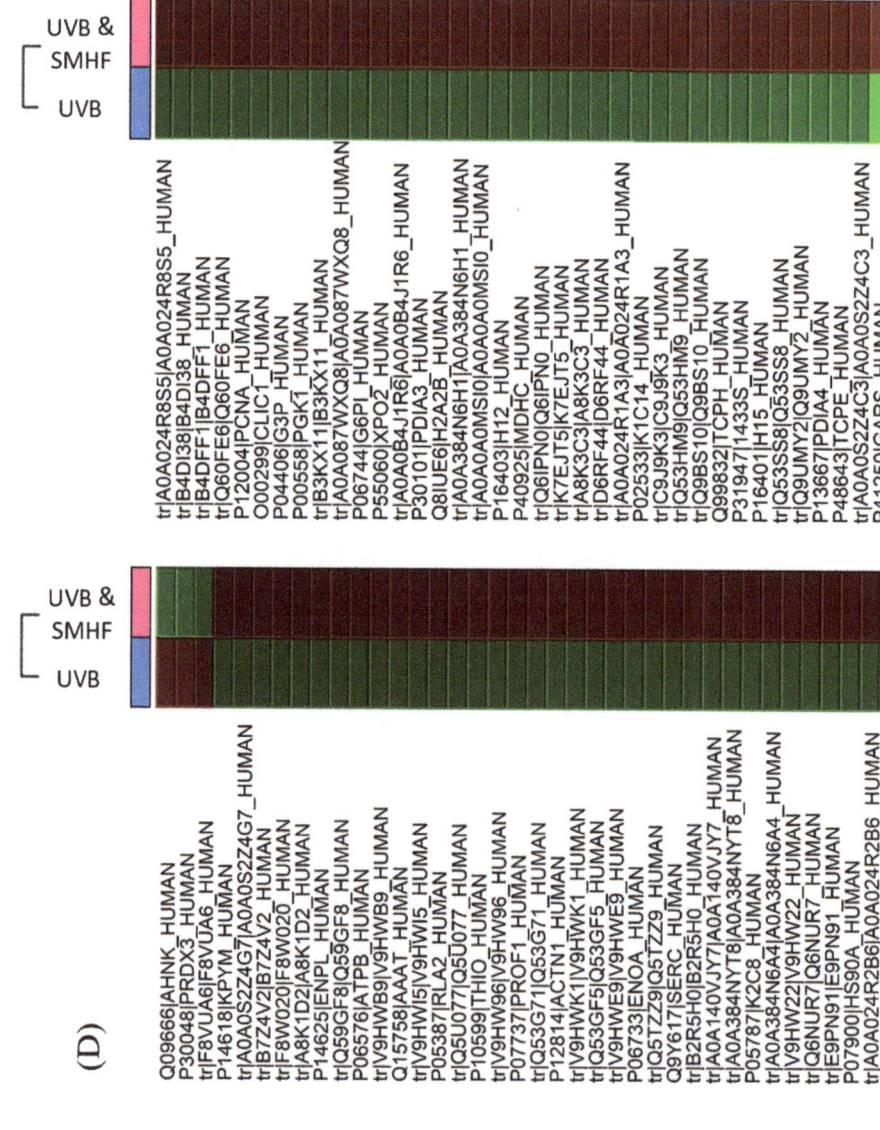

Figure 1. Cont.

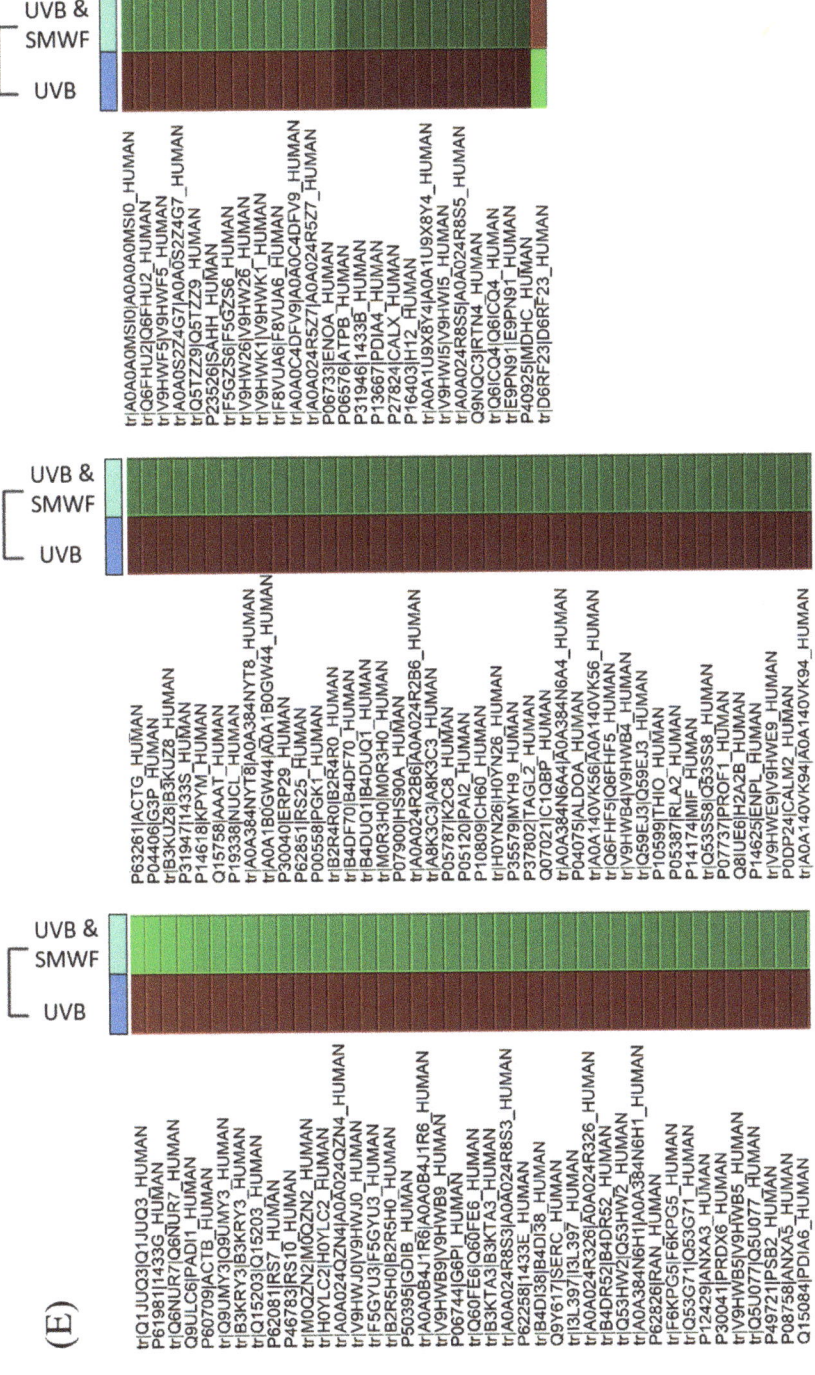

Figure 1. Heat maps obtained from Peaks Q displaying the changes in protein expression across controls and samples. The red color signifies the upregulation of protein expression while the green color signifies downregulation of protein expression. (**A**) The comparison of non-UVB and UVB control cells. (**B**–**E**) are the comparison of UVB control cells with cells that were simultaneously exposed to UVB while being treated with SMCE, SMEAF, SMHF, and SMWF, respectively.

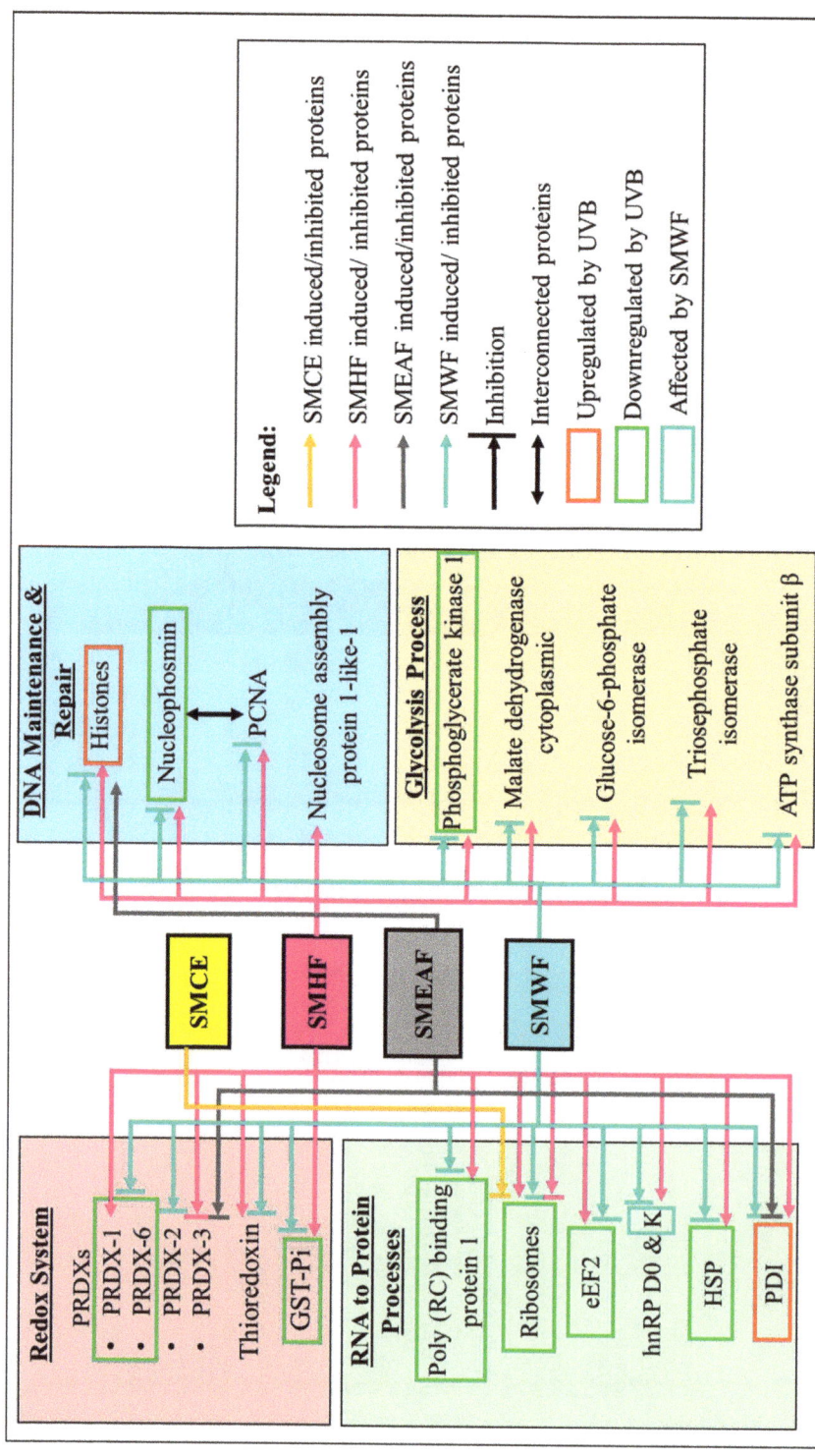

Figure 2. Dysregulation of proteins involved in the redox system, DNA maintenance and repair, RNA transcription to protein synthesis, and glycolysis process by UVB irradiation and the effect SMCE, SMHF, SMEAF, and SMWF has on the UVB irradiated cells.

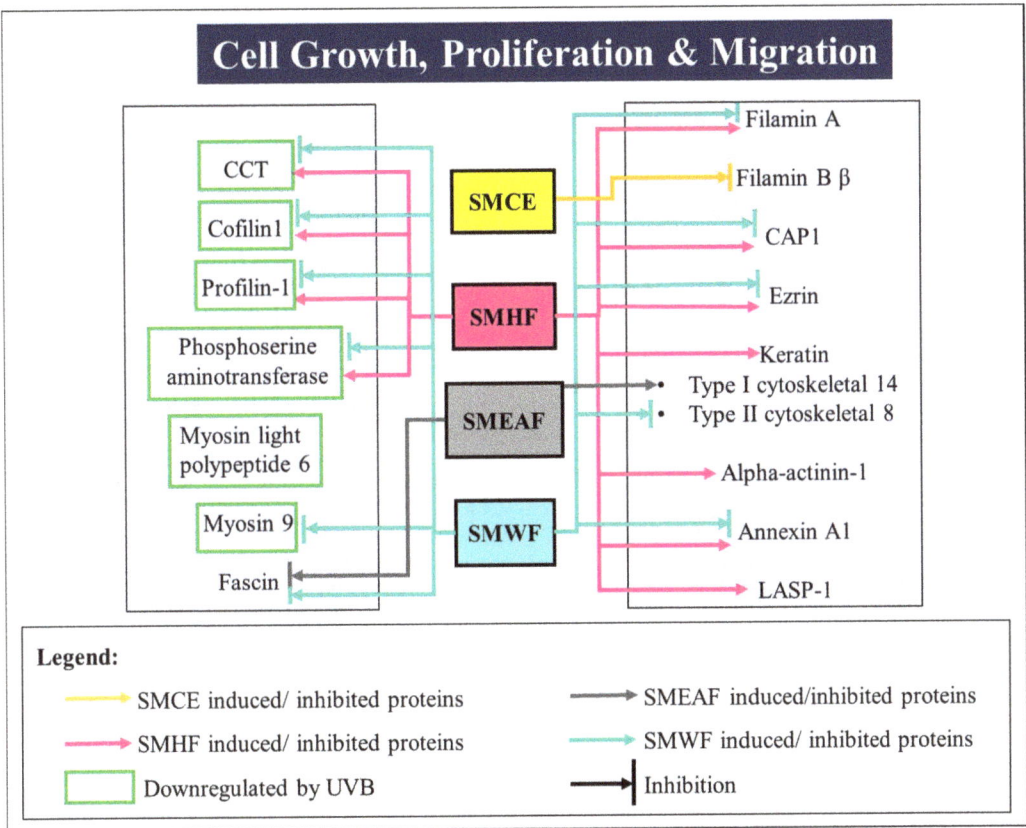

Figure 3. UVB-induced changes in the expression of proteins involved in cell growth, proliferation, and migration processes. These proteins then were either rescued or further suppressed after being treated with SMCE, SMHF, SMEAF, and SMWF.

2.2.2. Oxidative Damage Induced Activation of the Redox System

As UVB irradiation induces oxidative damage, it is expected that there would be changes in the molecular dynamics within the keratinocyte cells and one of them is the redox regulating PRDXs. PRDXs are a family of antioxidant enzymes with an aptitude for reducing alkyl hydroperoxide and hydrogen peroxide (H_2O_2) to alcohol and water, respectively [30]. In mammals, six distinct PRDXs have been discovered and can be found mostly in the cytosol, but some can also be found in the nucleus, mitochondria, lysosome, endoplasmic reticulum, or be secreted from the cell [31,32]. The secretion of PRDX like PRDX-2 triggers the production and secretion of TNF-α in macrophages, inducing a cascade of inflammatory response to the stimuli [33]. PRDXs also work together with thioredoxin and thioredoxin reductase in its redox reaction by obtaining electrons from thioredoxin/thioredoxin reductase systems [31]. It is through this relationship that H_2O_2 cell signaling is mediated and controlled, even in the event of PRDXs hyperoxidation [34,35]. The changes in PRDXs by UVB was previously reported in various studies on skin cells. In a study done by Liu et al. [36], PRDX 1 was upregulated after exposing HaCaT cells to 21 ± 1 mJ/cm^2 UVB for 18 h and then incubated for another 6 h before protein extraction. Another study by Wu et al. [37] on skin fibroblasts displayed that at low (17 mJ/cm^2) and middle (70 mJ/cm^2) doses of UVB irradiation, the human fibroblast cells exhibited an increase in PRDX-1, but for PRDX-4 and -6, this increase was only seen in the middle dose of UVB irradiation at the 24 h time point. At low doses of UVB, the PRDX-4 and -6 were negatively regulated as are all PRDXs at high (468 mJ/cm^2) dose of UVB. Although different results on PRDX expression levels were obtained due to different exposure levels, protein harvest times, and cell type, it is undeniable that UVB does affect the expression of PRDX. Knockdown of PRDX-3 and mutations in PRDX-6 in mice even showed increased keratinocyte apoptosis and tissue damage, respectively, due to increased susceptibility in oxidative damage [30,38], thus, indicating the importance of PRDX in cell survival against UVB damage.

In response to UVB irradiation, PRDX-1 and -6 were significantly downregulated but there were no significant changes in the expression of PRDX-2 and-3 as can be seen in Table 1 and Figure 2. No changes in any PRDX levels in comparison to the UVB control cells were observed after treating with SMCE. Despite that, its fraction, SMHF, initiated an upregulation in PRDX-1 and, subsequently, downregulated PRDX-3. No changes in PRDX-6 levels were detected, though, when the cells were treated with SMHF. Furthermore, SMHF also upregulated the expression of thioredoxin, suggesting that SMHF might have possibly activated the thioredoxin/thioredoxin reductase system along with PRDX-1. On another note, SMEAF also induced similar downregulation of PRDX-3 in the HaCaT cells as implicated by SMHF, but incited no changes in other PRDXs. As for SMWF, downregulation of all PRDX-1, -2, and, -6 were recorded.

The increase in PRDX-1 and thioredoxin in cells treated with SMHF suggests that the compounds in SMHF are not only able to counteract against UVB induced H_2O_2, but might also indirectly inhibit the activation of the apoptosis signal-regulating kinase-1 (ASK-1)-mediated apoptotic pathway. According to Kim et al. [39], PRDX-1 is able to negatively regulate ASK-1, a mitogen activated protein (MAP) kinase kinase, activity and it is through the inhibition of ASK-1 activation that PRDX-1 is able to inhibit MAP kinase kinase (MKK)3/6, Jun N-terminal kinase (JNK), and p38 phosphorylation, ultimately attenuating cell apoptosis. Another interesting point to note is that the expression of GST-pi, which was previously downregulated in UVB control cells compared to the non-UVB control cells, were now upregulated under the treatment of SMHF. This change in expression of GST-pi is intriguing because GST-pi is a known inhibitor of JNK. Under normal cell conditions, monomeric GST-pi forms a GST-pi-JNK complex with c-Jun, actively inhibiting the phosphorylation of JNK. However, under oxidative stress, monomeric GST-pi becomes detached from the complex and forms dimerization or multimerization of GST-pi. This enables JNK and c-Jun to be activated. Nevertheless, it is believed that under conditions where newly synthesized GST-pi is formed, the GST-pi-JNK complex will once again reform, effectively inhibiting the activation of JNK [40]. In short, with the increase in expression of PRDX-1, thioredoxin, and GST-pi protein, it is possible that SMHF might be able to improve cell survival in UVB irradiated keratinocytes. Nevertheless, this hypothesis needs to be investigated further as SMHF also suppresses PRDX-3, which also plays an important role as an antioxidant enzyme against cell apoptosis. As for SMCE, SMEAF, and SMHF, it can be seen that neither of these extracts and fractions can inhibit cell apoptosis via PRDXs.

2.2.3. Impact of UVB on DNA Maintenance and Repair in Keratinocyte Cells

Histones are indispensable in the regulation of all nuclear processes such as DNA replication, the progression of cell- cycle, transcription, etc. Assembled with two copies of histone H2A, H2B, H3, and H4 histone each, these histones form the nucleosome core particle, which wraps around 146 bp of DNA. To further stabilize the octameric core, H1 linker histone comes into play to form chromatin-specific high-order structures. Other additional histone variants and accessory proteins are also distributed along the chromatin fiber [41]. Due to the importance of its role in epigenetic regulation, any changes in histone modification or expression level will impact gene expression and determine cell fate. Based on the analysis, this study showed that the exposure of HaCaT cells to UVB upregulated the expression of histone H2A type 2-B and H2B (Table 1). This upregulation of histones is not uncommon under the phototoxicity of UVB. Both Sesto et al. [42] and Dazard et al. [43] had previously reported that UVB irradiation causes the expression of histones to be upregulated at the 24th hour in human keratinocytes after exposure. Dazard, Gal, Amariglio, Rechavi, Domany, and Givol [43] even suggested that the upregulation of histones could aid in DNA repair, but as of yet, the process has not been fully confirmed.

When the cells were treated with SMCE, there were no significant changes in expression level. However, after fractionation, SMHF not only further increased the expression of histone H2A type 1-B, but also H1.2 and H1.5. SMEAF also upregulated histone H1.2 and H2A type 1-A. In contrast, SMWF was the only fraction that downregulated histone H2A type 2-B, H2B, H1.2, and H4. Although, from this data it is possible that SMHF and SMEAF might be implicated in processes involving DNA repair and maintenance, however, the increased expression of H1.2 by both extracts is concerning. This is because, when histone H1.2 translocates from the nucleus to the cytosol, it has been reported to become pro-apoptotic and is able to induce cell death by interacting with the cytochrome C and proapoptotic Bcl-2 family in x-ray irradiated rat cells [44]. This is further supported by Ruiz-Vela and Korsmeyer [45], who showed that histone H1.2 promotes the activation of caspase 3 and 7 via apoptotic protease activating factor 1 (APAF-1) and caspase-9 in the UV induced apoptosis process. Hence, further studies must be done to better elucidate the effect SMHF and SMEAF have on histones.

On the other hand, nucleophosmin, a nucleolar phosphoprotein, plays an essential role in cell proliferation, ribosome biogenesis, and cell survival in DNA-damaged cells [46]. Some of the existing findings on the effect of UV on nucleophosmin showed that UV exposure upregulates nucleophosmin dose-dependently and the increase in nucleophosmin reduces cell death [47,48]. Similar observations on nucleophosmin upregulation were also seen when fibroblast cells were exposed separately to either UVA or UVB [37,49]. Subsequently, overexpression of nucleophosmin suppressed both p53 and p21 expression, although the limit of suppression on p21 stops at a certain dose of UV radiation. Higher doses of UV are unable to inhibit p21 expression, signifying that nucleophosmin may function as an early sensor mechanism against genotoxic stress [47]. The lack of nucleophosmin also arrested the cell cycle in the G2 phase following UV exposure, resulting in a delay of cell mitosis and proliferation [47]. Another study showed that nucleophosmin is able to increase PCNA expression by regulating the PCNA promoter, thus mediating DNA repair via the nucleotide excision repair pathway [48]. However, in this study, nucleophosmin was downregulated instead under UVB phototoxicity as can be seen in Figure 2. This suggests that the process of DNA repair had been inhibited by UVB at 24 h after exposure, thus encouraging the cells toward cell death. Although SMCE does not affect the expression levels of nucleophosmin, SMHF was able to inverse the effect UVB had on nucleophosmin. Furthermore, an increase in PCNA by SMHF was also observed, further emphasizing the potential of SMHF in initiating DNA repair in UV-damaged cells. Additionally, although these proteins had no changes between the UVB and non-UVB control, upregulation of these two proteins by SMHF might indirectly affect the DNA repair mechanism and even cell survival in keratinocyte cells. One of the proteins that were upregulated is the nucleosome assembly protein 1-like 1. This protein is known for its role in histones H2A and H2B transportation into the nucleus and also its ability to remove and replace H2A-H2B or other histone variant dimers, facilitating nucleosome sliding along the DNA in favor of thermodynamically better positions [50,51]. Another is the protein, GAPDH. Traditionally, this protein is commonly used as a housekeeping gene, but this protein is in fact involved in many different cellular processes such as DNA repair, membrane fusion and transport, tRNA export, cytoskeletal dynamics, and even cell death [52–57]. Therefore, more studies on SMHF are warranted. On the flip side, SMWF lowered the expression of nucleophosmin, PCNA, and GAPDH, implicating its inhibitory effect on DNA maintenance and repair in UVB damaged cells. Subsequently, SMEAF itself did not affect any of these proteins.

2.2.4. Modifications of Downstream Process from RNA to Protein

In addition to DNA damage, poly (RC) binding protein 1, which is responsible for RNA transcription, splicing, and translation were also downregulated by UVB exposure, affecting downstream processes of protein synthesis. Other proteins involved in protein synthesis are also dysregulated by UVB as depicted in Figure 2. These proteins are ribosomes and eEF2, a protein that acts as a catalyst to the ribosomes in shifting the mRNA-tRNA complex within ribosomes from the 5' end to 3' end in a process called translocation [58]. Apart from being involved in protein synthesis, ribosomes like ribosomal protein L12 are also involved in the translocation of newly synthesized ribosome protein from the cytoplasm to the nucleus with importin 11 [59]. Under the exposure of 100 mJ/cm^2 UVB, the ribosomal protein gene expression in primary human keratinocyte cells was reported to be significantly increased 6 h after exposure [60]. However, this increase in expression was significantly reduced at 24 h, along with the expression of eEF2. Following the treatment of cells with *S. macrophylla* extract and fractions, SMCE was seen to further reduce the expression of 60S ribosomal protein L18, but did not change the expression level of poly (RC) binding protein 1 or eEF2. Nonetheless, SMHF significantly elevated poly (RC) binding protein 1, several ribosomal proteins, and eEF2 levels. SMHF also increased the expression of hnRP D0 and K, in which both are of the same hnRP family as poly (RC) binding protein 1 and are also involved in RNA regulation either through DNA or RNA binding [61–63]. Again, SMEAF induced no changes in the expression of the proteins mentioned, while a completely opposite effect was obtained from cells treated with SMWF. The expression of various ribosomes, poly (RC) binding protein 1, eEF2, and hnRP K was diminished under the treatment of SMWF. Hence, it can be suggested that only SMHF might be able to augment the protein synthesis pathway after UVB.

On another note, under environmental stresses such as UVB exposure, the overexpression of ROS depletes Ca^{2+} ions from the endoplasmic reticulum (ER) lumen. Subsequently, this leads to the malfunction of ER chaperones and other proteins, which then leads to the accumulation of unfolded or misfolded proteins, promoting ER stress [64]. In the event this occurs, HSPs are expressed. HSPs are proteins that are expressed in response to cellular stress, for example, oxidative damage, chemical stress, hyperthermia, etc., and are commonly known for their maintenance of cellular proteostasis [65,66]. However, their location and physiological role vary according to their classification, which can be separated based on their molecular size that ranges from 10 to more than 100 kDa [67]. In addition, the expression of HSP has been correlated to cell survival [68]. As reported by Merwald et al. [69], increased expression of HSP-72 through prior induction demonstrated increased keratinocyte cell survival under UVB irradiation in comparison to the control. This suggests that the expression of HSP is key to stress tolerance and the survival of keratinocytes against phototoxicity.

The expression level of HSPs has been noted to change according to various time points after UVB exposure. In a study where the cellular protein was collected at the 5 h time point, HSP-60 and -70 kDa was increased in normal human epithelial keratinocyte cells [70]. However, at the 12th hour, Howell et al. [71] reported a decrease in the mRNA of heat shock cognate 71 kDa protein. Consecutively, in this study, the expression of two HSPs, HSP-70 kDa protein 8 and HSP-90α was significantly downregulated compared to the non-UVB control 24 h after exposure to 50 mJ/cm^2 UVB. It is possible that after the initial exposure of UVB, the expression of HSP is triggered and thus increases rapidly, but then the expression will slowly decrease with time as cell death occurs. Nevertheless, with the treatment of SMHF, not only was the expression of the two HSPs upregulated, but three other HSP, HSP-10, HSP-70 kDa protein 4, and HSP-70 kDa protein 9, were also upregulated. In contrast, SMWF was seen to significantly further downregulate the expression of various HSPs while no changes in HSP expression were seen in both SMCE and SMEAF.

The upregulation of HSP-10, HSP-70, and HSP-90 kDa by SMHF as seen in Table 1 is a good sign of cell survival as their functions are important in cell maintenance. HSP-

10 kDa, in conjuncture with HSP-60, was reported to inhibit cell apoptosis by modulating the Bcl-2 family and thereby the intrinsic apoptotic pathway in cardiac muscle cells [72]. On the other hand, HSP-70 kDa is involved in the folding, assembly, and refolding of aggregated and misfolded proteins. They also control the membrane translocation of proteins and regulate the activity of regulatory proteins [73]. Finally, HSP-90 kDa acts as an actin-binding protein, modulating cell migration, which is essential for wound healing [74,75]. Topical application of HSP90α infused in 10% carboxymethylcellulose cream on a 1×1 cm wound located on the back of nude mice for five days significantly accelerated wound closure. Further analysis of the mice skin revealed that the HSP90α cream instigated significant re-epithelialization and formed thicker and longer epidermis compared to the control [76]. Hence, the increase of these HSPs by SMHF is a good indicator of its potential photoprotective and wound healing abilities.

Besides HSPs, the presence of PDI under oxidative stress is also essential to cell survival. PDI plays an important role in adding disulfide bonds into proteins through the oxidase activity and via the isomerase activity, which aids in the rearrangement of incorrect disulfide bonds [77]. This means that through PDI, there are fewer misfolded proteins as disulfide bonds are vital in maintaining the structure, regulation, and function of proteins [78]. Although PDIs are anti-apoptotic biomarkers for tumor growth, metastasis, and angiogenesis, it may yet be an essential protein to look into and upregulate to promote cell survival and proliferation in skin cells after UVB irradiation [79,80]. In human fibroblast cells, both PDI and PDI-A3 were significantly upregulated 24 h after being exposed to 70 mJ/cm^2 UVB dosage [37]. Another study on the 3D cell culture of CDD 1102 KERTr human keratinocyte cell line also displayed a significant increase in PDI protein expression 24 h after exposure to 60 mJ/cm^2 UVB [81]. Similarly, in this study, PDI-A6 protein levels were significantly increased after UVB exposure while no changes in the other PDIs were seen (Table 1 and Figure 2). Treatment with SMCE induced no difference in the expression of PDI. Nevertheless, with the treatment of SMHF, not only were the increased levels of PDI-A6 maintained, but the PDI-A3 and PDI-A4 levels were also significantly elevated when compared to the UVB control. The increase in PDIs after UVB irradiation by SMHF is a good indication of the treatment in combat against UVB phototoxicity. On the other hand, SMEAF suppressed the expression of PDI-A3 while SMWF downregulated PDI-A4 and PDI-A6.

2.2.5. UVB Exposure Affects Cell Growth, Proliferation, and Migration

The dysregulation of actin cytoskeleton dynamics via UVB exposure could bring about the disruption of cell growth, division, migration, and proliferation necessary for wound healing, leading to cell death. In this experiment, proteins such as chaperonin-containing t-complex polypeptide 1 (CCT) -β and δ, cofilin 1, profilin 1, phosphoserine aminotransferase, myosin light polypeptide 6, and myosin 9 experienced significant downregulation 24 h after the exposure of UVB on HaCaT. These proteins are, in one way or another, connected to the regulation of actin rearrangement and cytoskeleton organization. For example, CCT, a cytosolic molecular chaperone, is known to assist in the folding of tubulin, actin, and many other cytosolic proteins [82]. It also acts as a modulator in the process of assembling the cell cytoskeleton [83]. Cofilin1 and profilin-1, on the other hand, work hand in hand with CAP1 and various other actin-associated proteins in actin filamin turn over [84,85]. Aberrant expressions of these proteins will result in the alteration of cell–cell adhesion, cell proliferation, and motility [85–87]. Next, phosphoserine aminotransferase, an enzyme involved in serine biosynthesis, has also been implicated in an increase in the proliferation of cells [88]. It was reported that when phosphoserine aminotransferase was suppressed, alterations in cell morphology and F-actin cytoskeletal arrangement was seen. Furthermore, the suppression also inhibited the migration and motility of triple negative breast cancer [89]. Finally, myosins are actin-based molecular motors that function to bind actin filaments together [90]. When the skin is wounded, myosin II, made partly of myosin

9, generates contractile forces necessary for cell motility and migration into the wounded area [91].

When the cells were treated with *S. macrophylla* extract and fractions, SMCE can be seen to decrease the expression of filamin B β. Filamins are actin-binding proteins that act as crosslinks to the actin cytoskeleton filaments to form a dynamic structure. They also aid in anchoring the structure to plasma adhesion receptors present on the membrane [92]. Hence, downregulation of filamin B β can possibly bring an impairment to cell structure and growth. Nevertheless, after fractionation, a big change in protein expression was seen for each fraction. This is especially true for SMHF as SMHF not only elevated CCT-β and -δ levels, but other CCT subunits like CCT- γ, ε, and η were also increased. SMHF also increased both the expression levels of cofilin 1, profilin 1, and phosphoserine aminotransferase. This could be an indication that SMHF might be able to counteract UVB damage by activating proteins needed in the wound healing process, despite not increasing myosin light polypeptide 6 and 9 expression levels. Further analysis on SMHF also showed that the fraction had additionally increased CAP1, filamin A, ezrin, keratin type I cytoskeletal 14, keratin II cytoskeletal 8, alpha-actinin 1, annexin A1, and LASP-1. These proteins that were elevated in expression have been ascribed as regulators of the actin cytoskeleton dynamics and some might potentially aid in the positive growth and proliferation of cells. In the case of filamin A, the lack of it was stated to cause a defect in the formation of cell junctions in filamin A null embryonic mice, while alpha-actinin 1 initiates cell motility through the regulation of focal adhesions, β4 integrin localization, and actin cytoskeleton organization [93,94]. Furthermore, the exposure of cell growth factors had shown stimulation of LASP 1 re-localization from the cell periphery to focal adhesion during cell migration, indicating its importance in cell migration [95]. On the other hand, even though annexin A1 is involved in many cellular processes, it was also reported to induce migration in fibroblast cells [96]. Finally, type I and II keratins are intermediate filaments and through post-translational modifications, they regulate cytoskeletal reorganization, while ezrin acts as linkers between the actin cytoskeleton and plasma membrane [97,98]. The other extract and fractions, besides SMEAF increasing the expression of keratin type I cytoskeletal 8, showed no upregulation in any of the proteins mentioned, suggesting that either they have very little or negative involvement in cell growth, proliferation, and migration after UVB irradiation. Although SMEAF increased the expression of keratin type I cytoskeletal 8, it also downregulated the expression of fascin, which is an actin bundling protein that has been reported to be involved in cell motility and invasion in cancer cell lines [99]. Therefore, due to the conflicting contradictory data, more studies are needed to truly understand the mechanism behind the molecular changes caused by SMEAF. SMWF, on the other hand, continues to downregulate many proteins involved in cell growth, proliferation, and migration, as can be seen in Figure 3, potentially suppressing the epidermis's ability to heal and repair itself after UVB irradiation.

2.2.6. SMHF Upregulates Proteins Involved in Cell Glycolysis

To maintain homeostasis and perform cellular maintenance such as DNA repair, protein turnover, transcription, and translation, etc., a considerable amount of energy is required [100]. Hence, changes in the glycolysis process are also important to take note as it could determine cell survival. In UVB control cells, phosphoglycerate kinase 1 was decreased, but this effect was attenuated by SMHF. In addition, SMHF also elevated the protein levels of malate dehydrogenase cytoplasmic, glucose-6-phosphate isomerase, triosephosphate isomerase, and ATP synthase subunit β. All these proteins listed are proteins that are involved in the glycolysis process with ATP synthase as the producer of ATP [101–106]. The upregulation of these proteins coincides with the data obtained for SMHF, whereby SMHF treated cells showed an increase in proteins involved in DNA repair and maintenance, RNA transcription and translation, protein processing after UVB irradiation. This further supports the possibility of SMHF attenuating the damage induced by UVB. Once again, no significant changes in these proteins were seen in SMCE and

SMEAF. However, SMWF displayed downregulation in phosphoglycerate kinase 1, malate dehydrogenase cytoplasmic, glucose-6-phosphate isomerase, triosephosphate isomerase, and ATP synthase subunit β, suggesting that the glycolysis process may be suppressed by SMWF.

2.2.7. Dysregulation of Gene Expression in HaCaT Cells after UVB Irradiation

Besides delving deep into UVB-induced dynamic proteomic dysregulation, the modulatory effect of UVB on inflammatory and collagen degrading genes was also investigated. Here, we investigated three inflammatory markers: TNF-α, NF-κB, and COX-2, and a collagen degrading marker, MMP-1. Skin exposure to UVB induces the upregulation of ROS, which stimulates the cascade of inflammatory response such as the expression of NF-κB [107]. In turn, NF-κB then activates the production of proinflammatory cytokines, like TNF-α [108]. According to Yeo et al. [109], the increase of TNF-α can upregulate the binding of early growth response-1 (EGR-1) to the MMP-1 promoter through activation of extracellular signal-regulated kinase (ERK)1/2, JNK, and p38 kinase, ultimately increasing the expression of the MMP-1 protein. Furthermore, the overexpression of TNF-α can also activate the ASK1-mediated pathway, which prolongs the prolonged activation of JNK and p38 [110], thus, suggesting an equally prolonged expression of MMP-1 in the cell. On another note, NF-κB can also directly affect the production of MMP-1 as it has been discovered that the MMP-1 promoter also contains NF-κB binding sites [111,112]. A study done by Elliott, Coon, Hays, Stadheim, and Vincenti [111] had shown that NF-κB1 homodimers, together with IL-1β and Bcl-3, can activate the MMP-1 transcription. Finally, the expression of COX-2 is mediated by NF-κB and its expression has been linked with UVB-induced cell inflammation, death, and skin carcinogenesis [113,114]. As per data procured, UVB control cells displayed a significant increase in TNF-α, NF-κB, COX-2, and MMP-1 expression level by 1.69, 1.44, 3.74, and 9.54-fold, respectively, compared to non-UVB control cells (Figure 4A–D). Nevertheless, SMCE was able to negatively regulate the expression of TNF-α to a 1.03-fold change and MMP-1 to a 5.07-fold change. On the other hand, SMHF was only able to downregulate NF-κB (0.64-fold), while SMEAF significantly decreased TNF-α, NF-κB, and MMP-1 levels by 0.72, 0.37, and 5.11-fold change, accordingly. Aside from that, SMWF displayed no significant changes in the expression of all four genes compared to the UVB control. This inability of SMWF to suppress all four genes coincides with the proteomics data obtained, which is that SMWF is unable to attenuate UVB-induced cell death. Regardless, none of the extract and fraction were able to significantly decrease COX-2 transcription levels, although SMCE and SMEAF showed slightly lowered expressions.

The implications of SMCE, SMEAF, and SMHF's ability to regulate these genes are important as these genes not only affect cell survival, but also the photoaging of the skin. Hence, the inhibition of SMEAF in the expression of NF-κB, TNF-α, and MMP-1 suggests that SMEAF can not only inhibit the UVB- induced inflammatory process, but is also able to suppress collagen degradation. SMCE also displayed the same aptitude as SMEAF, even though it was only able to slightly suppress the expression of NF-κB. However, in conjunction with the proteomics studies, even though SMHF significantly decreased the expression of NF-κB and significantly increased the expression of PRDX-1 (an inhibitor to the ASK-1 activity), it was not sufficient to suppress the expression of MMP-1. Based on this outcome, it might be suggested that the inability of SMHF to decrease MMP-1 expression could be due to its inability to significantly suppress the expression of TNF-α, showing the importance of TNF-α in the role of MMP-1 expression.

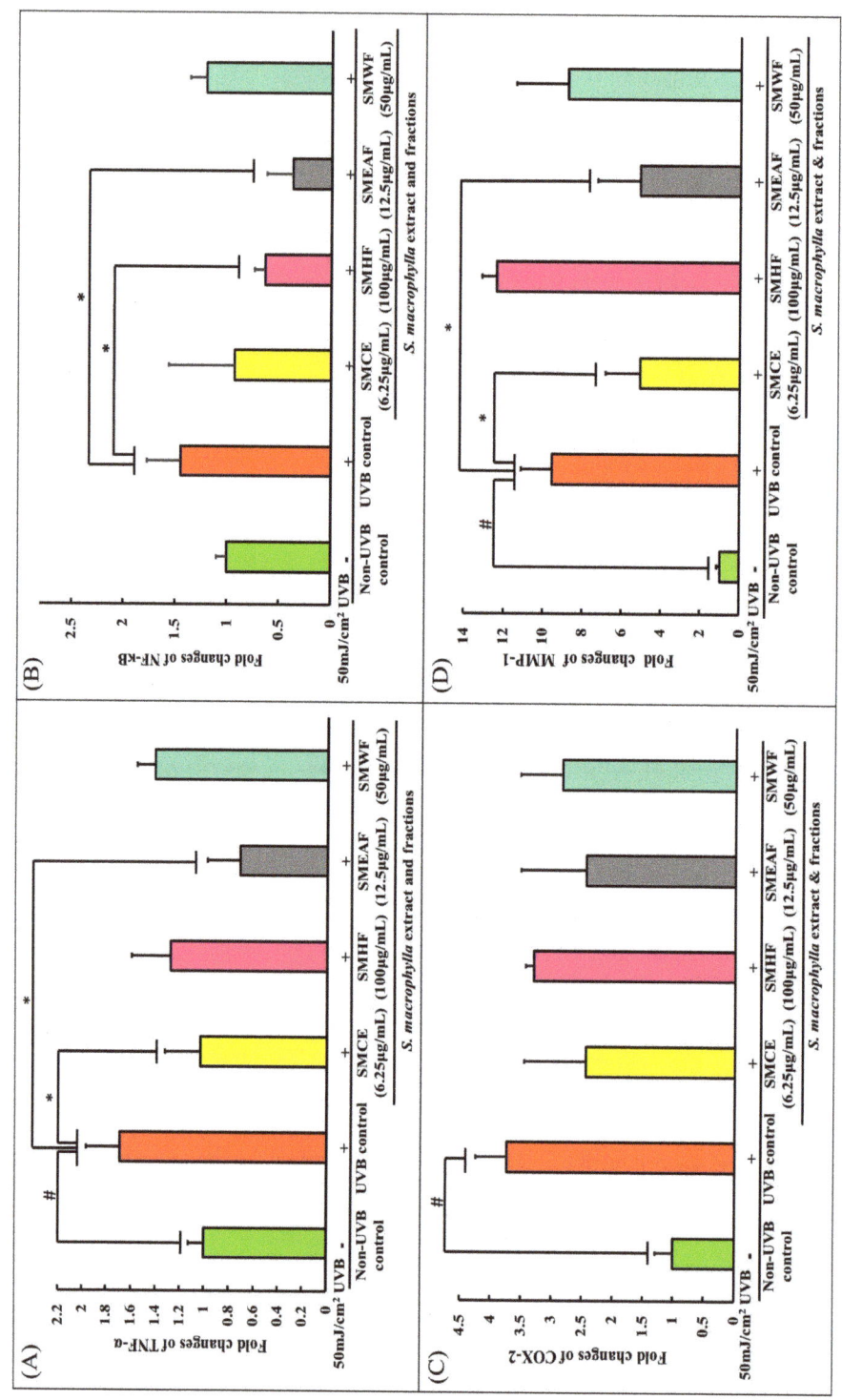

Figure 4. *Cont.*

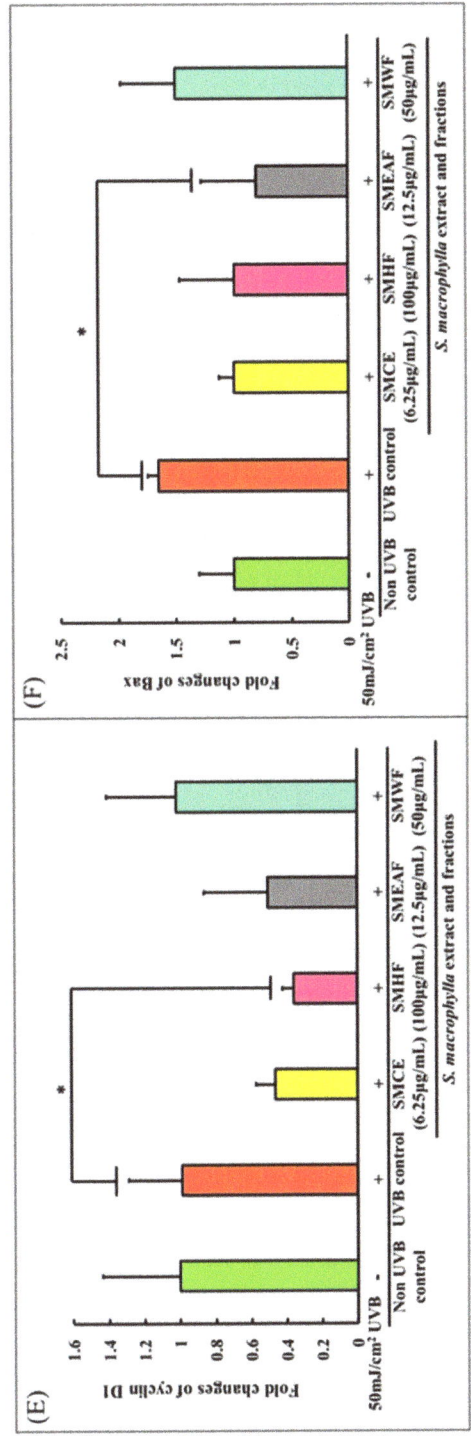

Figure 4. Gene expression changes in HaCaT cells after UVB irradiation and treatment with *S. macrophylla* extract and fractions. HaCaT cells were treated with SMCE, SMHF, SMEAF, and SMWF (at indicated concentrations) before being irradiated with 50 mJ/cm^2 UVB immediately. After irradiation, the extract was removed and the cells were left to incubate for 24 h before mRNA extraction. The fold changes in (**A**) TNF-α, (**B**) NF-κB, (**C**) COX-2, (**D**) MMP-1 (**E**) cyclin D1, and (**F**) Bax mRNA expression were analyzed using qPCR. Data were expressed as mean ± standard deviation. ($n \geq 3$; p-value ≤ 0.05; # symbolizes significant difference between non-UVB and UVB control; * symbolizes significant difference between UVB control and sample).

Other than the changes in inflammatory and collagen degradation, the effect of *S. macrophylla* against UVB-induced cell cycle arrest and apoptosis were also investigated. Subsequently, our study on cell cycle changes revealed that cyclin D1 experienced no changes in expression 24 h after 50 mJ/cm^2 UVB irradiation (Figure 4E). However, treatment with SMHF significantly decreased the expression of cyclin D1 compared to the UVB control cells. Both SMCE and SMEAF also decreased the expression of cyclin D1, although not significantly, while cells treated with SMWF showed no changes in the expression of cyclin D1. In the process of cell cycle, the coordination of CDK, CDK inhibitor, and cyclin expressions are essential to ensure continuous cell proliferation, and among the cell cycle related genes, both CDK2/cyclinE1 and CDK4/cyclinD1 complexes have a role to play in the G1/S phase transition, while CDK1 and cyclin B1 is responsible for the transition of cells from the G2 to M phase [115]. However, under environmental stresses, the progression of cell cycle can be inhibited, thus halting cell proliferation [116]. According to Han and He [117], cell cycle progression of keratinocytes is UVB dose-dependent. All cells exposed to UVB were significantly arrested in the S phase 18 h after exposure. They also displayed an increase in cyclin D1 expression at the 3 h time point and decreased expression at the 6 h timepoint during the S phase. This increase in cyclin D1 was reported to be due to the activation of the AKT, ERK, and EGFR pathways, which in turn are activated by MMP. After 48 h, cells exposed to lower doses of UVB were seen to successfully exit the S phase and returned to the G0-G1 phase, but those that were exposed with higher doses of UVB stayed longer in the S phase. The arrest of cells by UVB in the S phase is most likely due to the increase in cyclin D1 expression as Yang et al. [118] reported that it is necessary for cyclin D1 levels to be low in the S phase to allow for efficient DNA synthesis. As UVB has been known to cause damage to DNA, it could be possible that the cells were arrested to inhibit DNA synthesis. In the case of SMHF treated cells, the decrease in cyclin D1 expression coincides with the data obtained from LC-MS/MS as SMHF was shown to upregulate proteins that are involved in DNA repair and maintenance. With the combination of increased DNA repair proteins and decreased cyclin D1 expression, it could be possible that SMHF treated cells are able to actively repair and synthesize DNA. However, this claim needs to be further confirmed with further tests. As for SMCE and SMEAF, the decrease in cyclin D1 could be due to their ability to suppress MMP-1 expression, but this has yet to be confirmed.

In the apoptotic pathway, we focused on the expression of Bcl-2-associated X protein (Bax) as it plays an important role in forming the apoptotic pores at the mitochondrial outer membrane. When the apoptotic pores form, it is then at this stage where it is considered as the point of no return for the cells [119]. The permeabilization of the mitochondria outer membrane will then lead to the release of cytochrome C and the activation of caspases involved in the intrinsic apoptotic pathway [119]. In our study, an increase in Bax expression was seen at 24 h in UVB control cells, as shown in Figure 4F. This increase in Bax in UVB control cells is expected as it was previously reported in several studies, indicating that the exposure of keratinocytes to UVB would ultimately lead to the activation of apoptosis [120,121]. However, when treated with SMCE, SMHF, and SMEAF, the expression of Bax was decreased. In addition, the decrease in Bax mRNA expression was significant for cells treated with SMEAF. Thus, this suggests that SMCE contains bioactive compounds that may be able to suppress UVB-induced apoptosis and after fractionation, SMHF and, especially, SMEAF, are the two fractions that contain the compounds responsible for suppressing Bax. In contrast, there were no changes in the expression of Bax when the irradiated cells were treated with SMWF. The lack of change in both cyclin D1 and Bax expression together with suppression of DNA repair proteins in SMWF treated cells might suggest that the cells may be experiencing cell cycle arrest and might be entering the apoptosis pathway. Even so, further analysis on the apoptotic pathway, extract fractionation, and continuous bio-guided assays is still necessary to identify the compound that is able to reverse or inhibit the effect of UVB. The changes in gene expression and its related pathway can be seen in Figure 5.

Overall, these experiments showed that UVB does induce a massive change in both gene and protein expression, affecting their regulation and ultimately, cell survival. To provide an overview, Figure 6 depicts a summary of the gene and protein expression changes throughout the whole cell after UVB exposure. As the cells were irradiated with UVB, ROS and RNS were produced, while the redox system was inhibited. This leads to the production of inflammatory markers, which induces a cascade of reactions, leading to collagen degradation and subsequently, cell death. Furthermore, damage to the cell DNA also occurred, which in turn suppressed RNA transcription, translation, and protein processing. Subsequently, UVB exposure also decreased the expression of proteins involved in DNA maintenance and repair, inhibiting any form of repair on the UVB-induced CPD and pyrimidine-pyrimidone (6-4) photoproduct. Furthermore, the proteins involved in both the glycolysis process and cell growth, proliferation, and migration processes were also affected by UVB irradiation. All in all, the suppression of these genes and proteins will ultimately encourage the cell to begin cell death. However, this effect can be reversed with *S. macrophylla* as a treatment against UVB-induced photodamage. Although SMCE was only able to inhibit the production of inflammatory markers and subsequently collagen degradation itself, its fractions displayed a wider effect throughout the cells. SMHF exhibited a significant increase in a wide array of proteins that reverses the effect of UVB on many cell processes including the redox system, DNA synthesis and repair, glycolysis process, RNA to protein processes, and finally, it also induces cell growth, proliferation, and migration. On the other hand, SMEAF retained the same effect as SMCE but with the addition of being able to significantly inhibit Bax, indicating that the compounds responsible for the anti-inflammatory, anti-collagen degrading, and potentially anti-apoptotic activity may be within the SMEAF fraction. In comparison to the other fractions, SMWF was not only to be unable to inhibit the effect of UVB, but also additionally seems to further encourage cell death. Based on the data, it can be said that although SMCE can attenuate UVB-induced photodamage, its effect can be enhanced further via fractionation. This could be due to the removal of antagonistic compounds from the mixture, where in this case, SMHF showed an improved reversal effect against UVB damage while SMWF displayed the opposite effect. Hence, for future studies, further fractionation of the current existing fractions should be conducted to improve their efficacy against photoaging. Next, the creation of cosmetic formulation with *S. macrophylla* fractions as the active ingredient and in vivo human clinical testing of said formula on skin hydration, elasticity, sebum production, and pigmentation can also be conducted, as described by Tian et al. [122], Goh et al. [123], Adejokun and Dodou [124], and Mosquera et al. [125], to better study its anti-photoaging effect on the skin.

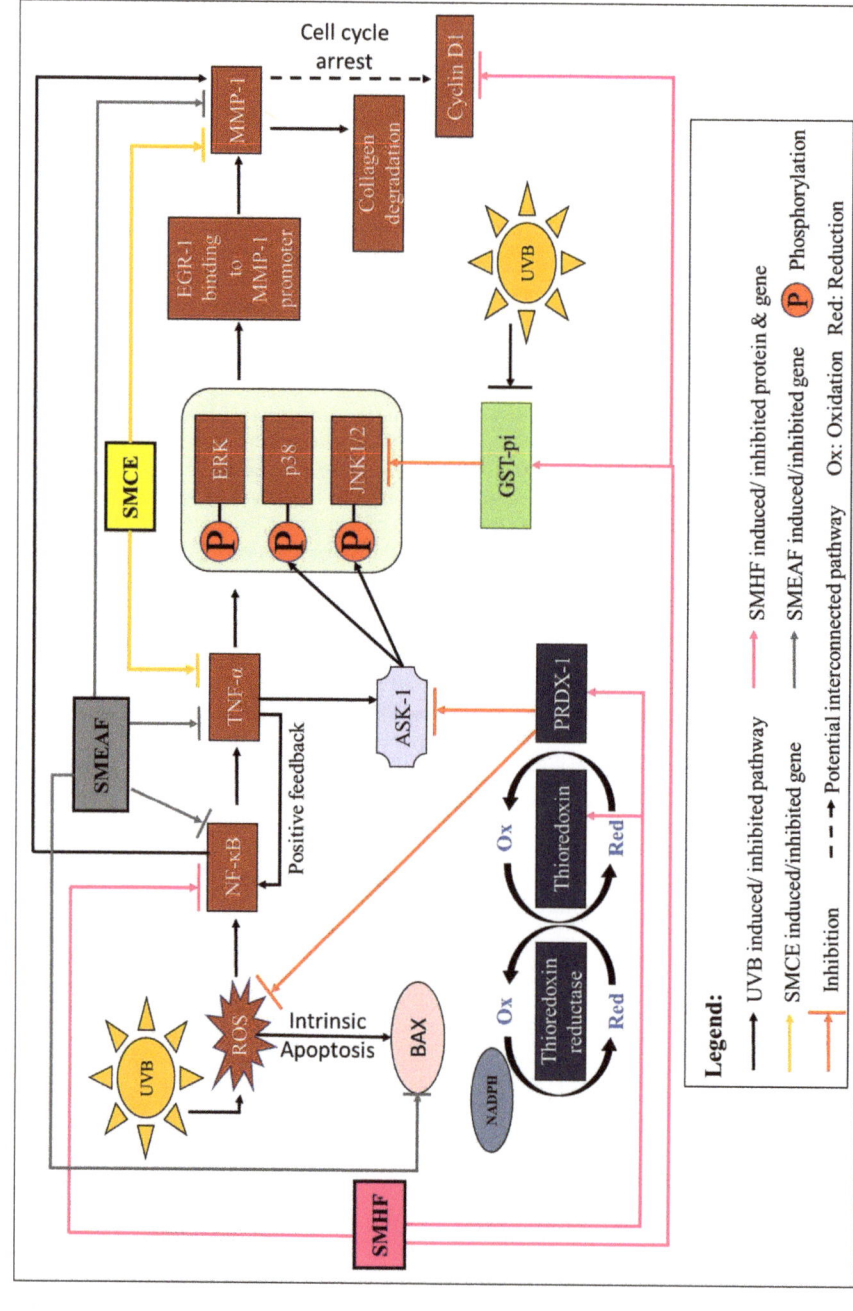

Figure 5. Activation of the redox system, cell inflammation, collagen degradation, cell cycle arrest and intrinsic apoptosis by UVB and the effect SMCE, SMHF, and SMEAF treatment has on the irradiated keratinocyte cells.

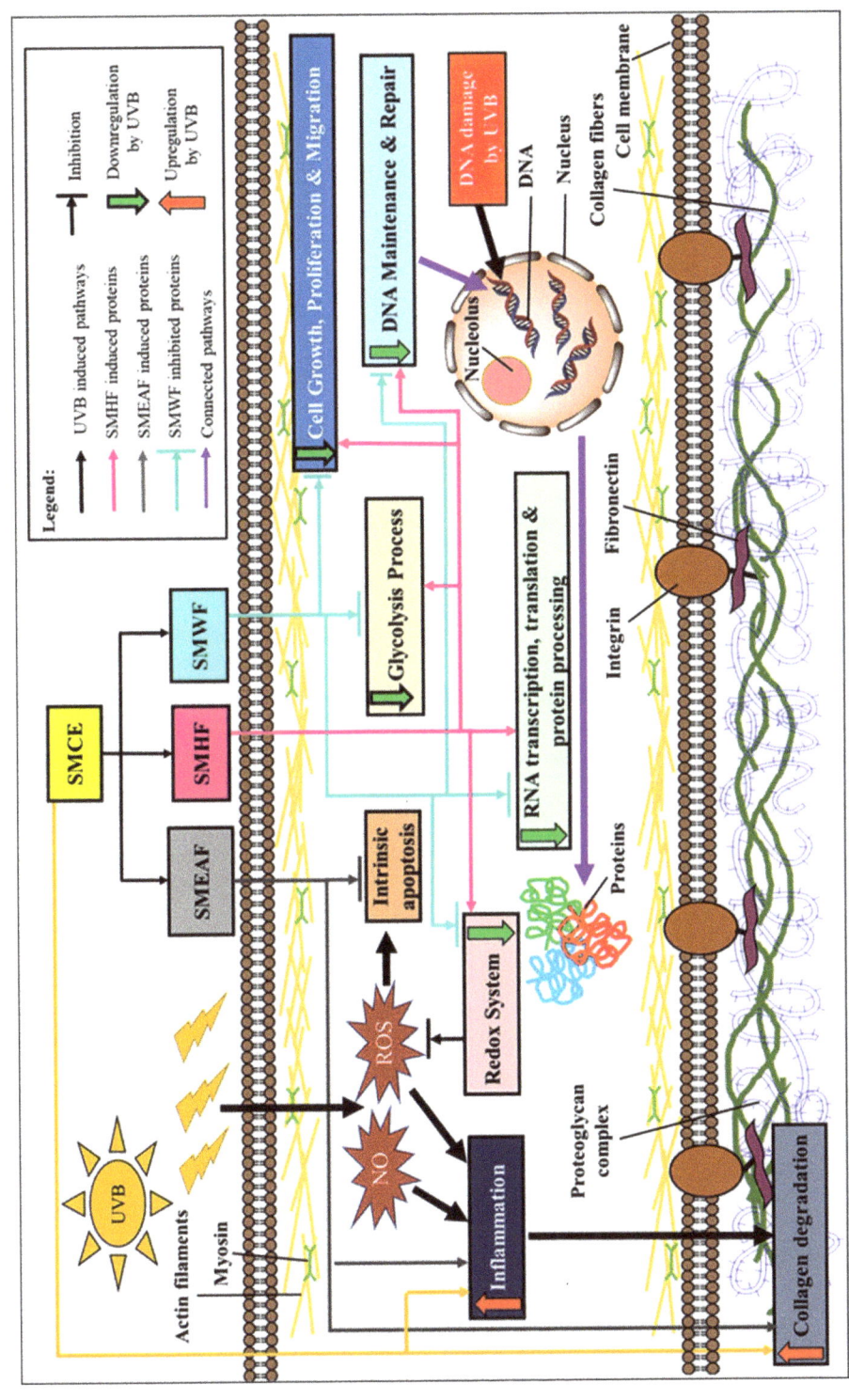

Figure 6. An overview of the effect of *S. macrophylla* extract and fractions against UVB-induced photodamage in keratinocyte cells.

3. Materials and Methods

3.1. Plant Material and Extraction

Seeds of *S. macrophylla* (3 kg) were purchased from a local market and a voucher of the specimen (No. KLU46901) was deposited at the Herbarium of Institute of Biological Sciences, Faculty of Science, University of Malaya, Malaysia. To extract the bioactive compounds, the seeds were first dried and then ground finely before being soaked in ethanol at room temperature for 72 h. After filtration, the mixture was concentrated in a rotary vacuum evaporator at 40 °C to yield SMCE. Some portions of SMCE were then further processed to obtain SMHF using the hexane solvent. The liquid hexane fraction was separated from its insoluble residues via filtration, dried using anhydrous sodium sulfate, and concentrated again with the rotary vacuum evaporator at 40 °C. The insoluble residues were then subjected to a solvent–solvent portioning of ethyl acetate and water at a 1:1 ratio. After complete separation of both layers, SMEAF was evaporated using rotary vacuum evaporation while SMWF was freeze-dried with a freeze dryer [126].

3.2. Cell Line and Maintenance

The human keratinocyte (HaCaT) cell line (American Tissue Culture Center, Chapel Hill, NC, USA) was used to emulate the epidermal skin cells. They were maintained with culture medium of 1× high-glucose Dulbecco's Modified Eagle Medium (DMEM), supplemented with GlutaMAX without HEPES (Gibco, Thermo Fisher, Waltham, MA, USA), 10.0% fetal bovine serum (Gibco, Thermo Fisher, Waltham, MA, USA), and 1.0% antibiotic/antimycotic solution (100 U/mL penicillin, 100 µg/mL streptomycin, and 25 µg/mL amphotericin B) (Gibco, Thermo Fisher, Waltham, MA, USA). The cells were incubated at 37 °C in 5% CO_2 atmospheric conditions [127].

3.3. Cytotoxicity Assay

The cytotoxicity assay was conducted to determine the non-cytotoxic concentration of each extract and fraction on HaCaT cells. The cells were incubated with 0, 1.56, 3.125, 6.25, 12.5, 25, 25, 50, and 100 µg/mL of each extract and fraction for 24 h before its cell viability was assessed using the 3-(4,5-dimethylthiazol-2-Y1)-2,5-diphenyltetrazolium bromide (MTT) assay. In each well, 20 µL of MTT was added and the plate was then incubated for 2 h. After that, the solution was removed and 100 µL of dimethyl sulfoxide (DMSO) was then added into each well. The absorbance was measured at 570 nm via a microplate reader and the percentage of cell viability was calculated after normalizing against the control cells.

3.4. UV Irradiation

To study the effect of UVB on keratinocyte cells, the cells were seeded in six well plates at a concentration of 300,000 cells/well. The following day, the cells were rinsed with phosphate buffered saline (PBS) and then treated with either 6.25 µg/mL SMCE, 100 µg/mL SMHF, 12.5 µg/mL SMEAF, or 50 µg/mL SMWF in 1.5 mL PBS per well, with 0.5% DMSO as the vehicle, before being irradiated with 50 mJ/cm^2 of UVB using a Philip UVB Broadband TL 20W/12 phototherapy lamp (Philip, Amsterdam, The Netherlands) according to Mahendra et al. [128] with slight modifications. The UVB dose was measured using a Lutron UV light meter UV-340A (Lutron, Taipei, Taiwan). After irradiation, the PBS solution was removed and replaced with 3 mL of media in each well. Non-UVB and UVB control plates (untreated cells that were either irradiated or non-irradiated) were treated the same way as the sample plates. This included rinsing and treating the cells with 0.5% DMSO in 1.5 mL PBS per well for a similar amount of time before the PBS was replaced with media. The cells were incubated for 24 h before RNA or protein extraction.

3.5. Protein Expression Studies

3.5.1. Preparation of Lysis Buffer

Fresh lysis buffer was prepared a day before protein extraction. The lysis buffer was prepared by combining 10 mM Tris solution with 0.1% Triton X. The pH of the solution was then adjusted to pH 7.4 using hydrochloric acid before being sterile filtered with a 0.22 µM cellulose acetate syringe filter (Sartorius, Göttingen, Germany) and stored at 4 °C.

3.5.2. Protein Extraction

Cells were harvested using Tryple E (Gibco, Thermo Fisher, Waltham, MA, USA) and pelleted via centrifugation at 1000 rpm for 5 min. To completely remove the presence of media, the cells were rinsed and pelleted twice. Next, the PBS was removed and the pellet was resuspended in 80 µL of ice-cold lysis buffer. The mixture was then freeze–thaw for three cycles under these conditions: −152 °C, 10 min; 37 °C for 2–3 min. Subsequently, the mixture was centrifuged at $10,000 \times g$ at 4 °C for 15 min and the supernatant was collected and stored at −80 °C in low protein binding microcentrifuge tubes (Eppendorf, Hamburg, Germany).

3.5.3. Bicinchoninic Acid (BCA) Protein Assay

Protein concentration was measured via a BCA Protein Assay Kit (Thermo Fisher, Waltham, MA, USA), according to the manufacturer's instructions. In short, reagent A and B were mixed at a 50:1 ratio to create the working reagent. After that, aa 10 µL protein sample was added to the well with 200 µL working reagent. The plate was then incubated at 37 °C for 30 min before allowing to cool and measured at 562 nm. Later, the absorbance obtained was subtracted with the blank absorbance prior to calculating the concentration based on the standard curve. The standard curve was built on the increasing concentrations (0–1500 µg/mL) of bovine serum albumin (BSA) (Sigma, St. Louis, MI, USA).

3.5.4. In-Solution Tryptic Digestion

Protein samples (10 mg/mL) were denatured in a solution containing 1 µL of 200 mM 1,4-dithiothreitol (DTT), 25 µL of 100 mM ammonium bicarbonate, and 25 µL trifluoroethanol (TFE). The mixture was mixed via vortexing and heated for 1 h at 60 °C. After incubation, 4 µL of 200 mM iodoacetamide (IAM) stock was added to the mixture to alkylate the proteins for 1 h in the dark at room temperature. Next, 1 µL of 200 mM DTT was added to quench the excess IAM and the mixture was again incubated for 1 h in the dark at room temperature. Subsequently, 300 µL of ultrapure water and 100 µL of 100 mM ammonium bicarbonate was added to the samples before following with 1 µL of 20 µg/mL MS grade trypsin (Thermoscientific, Waltham, MA, USA), which was reconstituted in 50 mM acetic acid. The mixture was incubated for 16 h at 37 °C for complete protein digestion. After that, 1 µL of formic acid was added to the mixture to stop the trypsin reaction. Finally, the samples were dried overnight in the MiniVac speed vacuum concentrator (Lobogene, Lillerød, Denmark) and stored at −20 °C.

3.5.5. Protein Sample Desalting and Cleanup

The samples were cleaned and desalted before LCMS/MS analysis using a Pierce® C18 Spin column (Thermoscientific, Waltham, MA, USA), according to the manufacturer's instructions. After clean up and desalting, the sample was once again dried overnight in the MiniVac speed vacuum concentrator (Lobogene, Lillerød, Denmark) and stored at −20 °C.

3.5.6. Analysis of Protein Samples with Nanoflow-Ultra High-Performance Chromatography-Tandem Mass Spectrometry (LC-MS/MS)

This method was done following the method described by Paudel et al. [129]. First, protein samples were dissolved in 30 µL 0.1% formic acid before centrifuging at 14,000 rpm for 10 min. Next, 1 µL of the sample was loaded into an Agilent C18, 300 Å large capacity chip (Agilent Technologies, Santa Clara, CA, USA), and the chip was mounted onto the Agilent 1200HPLC-Chip/MS interface, which was coupled with an Agilent 6500 iFunnel quadrupole-time of flight (Q-TOF) LC/MS system. The flow rate was set at 4 µL/min from an Agilent 1200 Series Capillary pump and 0.5 µL/min from an Agilent 1200 Series Nano Pump with solution A (0.1% formic acid in water) and solution B (90% acetonitrile and 0.1% formic acid in water). The samples were then eluted with multi-step gradients of 5–75% of solution B (30 min of 5–75% solution B, 9 min of 75% solution, and 8 min of 5–75% solution B). The ion polarity of Q-TOF was set at positive with a capillary voltage of 2050 V, the gas temperature at 325 °C, fragmentor voltage at 360 V, and finally, drying gas flow rate of 5 L/min. The spectra acquired were in auto MS/MS mode with an MS scan range of 110–3000 m/z and an MS/MS scan range of 50–3000 m/z. The precursor charge state selection and preference were fixed as doubly, triply, or more than triply charged state, with the exclusion of precursor 299/294457 m/z (Z = 1) and 1221.990637 m/z (Z = 1) (reference ions).

3.5.7. Protein Identification and Differential Expression Studies with PEAKS Bioinformatics Software

To determine the protein identification and differential expression, label-free quantification (LFQ) was conducted using PEAKS studio 7.5 (Bioinformatics Solution Inc., Waterloo, ON, Canada) using the method as described by Paudel et al. (2020) with slight modification. Homo sapiens (Uniprot database) was used for homology search and protein identification. The carbamidomethylation was preset as a fixed modification with maximum missed cleavages at 3. Parent mass and fragment mass error tolerance were set at 0.1 Da with monoisotopic as the precursor mass search type. Trypsin was then selected as the digestion enzyme. The data filtering parameters were set at 1% false discover rate (FDR) with unique peptides $\geq$1. The LFQ parameters used were: retention time shift tolerance of 6 min, mass error tolerance of 20 ppm, and FDR threshold of 1%. The protein expression of the UVB control cells was compared against the non-UVB control cells, while samples treated with SMCE, SMHF, SMEAF, SMWF were compared against the UVB control cells using hierarchical clustering. The heat map was generated by setting the protein significance $\geq$13, which was the p-value of 0.05, fold change $\geq$1, and had at least one unique peptide. To calculate the significance, PEAKS Q was used and experimental bias was taken into account via automatic normalization of protein ratios in accordance with the total ion chromatogram (TIC) [129].

3.6. Quantitative Polymerase Chain Reaction (PCR) Analysis of Gene Expression Changes in UVB Irradiated Cells

Total mRNA was then collected using 300 µL of Trizol. The Trizol was then processed with a 1:5 ratio of chloroform to Trizol. The solution was vortexed and incubated for 3 min at room temperature before centrifuging at 13,500 rpm at 4 °C for 15 min. The upper colorless aqueous phase was extracted and mixed with 100% isopropyl alcohol at a 1:2 ratio. The mixture was mixed gently and incubated for 10 min at room temperature before centrifuging again at 13,500 rpm at 4 °C for 15 min. The supernatant was discarded carefully and the pellet was rinsed with 75% ethanol before centrifuging at 10,500 rpm for 5 min at 4 °C. The supernatant was discarded and the rinsing process was repeated. Finally, the pellet was air-dried for 5 min and dissolved in 10 µL of DEPC ultrapure water. The RNA concentration was measured at A260/A280. Total mRNA that was collected was converted to cDNA using a High Capacity cDNA Reverse Transcription Kit (Applied Biosystem, San Francisco, CA, USA) and cDNA was prepared for gene expression reading using the Power SYBR Green PCR Master Mix. qPCR analysis was done using the Step One Plus Real-Time PCR System (Applied Biosystem, San Francisco, CA, USA). Primers for 18S ribosomal RNA, TNF-α, NF-κB, COX-2, Bax, cyclin D1, and MMP-1 were either obtained from journals or designed using NCBI Primer Blast. The sequences of each primer can be seen in Table 2.

Table 2. Forward and reverse primers that were used to study gene expression changes via qPCR.

Target Gene	Sequence	References
TNF-α	Forward: 5′-CCAGGCAGTCAGATCATCTTCTC-3′ Reverse: 5′-AGCTTGAGGGTTTGCTACAACAT-3′	[130]
NF-κB	Forward: 5′-GACGAGAACGGAGACACA-3′ Reverse: 5′-TGGTTGGTAGGTTGACAAC-3′	Designed with NCBI Primer Blast
COX-2	Forward 5′-TGCGCCTTTTCAAGGATGGA-3′ Reverse 5′-CCCCACAGCAAACCGTAGAT-3′	Designed with NCBI Primer Blast
MMP-1	Forward 5′-GGGAGATCATCGGGACAACTC-3′ Reverse 5′-TGAGCATCCCCTCCAATACC-3′	[131]
Cyclin D1	Forward 5′-TGCGCTGCTACCGTTGACT-3′ Reverse 5′-AGCGATGTGAATATTTCCAAACC-3′	[132]
Bax	Forward 5′-GTCGCCCTTTTCTACTTTGCCAG-3′ Reverse 5′-TCCAGCCCAACAGCCGCTCC-3′	[126]
18S ribosomal RNA	Forward: 5′-GGCCCTGTAATTGGAATGAGTC-3′ Reverse: 5′-CCAAGATCCAACTACGAGCTT-3′	[133]

3.7. Statistical Analysis

All quantitative data were analyzed using the SPSS statistical analysis software and the results were expressed as mean ± standard deviation (SD). One-way Analysis of Variance (ANOVA) and Tukey post-hoc were used to determine significant data. The significant value was set at $p \leq 0.05$.

4. Conclusions

In summary, a wide view of the adverse effect UVB has on a cellular level was portrayed through this research. The range of impact of UVB from the activation of the redox system and skin inflammation, to the suppression of protein synthesis, inhibition of cell growth and repair, induction of DNA damage, signaling of collagen degradation, and finally cell death are processes that cosmetic companies constantly battle with to keep their clients looking young. When tested with the *S. macrophylla* extract and fractions, two fractions, namely SMEAF and SMHF, exhibited potential photoprotective properties. Functioning via two completely different mechanisms, SMEAF showcased its ability to suppress inflammation, collagen degradation, and potentially the intrinsic apoptosis pathway on a cellular level while SMHF displayed its photoprotective properties through its involvement in the redox system, DNA repair, RNA transcription, protein maintenance and synthesis, cell growth, migration and proliferation, and cell glycolysis processes. Thus, as per the results, further analysis and fractionation of SMHF and SMEAF are warranted toward the making of anti-photoaging cosmetic active ingredients. Furthermore, in vivo human clinical studies with *S. macrophylla* cosmetic formulations can be conducted in the future to better evaluate the effect of *S. macrophylla* against photoaging.

Supplementary Materials: The following are available online. Table S1: Analysis of differentially expressed protein in UVB control cells and those irradiated but treated with *S. macrophylla* extract and fractions via LC-MS/MS.

Author Contributions: Conceptualization, B.H.G. and C.K.M.; Methodology, C.K.M., S.A.Z.A., and B.H.G.; Formal Analysis, C.K.M. and S.A.Z.A.; Investigation, C.K.M.; Resources, S.A.Z.A. and S.U.K.; Writing—Original Draft Preparation, C.K.M.; Writing—Review & Editing, L.C.M., S.A.Z.A., B.H.G., and P.P.; Supervision, B.H.G.; Project Administration, B.H.G. and C.K.M.; Funding Acquisition, T.T.H., L.-H.C., S.Y.T., and B.H.G. All authors have read and agreed to the published version of the manuscript.

Funding: This work was financially supported by External Industry Grants from Biotek Abadi Sdn Bhd (vote no. GBA-81811A), Monash Global Asia in the 21st Century (GA21) research grant (GA-HW-19-L01 & GA-HW-19-S02) and the Fundamental Research Grant Scheme (FRGS/1/2019/STG05/USM/02/10).

Institutional Review Board Statement: Not applicable.

Informed Consent Statement: Not applicable.

Data Availability Statement: The data used to support the findings of this study are available from the corresponding author upon request.

Conflicts of Interest: The authors declare that there are no conflict of interest.

References

1. Bogh, M.K.B.; Schmedes, A.V.; Philipsen, P.A.; Thieden, E.; Wulf, H.C. Vitamin D production after UVB exposure depends on baseline vitamin D and total cholesterol but not on skin pigmentation. *J. Investig. Dermatol.* **2010**, *130*, 546–553. [CrossRef]
2. Gathers, R.C.; Scherschun, L.; Malick, F.; Fivenson, D.P.; Lim, H.W. Narrowband UVB phototherapy for early-stage mycosis fungoides. *J. Am. Acad. Dermatol.* **2002**, *47*, 191–197. [CrossRef]
3. Kanwar, A.J.; Dogra, S.; Parsad, D.; Kumar, B. Narrow-band UVB for the treatment of vitiligo: An emerging effective and well-tolerated therapy. *Int. J. Dermatol.* **2005**, *44*, 57–60. [CrossRef]
4. Kirke, S.M.; Lowder, S.; Lloyd, J.J.; Diffey, B.L.; Matthews, J.N.S.; Farr, P.M. A randomized comparison of selective broadband UVB and narrowband UVB in the treatment of psoriasis. *J. Investig. Dermatol.* **2007**, *127*, 1641–1646. [CrossRef]
5. Scharffetter–Kochanek, K.; Brenneisen, P.; Wenk, J.; Herrmann, G.; Ma, W.; Kuhr, L.; Meewes, C.; Wlaschek, M. Photoaging of the skin from phenotype to mechanisms. *Exp. Gerontol.* **2000**, *35*, 307–316. [CrossRef]
6. Guinot, C.; Malvy, D.J.M.; Ambroisine, L.; Latreille, J.; Mauger, E.; Tenenhaus, M.; Morizot, F.; Lopez, S.; Le Fur, I.; Tschachler, E. Relative contribution of intrinsic vs extrinsic factors to skin aging as determined by a validated skin age score. *Arch. Dermatol.* **2002**, *138*, 1454–1460. [CrossRef]
7. Papakonstantinou, E.; Roth, M.; Karakiulakis, G. Hyaluronic acid: A key molecule in skin aging. *DermatoEndocrinol.* **2012**, *4*, 253–258. [CrossRef] [PubMed]
8. Addor, F.A.S.A. Beyond photoaging: Additional factors involved in the process of skin aging. *Clin. Cosmet. Investig. Dermatol.* **2018**, *11*, 437–443. [CrossRef]
9. Mahendra, C.K.; Tan, L.T.H.; Pusparajah, P.; Htar, T.T.; Chuah, L.-H.; Lee, V.S.; Low, L.E.; Tang, S.Y.; Chan, K.-G.; Goh, B.H. Detrimental effects of UVB on retinal pigment epithelial cells and its role in age-related macular degeneration. *Oxid. Med. Cell. Longev.* **2020**, *2020*, 1904178. [CrossRef] [PubMed]
10. Berkey, C.; Biniek, K.; Dauskardt, R.H. Screening sunscreens: Protecting the biomechanical barrier function of skin from solar ultraviolet radiation damage. *Int. J. Cosmet. Sci.* **2017**, *39*, 269–274. [CrossRef] [PubMed]
11. You, Y.-H.; Lee, D.-H.; Jung-Hoon, Y.; Nkajima, S.; Yasui, A.; Pfeifer, G.P. Cyclobutane pyrimidine dimers are responsible for the vast majority of mutations induced by UVB irradiation in mammalian cells. *J. Biol. Chem.* **2001**, *276*, 44688–44694. [CrossRef] [PubMed]
12. Kovacs, D.; Raffa, S.; Flori, E.; Aspite, N.; Briganti, S.; Cardinali, G.; Torrisi, M.R.; Picardo, M. Keratinocyte growth factor down-regulates intracellular ROS production induced by UVB. *J. Dermatol. Sci.* **2009**, *54*, 106–113. [CrossRef]
13. Filip, G.A.; Postescu, I.D.; Bolfa, P.; Catoi, C.; Muresan, A.; Clichici, S. Inhibition of UVB-induced skin phototoxicity by a grape seed extract as modulator of nitrosative stress, ERK/NF-kB signaling pathway and apoptosis, in SKH-1 mice. *Food Chem. Toxicol.* **2013**, *57*, 296–306. [CrossRef]
14. Bashir, M.M.; Sharma, M.R.; Werth, V.P. UVB and pro-inflammatory cytokines synergistically activate TNF-α production in keratinocytes through enhanced gene transcription. *J. Investig. Dermatol.* **2009**, *129*, 994–1001. [CrossRef] [PubMed]
15. Fagot, D.; Asselineau, D.; Bernerd, F. Direct role of human dermal fibroblasts and indirect participation of epidermal keratinocytes in MMP-1 production after UV-B irradiation. *Arch. Dermatol. Res.* **2002**, *293*, 576–583. [CrossRef]
16. Niwano, T.; Terazawa, S.; Nakajima, H.; Wakabayashi, Y.; Imokawa, G. Astaxanthin and withaferin A block paracrine cytokine interactions between UVB-exposed human keratinocytes and human melanocytes via the attenuation of endothelin-1 secretion and its downstream intracellular signaling. *Cytokine* **2015**, *73*, 184–197. [CrossRef]
17. Saito, M.; Tanaka, M.; Misawa, E.; Yao, R.; Nabeshima, K.; Yamauchi, K.; Abe, F.; Yamamoto, Y.; Furukawa, F. Oral administration of Aloe vera gel powder prevents UVB-induced decrease in skin elasticity via suppression of overexpression of MMPs in hairless mice. *Biosci. Biotechnol. Biochem.* **2016**, *80*, 1416–1424. [CrossRef]
18. Ewles, M.; Goodwin, L. Bioanalytical approaches to analyzing peptides and proteins by LC–MS/MS. *Bioanalysis* **2011**, *3*, 1379–1397. [CrossRef] [PubMed]
19. Goh, B.H.; Abdul Kadir, H.; Abdul Malek, S.N.; Ng, S.W. (αR,4R,4aR,6aS,7R,8S,10R,11S)-Methyl α-acet-oxy-4-(3-furan-yl)-10-hydroxy-4a,7,9,9-tetra-methyl-2,13-dioxo-1,4,4a,5,6,6a,7,8,9,10,11,12-dodeca-hydro-7,11-methano-2H-cyclo-octa-[f][2]benzopyran-8-acetate (6-O-acetyl-swietenolide) from the seeds of Swietenia macrophylla. *Acta Crystallogr. Sect. E Struct. Rep. Online* **2010**, *66*, o2802–o2803.

20. Wu, S.F.; Lin, C.K.; Chuang, Y.S.; Chang, F.R.; Tseng, C.K.; Wu, Y.C.; Lee, J.C. Anti–hepatitis C virus activity of 3–hydroxy caruilignan C from Swietenia macrophylla stems. *J. Viral Hepat.* **2012**, *19*, 364–370. [CrossRef] [PubMed]
21. Cheng, Y.-B.; Chien, Y.-T.; Lee, J.-C.; Tseng, C.-K.; Wang, H.-C.; Lo, I.-W.; Wu, Y.-H.; Wang, S.-Y.; Wu, Y.-C.; Chang, F.-R. Limonoids from the seeds of Swietenia macrophylla with inhibitory activity against dengue virus 2. *J. Nat. Prod.* **2014**, *77*, 2367–2374. [CrossRef] [PubMed]
22. Goh, B.H.; Abdul Kadir, H.; Abdul Malek, S.N.; Ng, S.W. Swietenolide diacetate from the seeds of Swietenia macrophylla. *Acta Crystallogr. Sect. E Struct. Rep. Online* **2010**, *66*, o1396. [CrossRef]
23. Goh, B.H.; Abdul Kadir, H. In vitro cytotoxic potential of *Swietenia macrophylla* King seeds against human carcinoma cell lines. *J. Med. Plant Res.* **2011**, *5*, 1395–1404.
24. Hashim, M.A.; Yam, M.F.; Hor, S.Y.; Lim, C.P.; Asmawi, M.Z.; Sadikun, A. Anti-hyperglycaemic activity of *Swietenia macrophylla* King (Meliaceae) seed extracts in normoglycaemic rats undergoing glucose tolerance tests. *Chin. Med.* **2013**, *8*, 11. [CrossRef] [PubMed]
25. Moghadamtousi, S.Z.; Goh, B.H.; Chan, C.K.; Shabab, T.; Kadir, H.A. Biological activities and phytochemicals of Swietenia macrophylla king. *Molecules* **2013**, *18*, 10465–10483. [CrossRef] [PubMed]
26. Sayyad, M.; Tiang, N.; Kumari, Y.; Goh, B.H.; Jaiswal, Y.; Rosli, R.; Williams, L.; Shaikh, M.F. Acute toxicity profiling of the ethyl acetate fraction of Swietenia macrophylla seeds and in-vitro neuroprotection studies. *Saudi Pharm. J.* **2017**, *25*, 196–205. [CrossRef]
27. Supriady, H.; Kamarudin, M.N.A.; Chan, C.K.; Goh, B.H.; Kadir, H.A. SMEAF attenuates the production of pro-inflammatory mediators through the inactivation of Akt-dependent NF-κB, p38 and ERK1/2 pathways in LPS-stimulated BV-2 microglial cells. *J. Funct. Foods* **2015**, *17*, 434–448. [CrossRef]
28. Mak, K.-K.; Shiming, Z.; Balijepalli, M.K.; Dinkova-Kostova, A.T.; Epemolu, O.; Mohd, Z.; Pichika, M.R. Studies on the mechanism of anti-inflammatory action of swietenine, a tetranortriterpenoid isolated from Swietenia macrophylla seeds. *Phytomedicine Plus* **2021**, *1*, 100018. [CrossRef]
29. Nilugal, K.C.; Fattepur, S.; Asmani, M.F.; Abdullah, I.; Vijendren, S.; Ugandar, R.E. Evaluation of wound healing acitivity of Swietenia macrophylla (Meliaceae) seed extract in albino rats. *Am. J. PharmTech Res.* **2017**, *7*, 113–124.
30. Wang, X.; Phelan, S.A.; Forsman-Semb, K.; Taylor, E.F.; Petros, C.; Brown, A.; Lerner, C.P.; Paigen, B. Mice with targeted mutation of peroxiredoxin 6 develop normally but are susceptible to oxidative stress. *J. Biol. Chem.* **2003**, *278*, 25179–25190. [CrossRef]
31. Fujii, J.; Ikeda, Y. Advances in our understanding of peroxiredoxin, a multifunctional, mammalian redox protein. *Redox Rep.* **2002**, *7*, 123–130. [CrossRef]
32. Hanschmann, E.-M.; Godoy, J.R.; Berndt, C.; Hudemann, C.; Lillig, C.H. Thioredoxins, glutaredoxins, and peroxiredoxins–molecular mechanisms and health significance: From cofactors to antioxidants to redox signaling. *Antioxid. Redox Signal.* **2013**, *19*, 1539–1605. [CrossRef] [PubMed]
33. Salzano, S.; Checconi, P.; Hanschmann, E.-M.; Lillig, C.H.; Bowler, L.D.; Chan, P.; Vaudry, D.; Mengozzi, M.; Coppo, L.; Sacre, S.; et al. Linkage of inflammation and oxidative stress via release of glutathionylated peroxiredoxin-2, which acts as a danger signal. *Proc. Natl. Acad. Sci. USA* **2014**, *111*, 12157–12162. [CrossRef]
34. Day, A.M.; Brown, J.D.; Taylor, S.R.; Rand, J.D.; Morgan, B.A.; Veal, E.A. Inactivation of a peroxiredoxin by hydrogen peroxide is critical for thioredoxin-mediated repair of oxidized proteins and cell survival. *Mol. Cell* **2012**, *45*, 398–408. [CrossRef]
35. Netto, L.E.S.; Antunes, F. The roles of peroxiredoxin and thioredoxin in hydrogen peroxide sensing and in signal transduction. *Mol. Cells* **2016**, *39*, 65–71.
36. Liu, S.; Guo, C.; Wu, D.; Ren, Y.; Sun, M.-Z.; Xu, P. Protein indicators for HaCaT cell damage induced by UVB irradiation. *J. Photochem. Photobiol. B* **2012**, *114*, 94–101. [CrossRef]
37. Wu, C.-L.; Chou, H.-C.; Cheng, C.-S.; Li, J.-M.; Lin, S.-T.; Chen, Y.-W.; Chan, H.-L. Proteomic analysis of UVB-induced protein expression- and redox-dependent changes in skin fibroblasts using lysine- and cysteine-labeling two-dimensional difference gel electrophoresis. *J. Proteomics* **2012**, *75*, 1991–2014. [CrossRef] [PubMed]
38. Baek, J.Y.; Park, S.; Park, J.; Jang, J.Y.; Wang, S.B.; Kim, S.R.; Woo, H.A.; Lim, K.M.; Chang, T.-S. Protective role of mitochondrial peroxiredoxin III against UVB-induced apoptosis of epidermal keratinocytes. *J. Investig. Dermatol.* **2017**, *137*, 1333–1342. [CrossRef] [PubMed]
39. Kim, S.Y.; Kim, T.J.; Lee, K.-Y. A novel function of peroxiredoxin 1 (Prx-1) in apoptosis signal-regulating kinase 1 (ASK1)-mediated signaling pathway. *FEBS Lett.* **2008**, *582*, 1913–1918. [CrossRef]
40. Adler, V.; Yin, Z.; Fuchs, S.Y.; Benezra, M.; Rosario, L.; Tew, K.D.; Pincus, M.R.; Sardana, M.; Henderson, C.J.; Wolf, C.R.; et al. Regulation of JNK signaling by GSTp. *EMBO J.* **1999**, *18*, 1321–1334. [CrossRef]
41. Verdone, L.; Agricola, E.; Caserta, M.; Di Mauro, E. Histone acetylation in gene regulation. *Brief. Funct. Genomic. Protemic.* **2006**, *5*, 209–221. [CrossRef]
42. Sesto, A.; Navarro, M.; Burslem, F.; Jorcano, J.L. Analysis of the ultraviolet B response in primary human keratinocytes using oligonucleotide microarrays. *Proc. Natl. Acad. Sci. USA* **2002**, *99*, 2965–2970. [CrossRef] [PubMed]
43. Dazard, J.-E.; Gal, H.; Amariglio, N.; Rechavi, G.; Domany, E.; Givol, D. Genome-wide comparison of human keratinocyte and squamous cell carcinoma responses to UVB irradiation: Implications for skin and epithelial cancer. *Oncogene* **2003**, *22*, 2993–3006. [CrossRef]
44. Zong, W.-X. Histone 1.2, another housekeeping protein that kills. *Cancer Biol. Ther.* **2004**, *3*, 42–43. [CrossRef]

45. Ruiz-Vela, A.; Korsmeyer, S.J. Proapoptotic histone H1.2 induces CASP-3 and -7 activation by forming a protein complex with CYT c, APAF-1 and CASP-9. *FEBS Lett.* **2007**, *581*, 3422–3428. [CrossRef]
46. Colombo, E.; Bonetti, P.; Lazzerini Denchi, E.; Martinelli, P.; Zamponi, R.; Marine, J.-C.; Helin, K.; Falini, B.; Pelicci, P.G. Nucleophosmin is required for DNA integrity and p19Arf protein stability. *Mol. Cell. Biol.* **2005**, *25*, 8874–8886. [CrossRef]
47. Maiguel, D.A.; Jones, L.; Chakravarty, D.; Yang, C.; Carrier, F. Nucleophosmin sets a threshold for p53 response to UV radiation. *Mol. Cell. Biol.* **2004**, *24*, 3703–3711. [CrossRef] [PubMed]
48. Wu, M.H.; Chang, J.H.; Yung, B.Y.M. Resistance to UV-induced cell-killing in nucleophosmin/B23 over-expressed NIH 3T3 fibroblasts: Enhancement of DNA repair and up-regulation of PCNA in association with nucleophosmin/B23 over-expression. *Carcinogenesis* **2002**, *23*, 93–100. [CrossRef] [PubMed]
49. Lamore, S.D.; Qiao, S.; Horn, D.; Wondrak, G.T. Proteomic identification of cathepsin B and nucleophosmin as novel UVA-targets in human skin fibroblasts. *Photochem. Photobiol.* **2010**, *86*, 1307–1317. [CrossRef] [PubMed]
50. Mosammaparast, N.; Ewart, C.S.; Pemberton, L.F. A role for nucleosome assembly protein 1 in the nuclear transport of histones H2A and H2B. *EMBO J.* **2002**, *21*, 6527–6538. [CrossRef]
51. Park, Y.J.; Chodaparambil, J.V.; Bao, Y.; McBryant, S.J.; Luger, K. Nucleosome assembly protein 1 exchanges histone H2A-H2B dimers and assists nucleosome sliding. *J. Biol. Chem.* **2005**, *280*, 1817–1825. [CrossRef]
52. Kumagai, H.; Sakai, H. A porcine brain protein (35 K protein) which bundles microtubules and its identification as glyceraldehyde 3-phosphate dehydrogenase. *J. Biochem.* **1983**, *93*, 1259–1269. [CrossRef]
53. Singh, R.; Green, M.R. Sequence-specific binding of transfer RNA by glyceraldehyde-3-phosphate dehydrogenase. *Science* **1993**, *259*, 365–368. [CrossRef] [PubMed]
54. Tisdale, E.J. Glyceraldehyde-3-phosphate dehydrogenase is required for vesicular transport in the early secretory pathway. *J. Biol. Chem.* **2001**, *276*, 2480–2486. [CrossRef] [PubMed]
55. Hara, M.R.; Agrawal, N.; Kim, S.F.; Cascio, M.B.; Fujimuro, M.; Ozeki, Y.; Takahashi, M.; Cheah, J.H.; Tankou, S.K.; Hester, L.D.; et al. S-nitrosylated GAPDH initiates apoptotic cell death by nuclear translocation following Siah1 binding. *Nat. Cell Biol.* **2005**, *7*, 665–674. [CrossRef] [PubMed]
56. Azam, S.; Jouvet, N.; Jilani, A.; Vongsamphanh, R.; Yang, X.; Yang, S.; Ramotar, D. Human glyceraldehyde-3-phosphate dehydrogenase plays a direct role in reactivating oxidized forms of the DNA repair enzyme APE1. *J. Biol. Chem.* **2008**, *283*, 30632–30641. [CrossRef]
57. Tristan, C.; Shahani, N.; Sedlak, T.W.; Sawa, A. The diverse functions of GAPDH: Views from different subcellular compartments. *Cell. Signal.* **2011**, *23*, 317–323. [CrossRef] [PubMed]
58. Susorov, D.; Zakharov, N.; Shuvalova, E.; Ivanov, A.; Egorova, T.; Shuvalov, A.; Shatsky, I.N.; Alkalaeva, E. Eukaryotic translation elongation factor 2 (eEF2) catalyzes reverse translocation of the eukaryotic ribosome. *J. Biol. Chem.* **2018**, *293*, 5220–5229. [CrossRef]
59. Plafker, S.M.; Macara, I.G. Ribosomal protein L12 uses a distinct nuclear import pathway mediated by importin 11. *Mol. Cell. Biol.* **2002**, *22*, 1266–1275. [CrossRef]
60. Takao, J.; Ariizumi, K.; Dougherty, I.I.; Cruz, P.D., Jr. Genomic scale analysis of the human keratinocyte response to broad-band ultraviolet-B irradiation. *Photodermatol. Photoimmunol. Photomed.* **2002**, *18*, 5–13. [CrossRef]
61. Kajita, Y.; Nakayama, J.; Aizawa, M.; Ishikawa, F. The UUAG-specific RNA binding protein, heterogeneous nuclear ribonucleoprotein D0. Common modular structure and binding properties of the 2xRBD-Gly family. *J. Biol. Chem.* **1995**, *270*, 22167–22175. [CrossRef]
62. Michelotti, E.F.; Michelotti, G.A.; Aronsohn, A.I.; Levens, D. Heterogeneous nuclear ribonucleoprotein K is a transcription factor. *Mol. Cell. Biol.* **1996**, *16*, 2350–2360. [CrossRef] [PubMed]
63. Guo, J.; Jia, R. Splicing factor poly(rC)-binding protein 1 is a novel and distinctive tumor suppressor. *J. Cell. Physiol.* **2018**, *234*, 33–41. [CrossRef]
64. Farrukh, M.R.; Nissar, U.A.; Afnan, Q.; Rafiq, R.A.; Sharma, L.; Amin, S.; Kaiser, P.; Sharma, P.R.; Tasduq, S.A. Oxidative stress mediated Ca2+ release manifests endoplasmic reticulum stress leading to unfolded protein response in UV-B irradiated human skin cells. *J. Dermatol. Sci.* **2014**, *75*, 24–35. [CrossRef]
65. Kalmar, B.; Greensmith, L. Induction of heat shock proteins for protection against oxidative stress. *Ad. Drug Deliv. Rev.* **2009**, *61*, 310–318. [CrossRef]
66. Edkins, A.L.; Price, J.T.; Pockley, A.G.; Blatch, G.L. Heat shock proteins as modulators and therapeutic targets of chronic disease: An integrated perspective. *Philos. Trans. R. Soc. Lond. B Biol. Sci.* **2018**, *373*, 20160521. [CrossRef] [PubMed]
67. Jee, H. Size dependent classification of heat shock proteins: A mini-review. *J. Exerc. Rehabil.* **2016**, *12*, 255–259. [CrossRef]
68. Beere, H.M. Death versus survival: Functional interaction between the apoptotic and stress-inducible heat shock protein pathways. *J. Clin. Investig.* **2005**, *115*, 2633–2639. [CrossRef] [PubMed]
69. Merwald, H.; Kokesch, C.; Klosner, G.; Matsui, M.; Trautinger, F. Induction of the 72-kilodalton heat shock protein and protection from ultraviolet B-induced cell death in human keratinocytes by repetitive exposure to heat shock or 15-deoxy-delta(12,14)-prostaglandin J2. *Cell Stress Chaperones* **2006**, *11*, 81–88. [CrossRef] [PubMed]
70. Perluigi, M.; Di Domenico, F.; Blarzino, C.; Foppoli, C.; Cini, C.; Giorgi, A.; Grillo, C.; De Marco, F.; Butterfield, D.A.; Schininà, M.E.; et al. Effects of UVB-induced oxidative stress on protein expression and specific protein oxidation in normal human epithelial keratinocytes: A proteomic approach. *Proteome Sci.* **2010**, *8*, 1–14. [CrossRef] [PubMed]

71. Howell, B.G.; Wang, B.; Freed, I.; Mamelak, A.J.; Watanabe, H.; Sauder, D.N. Microarray analysis of UVB-regulated genes in keratinocytes: Downregulation of angiogenesis inhibitor thrombospondin-1. *J. Dermatol. Sci.* **2004**, *34*, 185–194. [CrossRef]
72. Shan, Y.-X.; Liu, T.-J.; Su, H.-F.; Samsamshariat, A.; Mestril, R.; Wang, P.H. Hsp10 and Hsp60 modulate Bcl-2 family and mitochondria apoptosis signaling induced by doxorubicin in cardiac muscle cells. *J. Mol. Cell. Cardiol.* **2003**, *35*, 1135–1143. [CrossRef]
73. Mayer, M.P.; Bukau, B. Hsp70 chaperones: Cellular functions and molecular mechanism. *Cell. Mol. Life Sci.* **2005**, *62*, 670–684. [CrossRef]
74. Nishida, E.; Koyasu, S.; Sakai, H.; Yahara, I. Calmodulin-regulated binding of the 90-kDa heat shock protein to actin filaments. *J. Biol. Chem.* **1986**, *261*, 16033–16036. [CrossRef]
75. Taiyab, A.; Rao, C.M. HSP90 modulates actin dynamics: Inhibition of HSP90 leads to decreased cell motility and impairs invasion. *Biochim. Biophys. Acta* **2011**, *1813*, 213–221. [CrossRef]
76. Li, W.; Li, Y.; Guan, S.; Fan, J.; Cheng, C.-F.; Bright, A.M.; Chinn, C.; Chen, M.; Woodley, D.T. Extracellular heat shock protein-90α: Linking hypoxia to skin cell motility and wound healing. *EMBO J.* **2007**, *26*, 1221–1233. [CrossRef]
77. Wilkinson, B.; Gilbert, H.F. Protein disulfide isomerase. *Biochim. Biophys. Acta* **2004**, *1699*, 35–44. [CrossRef]
78. Wang, L.; Wang, X.; Wang, C.-C. Protein disulfide–isomerase, a folding catalyst and a redox-regulated chaperone. *Free Radic. Biol. Med.* **2015**, *83*, 305–313. [CrossRef]
79. Kuo, T.-F.; Chen, T.-Y.; Jiang, S.-T.; Chen, K.-W.; Chiang, Y.-M.; Hsu, Y.-J.; Liu, Y.-J.; Chen, H.-M.; Yokoyama, K.K.; Tsai, K.-C.; et al. Protein disulfide isomerase a4 acts as a novel regulator of cancer growth through the procaspase pathway. *Oncogene* **2017**, *36*, 5484–5496. [CrossRef]
80. Lin, C.-Y.; Hu, C.-T.; Cheng, C.-C.; Lee, M.-C.; Pan, S.-M.; Lin, T.-Y.; Wu, W.-S. Oxidation of heat shock protein 60 and protein disulfide isomerase activates ERK and migration of human hepatocellular carcinoma HepG2. *Oncotarget* **2016**, *7*, 11067–11082. [CrossRef] [PubMed]
81. Gęgotek, A.; Jarocka-Karpowicz, I.; Skrzydlewska, E. Synergistic cytoprotective effects of rutin and ascorbic acid on the proteomic profile of 3D-cultured keratinocytes exposed to UVA or UVB radiation. *Nutrients* **2019**, *11*, 2672. [CrossRef]
82. Yokota, S.; Yanagi, H.; Yura, T.; Kubota, H. Cytosolic chaperonin-containing t-complex polypeptide 1 changes the content of a particular subunit species concomitant with substrate binding and folding activities during the cell cycle. *Eur. J. Biochem.* **2001**, *268*, 4664–4673. [CrossRef]
83. Vallin, J.; Grantham, J. The role of the molecular chaperone CCT in protein folding and mediation of cytoskeleton-associated processes: Implications for cancer cell biology. *Cell Stress Chaperones* **2019**, *24*, 17–27. [CrossRef] [PubMed]
84. Balcer, H.I.; Goodman, A.L.; Rodal, A.A.; Smith, E.; Kugler, J.; Heuser, J.E.; Goode, B.L. Coordinated regulation of actin filament turnover by a high-molecular-weight Srv2/CAP complex, cofilin, profilin, and Aip1. *Curr. Biol.* **2003**, *13*, 2159–2169. [CrossRef]
85. Zhang, H.; Ghai, P.; Wu, H.; Wang, C.; Field, J.; Zhou, G.-L. Mammalian adenylyl cyclase-associated protein 1 (CAP1) regulates cofilin function, the actin cytoskeleton, and cell adhesion. *J. Biol. Chem.* **2013**, *288*, 20966–20977. [CrossRef] [PubMed]
86. Ding, Z.; Lambrechts, A.; Parepally, M.; Roy, P. Silencing profilin-1 inhibits endothelial cell proliferation, migration and cord morphogenesis. *J. Cell. Sci.* **2006**, *119*, 4127–4137. [CrossRef] [PubMed]
87. Yun, S.P.; Ryu, J.M.; Kim, M.O.; Park, J.H.; Han, H.J. Rapid actions of plasma membrane estrogen receptors regulate motility of mouse embryonic stem cells through a profilin-1/cofilin-1-directed kinase signaling pathway. *Mol. Endocrinol.* **2012**, *26*, 1291–1303. [CrossRef] [PubMed]
88. Vié, N.; Copois, V.; Bascoul-Mollevi, C.; Denis, V.; Bec, N.; Robert, B.; Fraslon, C.; Conseiller, E.; Molina, F.; Larroque, C.; et al. Overexpression of phosphoserine aminotransferase PSAT1 stimulates cell growth and increases chemoresistance of colon cancer cells. *Mol. Cancer* **2008**, *7*, 1–14. [CrossRef] [PubMed]
89. Metcalf, S.; Dougherty, S.; Kruer, T.; Hasan, N.; Biyik-Sit, R.; Reynolds, L.; Clem, B.F. Selective loss of phosphoserine aminotransferase 1 (PSAT1) suppresses migration, invasion, and experimental metastasis in triple negative breast cancer. *Clin. Exp. Metastasis* **2020**, *37*, 187–197. [CrossRef] [PubMed]
90. Chen, W.; Wang, W.; Sun, X.; Xie, S.; Xu, X.; Liu, M.; Yang, C.; Li, M.; Zhang, W.; Liu, W.; et al. NudCL2 regulates cell migration by stabilizing both myosin-9 and LIS1 with Hsp90. *Cell Death Dis.* **2020**, *11*, 1–15. [CrossRef]
91. Sarkar, S.; Egelhoff, T.; Baskaran, H. Insights into the roles of non-muscle myosin IIA in human keratinocyte migration. *Cell. Mol. Bioeng.* **2009**, *2*, 486–494. [CrossRef] [PubMed]
92. Xu, Q.; Wu, N.; Cui, L.; Wu, Z.; Qiu, G. Filamin B: The next hotspot in skeletal research? *J. Genet. Genom.* **2017**, *44*, 335–342. [CrossRef]
93. Feng, Y.; Chen, M.H.; Moskowitz, I.P.; Mendonza, A.M.; Vidali, L.; Nakamura, F.; Kwiatkowski, D.J.; Walsh, C.A. Filamin A (FLNA) is required for cell–cell contact in vascular development and cardiac morphogenesis. *Proc. Natl. Acad. Sci. USA* **2006**, *103*, 19836–19841. [CrossRef]
94. Hamill, K.J.; Hiroyasu, S.; Colburn, Z.T.; Ventrella, R.V.; Hopkinson, S.B.; Skalli, O.; Jones, J.C.R. Alpha actinin-1 regulates cell-matrix adhesion organization in keratinocytes: Consequences for skin cell motility. *J. Investig. Dermatol.* **2015**, *135*, 1043–1052. [CrossRef]
95. Lin, Y.H.; Park, Z.-Y.; Lin, D.; Brahmbhatt, A.A.; Rio, M.-C.; Yates, J.R., III; Klemke, R.L. Regulation of cell migration and survival by focal adhesion targeting of Lasp-1. *J. Cell. Biol.* **2004**, *165*, 421–432. [CrossRef] [PubMed]

96. Bizzarro, V.; Fontanella, B.; Carratù, A.; Belvedere, R.; Marfella, R.; Parente, L.; Petrella, A. Annexin A1 N-terminal derived peptide Ac2-26 stimulates fibroblast migration in high glucose conditions. *PLoS ONE* **2012**, *7*, e45639.
97. Tsukita, S.; Yonemura, S.; Tsukita, S. ERM (ezrin/radixin/moesin) family: From cytoskeleton to signal transduction. *Curr. Opin. Cell Biol.* **1997**, *9*, 70–75. [CrossRef]
98. Jacob, J.T.; Coulombe, P.A.; Kwan, R.; Omary, M.B. Types I and II Keratin Intermediate Filaments. *Cold Spring Harb. Perspect. Biol.* **2018**, *10*, a018275. [CrossRef]
99. Machesky, L.M.; Li, A. Fascin. *Commun. Integr. Biol.* **2010**, *3*, 263–270. [CrossRef] [PubMed]
100. Lunt, S.Y.; Vander Heiden, M.G. Aerobic glycolysis: Meeting the metabolic requirements of cell proliferation. *Annu. Rev. Cell Dev. Biol.* **2011**, *27*, 441–464. [CrossRef]
101. Yoshida, M.; Muneyuki, E.; Hisabori, T. ATP synthase—A marvellous rotary engine of the cell. *Nat. Rev. Mol. Cell Biol.* **2001**, *2*, 669–677. [CrossRef]
102. Grüning, N.-M.; Du, D.; Keller, M.A.; Luisi, B.F.; Ralser, M. Inhibition of triosephosphate isomerase by phosphoenolpyruvate in the feedback-regulation of glycolysis. *Open Biol.* **2014**, *4*, 130232. [CrossRef]
103. de Padua, M.C.; Delodi, G.; Vučetić, M.; Durivault, J.; Vial, V.; Bayer, P.; Noleto, G.R.; Mazure, N.M.; Ždralević, M.; Pouysségur, J. Disrupting glucose-6-phosphate isomerase fully suppresses the "Warburg effect" and activates OXPHOS with minimal impact on tumor growth except in hypoxia. *Oncotarget* **2017**, *8*, 87623–87637. [CrossRef] [PubMed]
104. Zong, M.; Lu, T.; Fan, S.; Zhang, H.; Gong, R.; Sun, L.; Fu, Z.; Fan, L. Glucose-6-phosphate isomerase promotes the proliferation and inhibits the apoptosis in fibroblast-like synoviocytes in rheumatoid arthritis. *Arthritis Res. Ther.* **2015**, *17*, 100. [CrossRef]
105. Jin, C.; Zhu, X.; Wu, H.; Wang, Y.; Hu, X. Perturbation of phosphoglycerate kinase 1 (PGK1) only marginally affects glycolysis in cancer cells. *J. Biol. Chem.* **2020**, *295*, 6425–6446. [CrossRef]
106. Hanse, E.A.; Ruan, C.; Kachman, M.; Wang, D.; Lowman, X.H.; Kelekar, A. Cytosolic malate dehydrogenase activity helps support glycolysis in actively proliferating cells and cancer. *Oncogene* **2017**, *36*, 3915–3924. [CrossRef]
107. Ding, M.; Li, J.; Leonard, S.S.; Shi, X.; Costa, M.; Castranova, V.; Vallyathan, V.; Huang, C. Differential role of hydrogen peroxide in UV-induced signal transduction. *Mol. Cell. Biochem.* **2002**, *234–235*, 81–90. [CrossRef]
108. Yamamoto, Y.; Gaynor, R.B. IκB kinases: Key regulators of the NF-κB pathway. *Trends Biochem. Sci.* **2004**, *29*, 72–79. [CrossRef]
109. Yeo, H.; Lee, J.Y.; Kim, J.; Ahn, S.S.; Jeong, J.Y.; Choi, J.H.; Lee, Y.H.; Shin, S.Y. Transcription factor EGR-1 transactivates the MMP1 gene promoter in response to TNFα in HaCaT keratinocytes. *BMB Rep.* **2020**, *53*, 323–328. [CrossRef] [PubMed]
110. Tobiume, K.; Matsuzawa, A.; Takahashi, T.; Nishitoh, H.; Morita, K.-I.; Takeda, K.; Minowa, O.; Miyazono, K.; Noda, T.; Ichijo, H. ASK1 is required for sustained activations of JNK/p38 MAP kinases and apoptosis. *EMBO Rep.* **2001**, *2*, 222–228. [CrossRef]
111. Elliott, S.F.; Coon, C.I.; Hays, E.; Stadheim, T.A.; Vincenti, M.P. Bcl-3 is an interleukin-1–responsive gene in chondrocytes and synovial fibroblasts that activates transcription of the matrix metalloproteinase 1 gene. *Arthritis Rheum.* **2002**, *46*, 3230–3239. [CrossRef]
112. Fanjul-Fernández, M.; Folgueras, A.R.; Cabrera, S.; López-Otín, C. Matrix metalloproteinases: Evolution, gene regulation and functional analysis in mouse models. *Biochim. Biophys. Acta* **2010**, *1803*, 3–19. [CrossRef]
113. Buckman, S.Y.; Gresham, A.; Hale, P.; Hruza, G.; Anast, J.; Masferrer, J.; Pentland, A.P. COX-2 expression is induced by UVB exposure in human skin: Implications for the development of skin cancer. *Carcinogenesis* **1998**, *19*, 723–729. [CrossRef] [PubMed]
114. He, Y.; Hu, Y.; Jiang, X.; Chen, T.; Ma, Y.; Wu, S.; Sun, J.; Jiao, R.; Li, X.; Deng, L.; et al. Cyanidin-3-O-glucoside inhibits the UVB-induced ROS/COX-2 pathway in HaCaT cells. *J. Photochem. Photobiol. B* **2017**, *177*, 24–31. [CrossRef] [PubMed]
115. Arora, S.; Tyagi, N.; Bhardwaj, A.; Rusu, L.; Palanki, R.; Vig, K.; Singh, S.R.; Singh, A.P.; Palanki, S.; Miller, M.E.; et al. Silver nanoparticles protect human keratinocytes against UVB radiation-induced DNA damage and apoptosis: Potential for prevention of skin carcinogenesis. *Nanomedicine* **2015**, *11*, 1265–1275. [CrossRef] [PubMed]
116. Pyo, C.-W.; Choi, J.H.; Oh, S.-M.; Choi, S.-Y. Oxidative stress-induced cyclin D1 depletion and its role in cell cycle processing. *Biochim. Biophys. Acta* **2013**, *1830*, 5316–5325. [CrossRef]
117. Han, W.; He, Y.Y. Requirement for metalloproteinase-dependent ERK and AKT activation in UVB-induced G1-S cell cycle progression of human keratinocytes. *Photochem. Photobiol.* **2009**, *85*, 997–1003. [CrossRef]
118. Yang, K.; Hitomi, M.; Stacey, D.W. Variations in cyclin D1 levels through the cell cycle determine the proliferative fate of a cell. *Cell Div.* **2006**, *1*, 1–8. [CrossRef]
119. Peña-Blanco, A.; García-Sáez, A.J. Bax, Bak and beyond—Mitochondrial performance in apoptosis. *FEBS J.* **2018**, *285*, 416–431. [CrossRef]
120. Ryu, B.; Ahn, B.-N.; Kang, K.-H.; Kim, Y.-S.; Li, Y.-X.; Kong, C.-S.; Kim, S.-K.; Kim, D.G. Dioxinodehydroeckol protects human keratinocyte cells from UVB-induced apoptosis modulated by related genes Bax/Bcl-2 and caspase pathway. *J. Photochem. Photobiol. B* **2015**, *153*, 352–357. [CrossRef]
121. Chen, F.; Tang, Y.; Sun, Y.; Veeraraghavan, V.P.; Mohan, S.K.; Cui, C. 6-shogaol, a active constiuents of ginger prevents UVB radiation mediated inflammation and oxidative stress through modulating NrF2 signaling in human epidermal keratinocytes (HaCaT cells). *J. Photochem. Photobiol. B.* **2019**, *197*, 111518. [CrossRef]
122. Tian, Y.; Hoshino, T.; Chen, C.J.; E, Y.; Yabe, S.; Liu, W. The evaluation of whitening efficacy of cosmetic products using a human skin pigmentation spot model. *Skin Res. Technol.* **2009**, *15*, 218–223. [CrossRef]
123. Goh, J.X.H.; Tan, L.T.-H.; Yew, H.C.; Pusparajah, P.; Lingham, P.; Long, C.M.; Lee, L.-H.; Goh, B.-H. Hydration effects of moisturizing gel on normal skin: A pilot study. *Prog. Drug Discov. Biomed. Sci.* **2019**, *2*, a0000023. [CrossRef]

124. Adejokun, D.A.; Dodou, K. A novel method for the evaluation of the long-term stability of cream formulations containing natural oils. *Cosmetics* **2020**, *7*, 86. [CrossRef]
125. Mosquera, T.; Peña, S.; Álvarez, P.; Noriega, P. Changes in skin elasticity and firmness caused by cosmetic formulas elaborated with essential oils of Aristeguietia glutinosa (matico) and Ocotea quixos (ishpingo). A statistical analysis. *Cosmetics* **2020**, *7*, 95. [CrossRef]
126. Goh, B.H.; Chan, C.K.; Kamarudin, M.N.A.; Abdul Kadir, H. Swietenia macrophylla King induces mitochondrial-mediated apoptosis through p53 upregulation in HCT116 colorectal carcinoma cells. *J. Ethnopharmacol.* **2014**, *153*, 375–385. [CrossRef]
127. Tan, L.T.-H.; Mahendra, C.K.; Yow, Y.-Y.; Chan, K.-G.; Khan, T.M.; Lee, L.-H.; Goh, B.-H. Streptomyces sp. MUM273b: A mangrove-derived potential source for antioxidant and UVB radiation protectants. *Microbiol. Open* **2019**, *8*, e859. [CrossRef]
128. Mahendra, C.K.; Tan, L.T.-H.; Yap, W.H.; Chan, C.K.; Pusparajah, P.; Goh, B.H. An optimized cosmetic screening assay for ultraviolet B (UVB) protective property of natural products. *Prog. Drug Discov. Biomed. Sci.* **2019**, *2*, 1–6. [CrossRef]
129. Paudel, Y.N.; Kumari, Y.; Abidin, S.A.Z.; Othman, I.; Shaikh, M.F. Pilocarpine induced behavioral and biochemical alterations in chronic seizure-like condition in adult zebrafish. *Int. J. Mol. Sci.* **2020**, *21*, 2492. [CrossRef]
130. Garcin, G.; Le Gallic, L.; Stoebner, P.-E.; Guezennec, A.; Guesnet, J.; Lavabre-Bertrand, T.; Martinez, J.; Meunier, L. Constitutive expression of MC1R in HaCaT keratinocytes inhibits basal and UVB-induced TNF-alpha production. *Photochem. Photobiol.* **2009**, *85*, 1440–1450. [CrossRef]
131. Adachi, H.; Murakami, Y.; Tanaka, H.; Nakata, S. Increase of stratifin triggered by ultraviolet irradiation is possibly related to premature aging of human skin. *Exp. Dermatol.* **2014**, *23*, 32–36. [CrossRef] [PubMed]
132. Tan, L.T.-H.; Chan, C.-K.; Chan, K.-G.; Pusparajah, P.; Khan, T.M.; Ser, H.-L.; Lee, L.-H.; Goh, B.-H. Streptomyces sp. MUM256: A Source for Apoptosis Inducing and Cell Cycle-Arresting Bioactive Compounds against Colon Cancer Cells. *Cancers* **2019**, *11*, 1742. [CrossRef] [PubMed]
133. Xie, S.-P.; Zhou, F.; Li, J.; Duan, S.-J. NEAT1 regulates MPP+-induced neuronal injury by targeting miR-124 in neuroblastoma cells. *Neurosci. Lett.* **2019**, *708*, 134340. [CrossRef]

Article

Simultaneous Determination of Polyamines and Steroids in Human Serum from Breast Cancer Patients Using Liquid Chromatography–Tandem Mass Spectrometry

Yu Ra Lee [1,2], Ji Won Lee [3], Jongki Hong [2,4,*] and Bong Chul Chung [1,2,*]

1. Molecular Recognition Research Center, Korea Institute of Science and Technology, Seoul 02792, Korea; T16627@kist.re.kr
2. KHU-KIST Department of Converging Science and Technology, Kyung Hee University, Seoul 02447, Korea
3. Department of Family Medicine, Gangnam Severance Hospital, Yonsei University College of Medicine, Seoul 06273, Korea; indi5645@yuhs.ac
4. College of Pharmacy, Kyung Hee University, Seoul 02447, Korea
* Correspondence: jhong@khu.ac.kr (J.H.); bcc0319@kist.re.kr (B.C.C.); Tel.: +82-2-961-9255 (J.H.); +82-2-958-5077 (B.C.C.)

Abstract: A simultaneous quantitative profiling method for polyamines and steroids using liquid chromatography–tandem mass spectrometry was developed and validated. We applied this method to human serum samples to simultaneously evaluate polyamine and steroid levels. Chemical derivatization was performed using isobutyl chloroformate to increase the sensitivity of polyamines. The method was validated, and the matrix effects were in the range of 78.7–126.3% and recoveries were in the range of 87.8–123.6%. Moreover, the intra-day accuracy and precision were in the ranges of 86.5–116.2% and 0.6–21.8%, respectively, whereas the inter-day accuracy and precision were in the ranges of 82.0–119.3% and 0.3–20.2%, respectively. The linearity was greater than 0.99. The validated method was used to investigate the differences in polyamine and steroid levels between treated breast cancer patients and normal controls. In our results, N-acetyl putrescine, N-acetyl spermidine, cadaverine, 1,3-diaminopropane, and epitestosterone were significantly higher in the breast cancer patient group. Through receiver operating characteristic curve analysis, all metabolites that were significantly increased in patient groups with areas under the curve >0.8 were shown. This mass spectrometry-based quantitative profiling method, used for the investigation of breast cancer, is also applicable to androgen-dependent diseases and polyamine-related diseases.

Keywords: polyamine; steroid; breast cancer; liquid chromatography–tandem mass spectrometry; serum

1. Introduction

Breast cancer is a common hormone-related cancer, which includes estrogen receptor-positive and progesterone receptor-positive disease. According to statistics from 2017, it is the fifth most common type of cancer in Korea and ranks fifth in cancer-related mortality. In addition, breast cancer ranks first among women as a cause of death from cancer [1]. Although the incidence of most cancers in Korean women has been declining since 2007, breast cancer continues to increase, with the highest incidence among other cancer types. Despite an increasing understanding of the molecular etiology of breast cancer over the past 20 years, there remains a lack of reliable biomarkers to monitor treatment efficacy associated with the disease. Currently, treatment of breast cancer commonly involves surgery, chemotherapy, radiotherapy, and hormone therapy [2]. However, it is important to confirm the effect of breast cancer treatment with a simple experimental method.

Polyamines are aliphatic amines composed of straight chains of carbon atoms. Amines have several biological implications, particularly polyamines, which mainly act as proliferation factors in cells. Cancer cells are highly proliferative; therefore, polyamines are one of the most important biomarkers in cancer research. The increase in polyamine concentrations in urine samples from patients with malignant cancer was first reported by D. H. Russell in 1971 [3]. Since then, polyamines have been analyzed in various biological fluids to investigate their potential as markers for the early diagnosis of cancer, evaluation of progression, and prediction of disease recurrence. It can also be seen that the number of polyamines is distributed differently depending on the type of lung and liver cancer [4]. In breast cancer patients, it has been reported that acetylated polyamines are present at a much higher concentration compared to that in normal human breast tissue [5,6]. Moreover, altered polyamines can be useful markers for the evaluation of breast cancer treatment efficacy [7].

It has been suggested that hormones such as progesterone, estrogens, and androgens are implicated in the development and/or growth of normal and neoplastic mammary tissue. Androgens, which play an important role in the development of prostate cancer, have also been shown to be associated with breast cancer, attracting academic attention. This is because the growth of estrogen receptor-positive breast cancer is decreased after blocking the androgen receptor [8]. In addition, androgens have been proposed to control tumor growth rates [9]. Free testosterone, an androgen, might play an important role in the development of breast cancer in women [10]. One study found that young women with high levels of a male hormone, like androgens, have a higher risk of developing breast cancer [11]. Progesterone, a female hormone, is also associated with breast cancer. In particular, progesterone receptors in breast cancer cells interact with estrogen receptors to change their mode of action and delay tumor growth [12]. This study suggests that an overall evaluation of polyamines and steroids will provide information on breast cancer treatment. Therefore, profiling the combined metabolism of polyamines and steroids is required in breast cancer patients.

Metabolic approaches can monitor an individual's status and help detect potential cancer biomarkers [13]. Gas chromatography–mass spectrometry and liquid chromatography–mass spectrometry (LC-MS) are used for quantitative metabolic profiling, as mass spectrometry delivers high sensitivity and is capable of characterizing complex biological samples [14,15]. In particular, in the case of cancer patients, since normal cells are mutated, not only free polyamines but also acetylated polyamines must be analyzed in relation to intracellular metabolic pathways. However, while analyzing acetylated polyamines using gas chromatography–mass spectrometry, additional active hydrogen must be removed, which requires additional derivatization [16]. One disadvantage of this is that the experiment time is increased due to the use of two or more derivatizations to remove active hydrogen sites. This process makes it less suitable for the analysis of acetylated polyamines. In addition, more specific and sensitive results can be obtained when using liquid chromatography–tandem mass spectrometry (LC-MS/MS) than LC-MS, which can be carried out using one derivatization, and thus, we conducted an experiment using the LC-MS/MS condition.

Therefore, in this study, an analytical method was validated for the simultaneous quantitative profiling of serum polyamines and steroids. To the best of our knowledge, this study is the first to simultaneously analyze polyamines and steroids in human serum samples from breast cancer patients. The developed and validated method was applied to analyze the serum concentrations of nine polyamines and eight steroids in patients with breast cancer after treatment and in normal female subjects using LC-MS/MS system.

2. Results and Discussion

2.1. Sample Preparation and Optimization

Polyamines have a positive charge at physiological pH [17] due to amino groups in their molecular structure; therefore, they easily bind with other substances. However, they are low-molecular weight molecules that rapidly elute from the chromatogram and therefore hinder accurate analysis. To solve this problem, we used an amine-carbamylated derivatization agent, isobutyl chloroformate. Reaction with a derivatization reagent prioritizes the binding of polyamines to other substances, and, therefore, it is possible to selectively extract polyamines.

Nine polyamines that were detected in the biological sample were subjected to an amine-carbamylated derivatization reaction. After polyamine derivatization, deprotonation was difficult because the amine and carbamoyl groups formed a stable bond. In addition, acetyl polyamine was substituted with nitrogen to which no acetyl group was attached. As a result, one carbamoyl group was substituted in N-acetyl putrescine (N-PUT) and N-acetyl cadaverine (N-CAD), two carbamoyl groups in N-acetyl spermidine (N-SPD), and three carbamoyl groups in N-acetyl spermine (N-SPM).

The carbamylation procedure that has been routinely used in our laboratory for polyamine analysis [18] had to be optimized. Serum samples were precipitated by reactions at high temperature (60 °C) for 20 min, derivatized using isobutyl chloroformate, and extracted using the liquid–liquid extraction (LLE) method with diethyl ether. Derivatization conditions were optimized to improve analyte sensitivity. Optimum conditions were selected by comparing the peak areas. A reaction time of 15 min (Figure 1A) and a reaction temperature of 35 °C were selected (Figure 1B). The LLE method was used to clean the sample and minimize interference. Three different eluents, diethyl ether, ethyl acetate, and methyl tert-butyl ether (MTBE), were tested to enhance the extraction efficacy. The extraction efficiencies of polyamines were relatively higher in MTBE; however, those of cadaverine (CAD), PUT, and 1,3-diaminopropane (DAP) were outside the acceptable range. The extraction efficiencies of all analytes were acceptable when diethyl ether was used (Figure 1C). This optimization condition was the same as that in our previous experimental conditions [18].

2.2. Liquid Chromatography–Tandem Mass Spectrometry

After the derivatization reaction, the analyte in the chromatogram can be effectively separated and analyzed simultaneously (Figure 2). In addition, when isobutyl chloroformate is used, the reaction in the aqueous solution is easy and exhibits high efficiency. This reaction can occur even at room temperature (20 to 25 °C) or with a little warming, with the completion of the derivatization reaction occurring in a short time (5 to 15 min).

All polyamines and steroids produced protonated precursor ions $[M + H]^+$ in the positive ion mode. All polyamines were derivatized with isobutyl chloroformate, whereas the steroids were not derivatized. Therefore, all steroids were detected in the free form, and steroid analysis revealed good intensity even without derivatization. All polyamines showed the $[M + H - OCH_2C_3H_7]^+$ ion as the base peak for quantitation (Figure 3). All steroids showed the $[M + H - H_2O]^+$ ion as the base peak of a fragment of high intensity for quantitation. In principle, LC-MS/MS using a stable internal standard (IS) is an optimal method for quantitative analysis. In particular, among our analytes, epimer-type substances (epitestosterone, testosterone, 17α-hydroxyprogesterone, and 11β-hydroxyprogesterone) might show false positives. However, we tried to separate the substances in the form of epimers as much as possible by adjusting the retention time, and there was a difference of approximately 0.6–1 min. Using the present method with aqueous extracts of serum, we achieved excellent separation of nine polyamines and eight steroids with no significantly interfering background peaks.

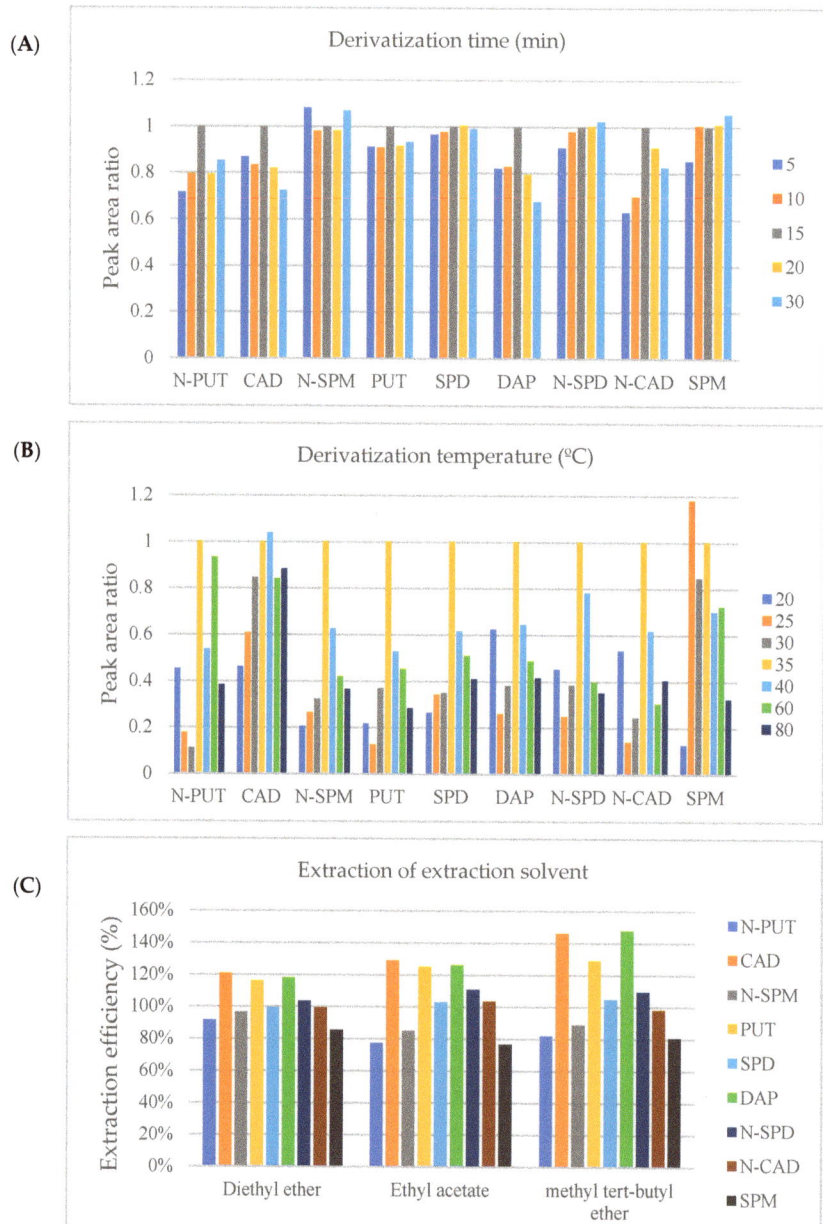

Figure 1. Comparison of the optimization procedures. (**A**) Isobutyl chloroformate derivatization under varying conditions including reaction time, (**B**) isobutyl chloroformate derivatization under varying reaction temperatures, and (**C**) extraction of liquid–liquid extraction solvents (N-PUT: *N*-acetyl putrescine; CAD: cadaverine; N-SPM: *N*-acetyl spermine; PUT: putrescine; SPD: spermidine; DAP: 1,3-diaminopropane; N-SPD: *N*-acetyl spermidine; N-CAD: *N*-acetyl cadaverine; SPM: spermine). The peak area ratios were expressed by dividing (**A**) 15 min and (**B**) 35 °C as standard values that we used for the conditions of this experiment.

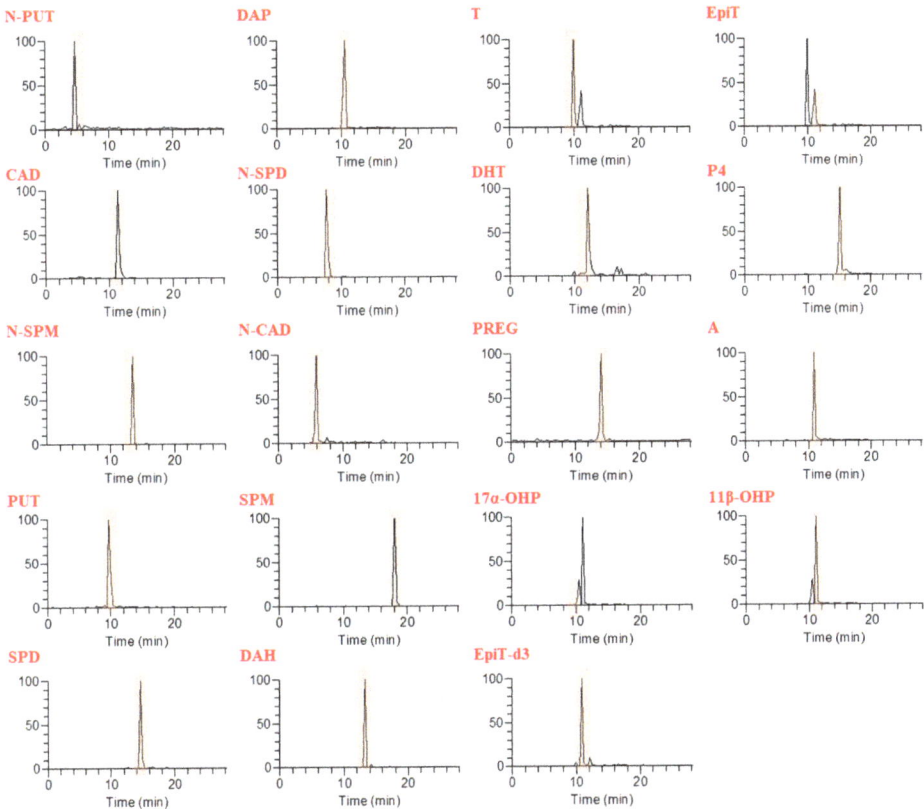

Figure 2. Chromatograms of nine polyamines, eight steroids, and two internal standards (IS) in the selected reaction monitoring mode, with target standard pretreatment (500 ng/mL). N-PUT: *N*-acetyl putrescine; CAD: cadaverine; N-SPM: *N*-acetyl spermine; PUT: putrescine; SPD: spermidine; DAP: 1,3-diaminopropane; N-SPD: *N*-acetyl spermidine; N-CAD: *N*-acetyl cadaverine; SPM: spermine; T: testosterone; EpiT: epitestosterone; DHT: dihydrotestosterone; PREG: pregnenolone; 17α-OHP:17α-hydroxyprogesterone; 11β-OHP: 11β-hydroxyprogesterone; A: androstenedione; P4: progesterone.

Figure 3. Fragmentation pattern of putrescine upon isobutyl chloroformate derivatization.

2.3. Method Validation

The developed method was validated by assessing the accuracy and precision of the quality control (QC) samples with four different concentrations. Moreover, linearity was performed with 10 calibration points (0.1, 1, 5, 10, 50, 100, 500, 1000, 2000, and 5000 ng/mL), excluding 5000 ng/mL and adding 0.5 ng/mL for 17α-hydroxyprogesterone (17α-OHP), 11β-hydroxyprogesterone (11β-OHP), androstenedione (A), and progesterone (P4). The regression equation was found to be linear over the dynamic ranges of all analytes (with correlation coefficient, R^2 > 0.99). The limit of quantification (LOQ) values were determined for most polyamines at 1 ng/mL, except N-acetyl spermine (N-SPM; 0.1 ng/mL). In contrast, the LOQ values were determined for testosterone (T), epitestosterone (EpiT), dihydrotestosterone (DHT), and pregnenolone (PREG) at 1 ng/mL, and the LOQ values were determined for 17α-OHP, 11β-OHP, A, and P4 at 0.1 ng/mL. As shown in Table 1, we checked the matrix effect and overall recovery. Most analytes did not present any significant matrix effect, ranging from 78.7 to 126.3%, and the recovery of analytes was 87.8–123.6%. Precision was assessed based on the coefficient of variation (% CV), and accuracy was assessed through the relative error rate (% bias). For polyamine analysis, the intra-day (n = 3) precision (coefficient of variation (% CV)) and accuracy (% bias) were in the ranges of 1.6–21.2% and 86.5–116.2%, respectively, whereas the inter-day (n = 3) precision and accuracy were in the ranges of 0.3–20.2% and 87.8–119.3%, respectively. For steroid analysis, the intra-day (n = 3) precision (% CV) and accuracy (% bias) were in the ranges of 0.6–21.8% and 91–114.3%, respectively, whereas the inter-day (n = 3) precision and accuracy were in the ranges of 1.2–18.5% and 82–108.0%, respectively (Table 2).

Table 1. Calibration range, linear regression equation, limit of quantification (LOQ), matrix effect, and recovery of polyamines and steroids.

Analytes	Calibration Range	Linear Regression Equation	Standard Errors of the Slope	Standard Errors of the Intercept	R^2	LOQ	Matrix Effect (%)	Recovery (%)
N-PUT	1–5000	y = 0.0008x + 0.0986	1.42^{-5}	0.03	0.998	1	101.4	107.4
CAD	1–5000	y = 0.0001x + 0.0718	6.07^{-6}	0.01	0.995	1	83.1	87.9
N-SPM	0.1–5000	y = 0.084x − 0.6771	1.76^{-4}	0.34	0.998	0.1	86.8	101.8
PUT	1–5000	y = 0.001x + 0.0949	7.90^{-5}	0.15	0.992	1	115.5	98.2
SPD	1–5000	y = 0.01x − 0.6062	1.59^{-3}	0.31	0.999	1	126.3	90.1
DAP	1–5000	y = 0.0013x + 0.0449	5.89^{-5}	0.11	0.996	1	89.7	101.7
N-SPD	1–5000	y = 0.0067x + 0.419	3.76^{-4}	0.73	0.990	1	118.4	123.6
N-CAD	1–5000	y = 0.0003x + 0.13	1.26^{-5}	0.02	0.992	1	87.5	100.7
SPM	1–5000	y = 6E-05x + 0.0088	1.79^{-5}	0.03	0.995	1	86.4	107.8
T	1–2000	y = 0.0032x + 0.1867	7.92^{-5}	0.15	0.996	1	106.7	89.9
EpiT	1–2000	y = 0.0045x + 0.4609	1.59^{-4}	0.31	0.993	1	103.4	100.4
DHT	1–2000	y = 0.0019x + 0.0017	6.13^{-5}	0.12	0.994	1	82.1	104.7
PREG	1–2000	y = 0.0003x − 0.0244	1.11^{-5}	0.02	0.991	1	79.3	97.7
17α-OHP	0.1–2000	y = 0.0115x + 0.1561	3.94^{-4}	0.77	0.993	0.1	84.6	87.8
11β-OHP	0.1–2000	y = 0.0049x + 0.1846	1.83^{-4}	0.36	0.992	0.1	88.6	91.8
A	0.1–2000	y = 0.0017x + 0.038	3.38^{-5}	0.07	0.998	0.1	78.7	101.0
P4	0.1–2000	y = 0.0062x − 0.2893	3.01^{-4}	0.58	0.994	0.1	100.9	96.7

N-PUT: *N*-acetyl putrescine; CAD: cadaverine; N-SPM: *N*-acetyl spermine; PUT: putrescine; SPD: spermidine; DAP: 1,3-diaminopropane; N-SPD: *N*-acetyl spermidine; N-CAD: *N*-acetyl cadaverine; SPM: spermine; T: testosterone; EpiT: epitestosterone; DHT: dihydrotestosterone; PREG: pregnenolone; 17α-OHP: 17α-hydroxyprogesterone; 11β-OHP: 11β-hydroxyprogesterone; A: androstenedione; P4: progesterone.

Table 2. Intra-day and inter-day validation of polyamines and steroids

Analytes	Spiked Concentration (ng/mL)	Intra-Day (n = 3)		Inter-Day (n = 3)	
		Accuracy (%Bias)	Precision (%CV)	Accuracy (%Bias)	Precision (%CV)
N-PUT	10	91.2	8.1	88.1	11.3
	50	96.8	5.4	96.4	13.6
	500	110.4	7.0	92.9	18.5
	1000	104.1	7.0	106.7	2.6
CAD	10	86.5	13.0	93.8	13.4
	50	107.4	5.9	100.0	7.9
	500	95.4	11.1	103.2	5.2
	1000	92.1	9.2	102.5	9.8
N-SPM	1	104.0	16.9	104.6	19.2
	50	88.2	15.3	101.3	0.3
	500	113.6	6.2	110.5	3.1
	1000	112.8	4.8	118.3	2.9
PUT	10	99.0	10.5	107.3	18.3
	50	103.5	8.6	105.5	5.8
	500	99.1	15.3	115.1	3.2
	1000	103.4	15.4	102.0	17.0
SPD	10	105.8	16.3	96.2	19.7
	50	116.2	21.2	109.6	17.5
	500	93.8	18.3	92.0	6.8
	1000	98.6	19.6	104.7	11.5
DAP	10	108.6	17.1	94.3	16.4
	50	105.4	8.7	94.7	20.2
	500	101.2	1.7	106.6	10.1
	1000	103.9	6.2	93.6	6.6
N-SPD	10	108.5	1.6	106.5	18.3
	50	110.6	11.5	104.8	11.5
	500	104.3	6.5	105.6	13.6
	1000	114.5	2.7	106.3	16.5
N-CAD	10	95.5	13.9	115.6	4.1
	50	109.6	15.1	111.0	14.3
	500	110.5	9.6	90.7	12.4
	1000	97.5	13.5	90.3	9.5
SPM	10	91.5	14.9	92.4	0.7
	50	104.2	13.3	119.3	7.8
	500	104.3	10.6	87.8	6.9
	1000	103.5	12.2	98.4	16.4
T	10	104.9	12.5	106.8	17.6
	50	94.8	14.0	106.1	11.2
	500	104.0	2.4	94.6	2.5
	1000	100.7	9.0	100.5	8.1
EpiT	10	100.9	7.0	101.7	5.9
	50	106.8	9.0	102.1	6.4
	500	108.6	10.0	104.0	4.3
	1000	105.6	6.9	99.9	12.3
DHT	10	106.6	8.7	97.7	9.5
	50	114.3	6.5	90.2	11.9
	500	111.5	3.9	97.6	16.0
	1000	100.6	6.8	95.5	18.5
PREG	10	111.1	17.8	82.0	17.7
	50	92.4	0.6	96.6	7.5
	500	98.9	16.0	96.1	11.4
	1000	103.7	7.4	101.7	8.6

Table 2. Cont.

Analytes	Spiked Concentration (ng/mL)	Intra-Day (n = 3)		Inter-Day (n = 3)	
		Accuracy (%Bias)	Precision (%CV)	Accuracy (%Bias)	Precision (%CV)
17α-OHP	1	101.5	17.1	107.9	8.4
	50	102.2	15.4	106.8	7.9
	500	109.1	13.2	100.4	3.8
	1000	96.4	1.2	95.0	1.2
11β-OHP	1	98.9	3.5	99.4	13.5
	50	95.0	21.8	108.0	4.7
	500	109.3	6.0	92.0	7.7
	1000	109.9	7.1	106.1	7.3
A	1	106.1	5.7	107.0	10.0
	50	96.9	6.9	94.8	10.2
	500	101.0	1.0	107.8	4.2
	1000	93.3	8.5	91.2	9.0
P4	1	99.9	19.7	101.0	1.7
	50	101.0	3.5	97.1	9.4
	500	91.0	13.1	98.3	8.6
	1000	110.8	5.3	103.0	18.1

N-PUT: N-acetyl putrescine; CAD: cadaverine; N-SPM: N-acetyl spermine; PUT: putrescine; SPD: spermidine; DAP: 1,3-diaminopropane; N-SPD: N-acetyl spermidine; N-CAD: N-acetyl cadaverine; SPM: spermine; T: testosterone; EpiT: epitestosterone; DHT: dihydrotestosterone; PREG: pregnenolone; 17α-OHP: 17α-hydroxyprogesterone; 11β-OHP: 11β-hydroxyprogesterone; A: androstenedione; P4: progesterone.

2.4. Application of Serum Polyamine and Steroid Profiles to Patients with Breast Cancer and Normal Controls

In this study, we quantitatively analyzed polyamines and steroids in human serum from treated cancer patients and normal controls. Using 200 μL aliquots of the serum samples, we detected nine polyamines and eight steroids for which the concentrations varied in the ranges of 0.14–632.22 ng/mL with polyamines and 0.12–62.74 ng/mL with steroids. Large variations in both polyamine and steroid levels were observed between patients with breast cancer after treatment and normal controls. As shown in Table 3, all polyamines were higher in the patient groups, whereas most of the steroids were higher in the patient groups, except 11β-OHP. Among the metabolites we analyzed, it was confirmed that polyamines, androgen, and progesterone levels were similar to those of reference ranges [19–22].

2.5. Receiver Operating Characteristic Curve

Receiver operating characteristic (ROC) curve analysis is generally performed to derive potential biomarkers. The results of ROC curve analysis were based on the results of univariate and multivariate statistical analyses that ensured the reliability of potential biomarkers for independent validation. Through t-test analysis, N-PUT, N-SPD, DAP, CAD, and EpiT were found to be significant markers, and the ROC curves for these five compounds were plotted (Figure 4). The area under the curve (AUC) for the remaining polyamines, namely N-SPD, DAP, CAD, and EpiT, was higher than 0.8, whereas the AUC for N-PUT and EpiT was higher than 0.9; therefore, it was possible to predict the potential markers.

Table 3. Concentrations of nine polyamines and eight steroids in human serum samples of patients with breast cancer after treatment and normal controls (ng/mL).

	Normal Controls (n = 10)		Patients (n = 10)		p Value
	Mean ± SD	Median, Range	Mean ± SD	Median, Range	
Polyamines					
N-PUT	11.12 ± 4.34	11.09, 1.6–16.53	61.64 ± 58.88	46.57, 15.88–183.08	0.021
CAD	60.3 ± 27.33	66.35, 12.17–96.11	208.21 ± 169.71	174.62, 31.9–545.38	0.027
N-SPM	0.53 ± 0.45	0.45, 0.14–1.81	1.1 ± 1.13	0.82, 0.16–3.6	0.12
PUT	54.88 ± 70.74	16.3, 2.65–232.98	120.94 ± 170.91	38.52, 2.38–586.46	0.146
SPD	15.15 ± 20.73	4.88, 1.3–76.45	19.92 ± 22.53	13.62, 1.06–77.13	0.551
DAP	14.38 ± 6.08	14.99, 2.5–22.12	39.48 ± 21.64	36.79, 12.02–67.47	0.003
N-SPD	31.06 ± 15.91	31.05, 4.42–58.37	57.75 ± 25.66	58.35, 15.51–95.85	0.016
N-CAD	20.04 ± 16.63	16.45, 2.45–50.16	76.8 ± 84.72	39.11, 15.28–236.6	0.109
SPM	63 ± 48.54	60, 13.91–172.96	179.37 ± 207.84	76.83, 7.58–632.22	0.121
Steroids					
T	4.86 ± 3.44	3.52, 1.3–10.09	8.06 ± 6.24	6.52, 1.78–18.88	0.095
EpiT	12.64 ± 5.36	10.79, 7.51–24.55	28.74 ± 13.93	24.85, 12.39–53.79	0.004
DHT	2.35 ± 1.22	1.7, 1.6–3.76	2.71 ± 1.94	2.71, 1.34–4.08	0.809
PREG	32.98 ± 11.37	31.97, 16.16–47.4	44.87 ± 15.27	44.47, 23.56–62.74	0.106
17α-OHP	1.15 ± 1.21	0.73, 0.14–3.76	1.15 ± 1.35	0.71, 0.12–4.08	0.999
11β-OHP	2.12 ± 1.7	1.9, 0.62–5.08	1.78 ± 1.47	1.42, 0.42–4.6	0.698
A	1.41 ± 1.07	1.01, 0.29–3.29	2.73 ± 2.03	2.02, 0.68–6.73	0.09
P4	1.3 ± 0.87	1.05, 0.18–3.64	1.47 ± 0.82	1.34, 0.27–3.06	0.569

N-PUT: *N*-acetyl putrescine; CAD: cadaverine; N-SPM: *N*-acetyl spermine; PUT: putrescine; SPD: spermidine; DAP: 1,3-diaminopropane; N-SPD: *N*-acetyl spermidine; N-CAD: *N*-acetyl cadaverine; SPM: spermine; T: testosterone; EpiT: epitestosterone; DHT: dihydrotestosterone; PREG: pregnenolone; 17α-OHP: 17α-hydroxyprogesterone; 11β-OHP: 11β-hydroxyprogesterone; A: androstenedione; P4: progesterone.

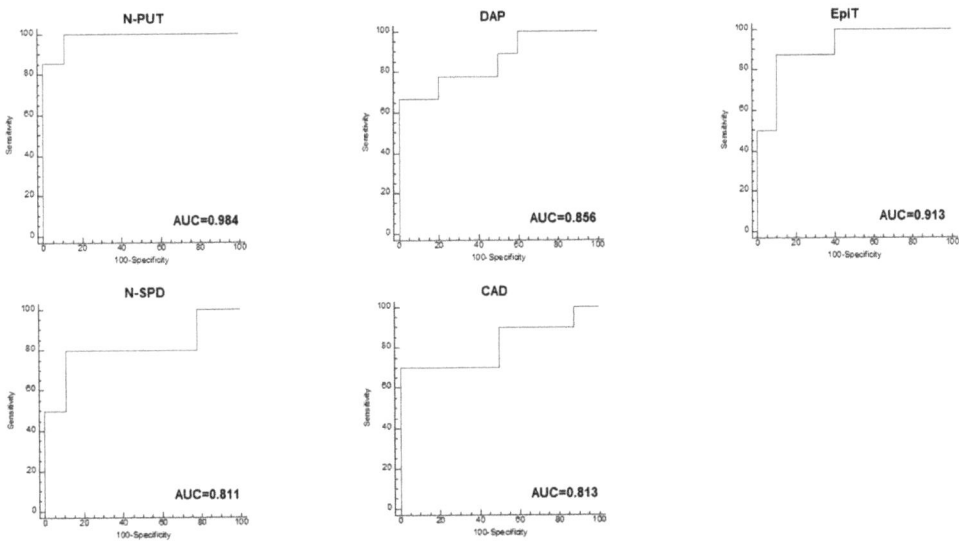

Figure 4. Univariate receiver operator characteristic (ROC) curve analyses for predicting biomarker performance in serum samples from treated breast cancer patients. Typical ROC curve plots of potential biomarkers with high-performance prediction are shown; several metabolites with AUC > 0.8 were observed. The ROC value of each metabolite was as follows: N-PUT, 0.984; N-SPD, 0.811; DAP, 0.856; CAD, 0.813; EpiT, 0.913 (N-PUT: *N*-acetyl putrescine; N-SPD: *N*-acetyl spermidine; DAP: 1,3-diaminopropane; CAD: cadaverine; EpiT: epitestosterone).

All polyamines were higher in treated cancer patients than in normal controls. Since polyamines are involved in cell proliferation, polyamine levels increase in cancer patients. The results corroborate several studies conducted in human urine [23], serum [7], and saliva [24]. In particular, N-PUT, N-SPD, and DAP were significantly higher in patients, which correlated with other breast cancer research [7]. DAP has been reported to be associated with cell proliferation [25]. In addition, N-SPD and N-SPM were not detected in tissues of normal humans, whereas N-SPD was reported to increase rapidly in tissues of breast cancer patients [6]. In particular, acetylpolyamine is an important factor in breast cancer research, and our results are consistent with studies that revealed a correlation between intracellular polyamine levels and alterations in histone acetylation and deacetylation in normal and cancer cells [26,27]. Most of the steroids, except 11β-OHP, were increased in breast cancer patients. In particular, EpiT was significantly higher in the patient group. Although there are few studies on the correlation between breast cancer and EpiT, certain studies have reported that epitestosterone levels are five times higher in breast cysts [28].

3. Materials and Methods

3.1. Chemicals

Reference standards for nine polyamines and steroids were obtained from Sigma-Aldrich (St. Louis, MO, USA), Steraloids (Newport, RI, USA), and Tokyo Chemical Industry (Tokyo, Japan). The IS, d_3-epitestosterone, and 1,6-diaminohexane were obtained from Sigma-Aldrich for calibration. For both calibration and QC, a commercially available blank serum sample (UTAK laboratories Inc, Valencia, CA, USA) was used. Sodium carbonate, sodium bicarbonate, formic acid (ACS reagent), and isobutyl chloroformate (derivatization solution) were obtained from Sigma-Aldrich. All high-performance liquid chromatography (HPLC)-grade organic solvents, including diethyl ether, acetonitrile, and methanol, were acquired from Burdick and Jackson (Muskegon, MI, USA). Deionized water (DW) was obtained using a Milli-Q purification system (Millipore, Billerica, MA, USA).

3.2. Preparation of Standard Solution

To prepare the reference standard stock solutions, polyamines and steroids were dissolved in methanol at a concentration of 1000 μg/mL. Working solutions were prepared using serial dilution with methanol at concentrations ranging from 100 μg/mL to 1 ng/mL. Moreover, 1,6-diaminohexane (DAH), one of the ISs, was dissolved in methanol at a concentration of 100 μg/mL. Working solutions were prepared using serial dilution with methanol at a concentration of 1 μg/mL. Further, EpiT-d3, another IS, was dissolved in acetonitrile at a concentration of 10 μg/mL. Working solutions were prepared using dilution with acetonitrile at a concentration of 1 μg/mL. All standard solutions were stored at −20 °C until further use.

3.3. Sample Information and Ethics Statement

Quantitative profiling of polyamines and steroids was conducted with 20 human serum samples from treated breast cancer patients (n = 10; aged 27 to 60 years, mean 51.5) and normal controls (n = 10; aged 27 to 59 years, mean 45.7). Serum samples were collected at the Yonsei University College of Medicine, and informed consent was obtained from each participant before collection. The study was approved by the institutional review board of Yonsei University College of Medicine (IRB No. 3-2017-0097). Participants were diagnosed with breast cancer of stage I–III and completed cancer treatment. Moreover, breast cancer patients completed cancer treatment, including surgery, adjuvant chemotherapy, and radiotherapy.

3.4. Sample Preparation

To remove proteins, the serum samples (200 μL) were added to 1 mL of DW at 60 °C for 20 min. After heating, 50 μL of the IS (1 μg/mL of DAH and EpiT-d$_3$) was added. Thereafter, the pH values of the samples were adjusted to 9.0 with 1.0 M sodium carbonate buffer (25 µL), and the derivatization reagent (20 µL) was added. Then, the mixture was incubated at 35 °C for 15 min. After cooling, the solution was extracted twice with 2 mL of diethyl ether for 15 min along with shaking and centrifuged for 5 min at 1300× g using a Heraeus SEPATECH Varifuge 3.0, and the organic solvent was transferred to a new test tube. The entire organic layer was evaporated. The method followed in this study was the same as that described by Byun et al. [18]. The residue was reconstituted with 100 µL of MeOH, and a 5 µL aliquot was injected into the liquid chromatography–tandem mass spectrometry system.

3.5. Liquid Chromatography–Tandem Mass Spectrometry

LC-MS/MS analysis was performed using a Shiseido nanospace SI-2 HPLC system (OSAKA SODA, Osaka, Japan) coupled with a Thermo LTQ XL ion trap MS capable of electrospray ionization (Thermo, San Jose, CA, USA). Chromatographic separation was achieved using a Thermo Hypersil GOLD C18 column (150 × 2.1 mm, particle size: 3 µm) at a flow rate of 100 µL/min. A gradient eluent (A: 0.1% formic acid in 5% acetonitrile; B: 0.1% formic acid in 95% acetonitrile) was used. The gradient elution system was controlled as follows: 0 min, 50% B; 0–12 min, 50–95% B (hold for 5 min); 17–18 min, 95%–50% B. The gradient was then returned to the initial condition (50% B) and held for 10 min before running the next sample. The column and autosampler temperatures were maintained at 35 °C and 4 °C, respectively. The use of tandem MS systems is robust and sensitive. First, we acquired the full spectrum of the selected precursor. Moreover, we acquired the MS spectrum using a full scan in selected reaction monitoring (SRM) mode, and we chose the product ion value with a higher ion intensity parameter. For each analyte, the most abundant ion product was selected as the quantitation ion in SRM mode analysis. MS was performed under the following conditions: spray voltage = 5.0 kV; capillary temperature = 350 °C; sheath gas flow rate = 20 arb, auxiliary gas flow rate = 5 arb; capillary voltage = 32 V; tube lens voltage = 85 V; multipole 00 offset = −4.5V. All analytes were detected in positive ion SRM mode. The analytical conditions were optimized for each analyte (Table 4).

Table 4. Optimized MS information for polyamines and steroids.

Compound	Abbreviation	Precursor Ion (m/z)	Product Ion (m/z)	Normalized Collision Energy (%)	Retention Time (min)
1,3-diaminopropane	DAP	275.0	201.1	22	9.3
Putrescine	PUT	289.0	215.0	28	10
Spermidine	SPD	446.1	372.3	35	14.6
Spermine	SPM	603.0	529.2	48	17.9
1,6-diaminohexane	DAH	317.0	243.0	27	12.45
Cadaverine	CAD	303.0	229.0	45	11.2
N-acetyl putrescine	N-PUT	231.0	157.0	28	4.8
N-acetyl spermidine	N-SPD	388.0	314.2	24	7.6
N-acetyl spermine	N-SPM	545.4	471.3	37	12.5
N-acetyl cadaverine	N-CAD	245.0	171.1	51	5.1
Testosterone	T	289.2	271.3	56	8.7
Dihydrotestosterone	DHT	291.2	255.3	24	11.1
Epitestosterone	EpiT	289.2	271.3	26	9.8
Epitestosterone-d$_3$	EpiT-d3	292.2	256.4	35	9.7
Androstenedione	A	287.2	269.3	48	9.7
Pregnenolone	PREG	317.2	299.3	29	14.1
Progesterone	P4	315.2	297.3	30	13.9
17α-Hydroxyprogesterone	17α-OHP	331.2	313.3	22	9.8
11β-Hydroxyprogesterone	11β-OHP	331.2	313.3	29	9.2

3.6. Validation

We carried out validation according the ICH guideline on bioanalytical method validation [29]. For validating the method, QC samples were prepared in commercial blank serum by spiking all target analytes at four different concentrations within their respective calibration ranges. Calibration standards and QC samples were prepared by adding a dilution of the stock analyte solution to the blank serum samples on every validation day. Further, the ion peak area of each metabolite was normalized by dividing it by the IS (e.g., polyamine groups-DAH; steroid groups-EpiT-d_3).

Calibration curves were plotted with concentrations ranging from 1 to 5000 ng/mL for most polyamines (except N-SPM; range = 0.1 to 5000 ng/mL); further, T, EpiT, DHT, and PREG were plotted from 1 to 2000 ng/mL and 17α-OHP, 11β-OHP, A, and P4 were plotted from 0.1 to 2000 ng/mL. The linearity was evaluated using the correlation coefficient (R^2) of the calibration curves. LOQ was defined as the lowest concentration with a signal-to-noise ratio greater than 10. Recovery was assessed to determine whether analyte compounds were lost during sample pretreatment. Overall recoveries were calculated by comparing the peak area ratios of analytes with the IS from all pretreatment steps versus those of only their derivatization steps. The matrix effect of urine samples due to endogenous substances compared to a standard solution was calculated as follows: (the ratio of spiked analyte standards and ISs in urine—the ratio of analytes and internal standards in blank urine)/(the ratio of the corresponding standard and IS in the standard solution) × 100 (%) [30]. Accuracy and precision were determined using QC samples at four concentrations, 10, 50, 500, and 1000 ng/mL, for most analytes, whereas analytes with an LOQ of 0.1 were measured at four concentrations of 1, 50, 500, and 1000 ng/mL. Accuracy was evaluated as the percent relative error (% bias), and precision was evaluated as the coefficient of variation (% CV). Intra-day validation was confirmed by analyzing three replicate samples, whereas inter-day validation was confirmed by analyzing the sample on three different days.

3.7. Statistical Testing and Data Processing

Polyamine and steroid concentrations were obtained from the calibration curves. Polyamines and steroid levels are expressed as the mean ± standard deviation (SD). We used Tune plus and Xcalibur for data analysis. Comparisons between normal controls and patients with breast cancer were performed using an independent samples Student's t-test. For each of the measurements, the difference between groups was analyzed using a t-test followed by Bonferroni correction. The threshold of significance was set at $p < 0.05$. ROC curves to find the possible candidate biomarkers were plotted using MedCalc software (MedCalc Software, Mariakerke, Belgium). The threshold of significant markers was set to an AUC value of > 0.8.

4. Conclusions

In this study, a simultaneous analysis method using LC-MS/MS with isobutyl chloroformate derivatization was validated for serum polyamines and steroids. Polyamines and steroids have different functional groups; therefore, we derivatized polyamines with isobutyl chloroformate but not steroids. Combined quantitative profiling was performed, and sample preparation involved derivatization and an LLE step. This method was successful and reliable with satisfactory peak separation. Thus, we applied quantitative profiling of serum polyamines and steroids in age-matched breast cancer patients and normal controls and observed certain significant differences. Since we conducted the experiment with 10 pairs, it is not accurate, but through our simultaneous analysis method, we might be able to confirm the treatment effect for breast cancer patients. In our results, the clinical impact of several significant compounds showing clinical relevance is expected to be confirmed by extending the number of patients in the future. Although there are several studies on polyamines and steroids separately, our study is the first to analyze polyamines and steroids simultaneously. This method is advantageous as it makes use of a small quantity of samples and reduces the analysis time. In addition, this MS-based quantitative profiling method will be applicable to not only polyamine-related diseases, such as cancer, but also androgen-dependent diseases, such as male pattern baldness and benign prostatic hyperplasia.

Author Contributions: Conceptualization, J.H.; methodology, Y.R.L.; software, Y.R.L.; validation, Y.R.L.; investigation, J.W.L.; resources, J.W.L.; data curation, Y.R.L.; writing—original draft preparation, Y.R.L.; writing—review and editing, J.H. and B.C.C.; supervision, B.C.C.; project administration, B.C.C. All authors have read and agreed to the published version of the manuscript.

Funding: This study was supported by a grant from the Korea Institute of Science and Technology Institutional Program (Project No. 2E31093).

Institutional Review Board Statement: The study was approved by the institutional review board of Yonsei University College of Medicine (IRB No. 3-2017-0097).

Informed Consent Statement: Informed consent was obtained from all subjects involved in the study.

Data Availability Statement: Not applicable.

Conflicts of Interest: The authors declare no conflict of interest.

Sample Availability: Not available.

References

1. Kang, S.Y.; Kim, Y.S.; Kim, Z.; Kim, H.Y.; Kim, H.J.; Park, S.; Bae, S.Y.; Yoon, K.H.; Lee, S.B.; Lee, S.K.; et al. Breast Cancer Statistics in Korea in 2017: Data from a Breast Cancer Registry. *J. Breast Cancer* **2020**, *23*, 115–128. [CrossRef] [PubMed]
2. Harwansh, R.K.; Deshmukh, R. Breast cancer: An insight into its inflammatory, molecular, pathological and targeted facets with update on investigational drugs. *Crit. Rev. Oncol./Hematol.* **2020**, *154*, 103070. [CrossRef] [PubMed]
3. Russell, D.H. Increased polyamine concentrations in the urine of human cancer patients. *Nat. New Biol.* **1971**, *233*, 144–145. [CrossRef] [PubMed]
4. Xu, H.; Liu, R.; He, B.; Bi, C.W.; Bi, K.; Li, Q. Polyamine Metabolites Profiling for Characterization of Lung and Liver Cancer Using an LC-Tandem MS Method with Multiple Statistical Data Mining Strategies: Discovering Potential Cancer Biomarkers in Human Plasma and Urine. *Molecules* **2016**, *21*, 1040. [CrossRef] [PubMed]
5. Kingsnorth, A.N.; Wallace, H.M. Elevation of monoacetylated polyamines in human breast cancers. *Eur. J. Cancer Clin. Oncol.* **1985**, *21*, 1057–1062. [CrossRef]
6. Persson, L.; Rosengren, E. Increased formation of N1-acetylspermidine in human breast cancer. *Cancer Lett.* **1989**, *45*, 83–86. [CrossRef]
7. Byun, J.A.; Choi, M.H.; Moon, M.H.; Kong, G.; Chul Chung, B. Serum polyamines in pre- and post-operative patients with breast cancer corrected by menopausal status. *Cancer Lett* **2009**, *273*, 300–304. [CrossRef] [PubMed]
8. Barton, V.N.; D'Amato, N.C.; Gordon, M.A.; Lind, H.T.; Spoelstra, N.S.; Babbs, B.L.; Heinz, R.E.; Elias, A.; Jedlicka, P.; Jacobsen, B.M.; et al. Multiple molecular subtypes of triple-negative breast cancer critically rely on androgen receptor and respond to enzalutamide in vivo. *Mol. Cancer Ther.* **2015**, *14*, 769–778. [CrossRef]
9. Bulbrook, R.D.; Thomas, B.S. Hormones are ambiguous risk factors for breast cancer. *Acta Oncol.* **1989**, *28*, 841–847. [CrossRef]
10. Cauley, J.A.; Lucas, F.L.; Kuller, L.H.; Stone, K.; Browner, W.; Cummings, S.R. Elevated serum estradiol and testosterone concentrations are associated with a high risk for breast cancer. Study of Osteoporotic Fractures Research Group. *Ann. Intern. Med.* **1999**, *130*, 270–277. [CrossRef] [PubMed]
11. Kaaks, R.; Berrino, F.; Key, T.; Rinaldi, S.; Dossus, L.; Biessy, C.; Secreto, G.; Amiano, P.; Bingham, S.; Boeing, H.; et al. Serum sex steroids in premenopausal women and breast cancer risk within the European Prospective Investigation into Cancer and Nutrition (EPIC). *J. Natl. Cancer Inst.* **2005**, *97*, 755–765. [CrossRef]
12. Mohammed, H.; Russell, I.A.; Stark, R.; Rueda, O.M.; Hickey, T.E.; Tarulli, G.A.; Serandour, A.A.; Birrell, S.N.; Bruna, A.; Saadi, A.; et al. Progesterone receptor modulates ERalpha action in breast cancer. *Nature* **2015**, *523*, 313–317. [CrossRef] [PubMed]
13. Silva, C.; Perestrelo, R.; Silva, P.; Tomas, H.; Camara, J.S. Breast Cancer Metabolomics: From Analytical Platforms to Multivariate Data Analysis. A Review. *Metabolites* **2019**, *9*, 102. [CrossRef] [PubMed]
14. Fiehn, O.; Kopka, J.; Trethewey, R.N.; Willmitzer, L. Identification of uncommon plant metabolites based on calculation of elemental compositions using gas chromatography and quadrupole mass spectrometry. *Anal. Chem.* **2000**, *72*, 3573–3580. [CrossRef]
15. Liu, X.; Locasale, J.W. Metabolomics: A Primer. *Trends Biochem Sci* **2017**, *42*, 274–284. [CrossRef] [PubMed]
16. Choi, M.H.; Kim, K.R.; Chung, B.C. Determination of hair polyamines as N-ethoxycarbonyl-N-pentafluoropropionyl derivatives by gas chromatography-mass spectrometry. *J. Chromatogr. A* **2000**, *897*, 295–305. [CrossRef]
17. Schuber, F. Influence of polyamines on membrane functions. *Biochem. J* **1989**, *260*, 1–10. [CrossRef]

18. Byun, J.A.; Lee, S.H.; Jung, B.H.; Choi, M.H.; Moon, M.H.; Chung, B.C. Analysis of polyamines as carbamoyl derivatives in urine and serum by liquid chromatography-tandem mass spectrometry. *Biomed. Chromatogr.* **2008**, *22*, 73–80. [CrossRef] [PubMed]
19. Loser, C.; Wunderlich, U.; Folsch, U.R. Reversed-phase liquid chromatographic separation and simultaneous fluorimetric detection of polyamines and their monoacetyl derivatives in human and animal urine, serum and tissue samples: An improved, rapid and sensitive method for routine application. *J. Chromatogr.* **1988**, *430*, 249–262. [CrossRef]
20. Starka, L. Epitestosterone. *J. Steroid Biochem. Mol. Biol.* **2003**, *87*, 27–34. [CrossRef]
21. Swerdloff, R.S.; Dudley, R.E.; Page, S.T.; Wang, C.; Salameh, W.A. Dihydrotestosterone: Biochemistry, Physiology, and Clinical Implications of Elevated Blood Levels. *Endocr. Rev.* **2017**, *38*, 220–254. [CrossRef]
22. Concannon, P.W.; Butler, W.R.; Hansel, W.; Knight, P.J.; Hamilton, J.M. Parturition and Lactation in the Bitch: Serum Progesterone, Cortisol and Prolactin. *Biol. Reprod.* **1978**, *19*, 1113–1118. [CrossRef] [PubMed]
23. Lee, S.H.; Kim, S.O.; Lee, H.D.; Chung, B.C. Estrogens and polyamines in breast cancer: Their profiles and values in disease staging. *Cancer Lett* **1998**, *133*, 47–56. [CrossRef]
24. Takayama, T.; Tsutsui, H.; Shimizu, I.; Toyama, T.; Yoshimoto, N.; Endo, Y.; Inoue, K.; Todoroki, K.; Min, J.Z.; Mizuno, H.; et al. Diagnostic approach to breast cancer patients based on target metabolomics in saliva by liquid chromatography with tandem mass spectrometry. *Clin. Chim. Acta* **2016**, *452*, 18–26. [CrossRef]
25. Rupniak, H.T.; Gladden, J.G.; Paul, D. The in vivo effects of a polyamine analogue on tissue stem cell proliferation. *Eur. J. Cancer Clin. Oncol.* **1982**, *18*, 1353–1359. [CrossRef]
26. Korzus, E.; Torchia, J.; Rose, D.W.; Xu, L.; Kurokawa, R.; McInerney, E.M.; Mullen, T.M.; Glass, C.K.; Rosenfeld, M.G. Transcription factor-specific requirements for coactivators and their acetyltransferase functions. *Science* **1998**, *279*, 703–707. [CrossRef] [PubMed]
27. D'Agostino, L.; Di Luccia, A. Polyamines interact with DNA as molecular aggregates. *Eur. J. Biochem.* **2002**, *269*, 4317–4325. [CrossRef]
28. Bicikova, M.; Szamel, I.; Hill, M.; Tallova, J.; Starka, L. Allopregnanolone, pregnenolone sulfate, and epitestosterone in breast cyst fluid. *Steroids* **2001**, *66*, 55–57. [CrossRef]
29. Ich Harmonised Guideline. Bioanalytical Method Validation M10. 2019. Available online: https://www.ema.europa.eu/en/documents/scientific-guideline/draft-ich-guideline-m10-bioanalytical-method-validation-step-2b_en.pdf (accessed on 14 March 2019).
30. Lee, W.; Park, N.H.; Ahn, T.B.; Chung, B.C.; Hong, J. Profiling of a wide range of neurochemicals in human urine by very-high-performance liquid chromatography-tandem mass spectrometry combined with in situ selective derivatization. *J. Chromatogr. A* **2017**, *1526*, 47–57. [CrossRef]

MDPI
St. Alban-Anlage 66
4052 Basel
Switzerland
Tel. +41 61 683 77 34
Fax +41 61 302 89 18
www.mdpi.com

Molecules Editorial Office
E-mail: molecules@mdpi.com
www.mdpi.com/journal/molecules

www.ingramcontent.com/pod-product-compliance
Lightning Source LLC
LaVergne TN
LVHW070724100526
838202LV00013B/1167